Bats of Texas

Number Forty-three

W. L. MOODY JR.

NATURAL HISTORY SERIES

Bats of Texas

Loren K. Ammerman, Christine L. Hice,

and David J. Schmidly

ILLUSTRATED BY CARSON M. BROWN

PHOTOGRAPHS BY J. SCOTT ALTENBACH

TEXAS A&M UNIVERSITY PRESS   College Station

Manufactured in China by Everbest Printing Co. through

FCI Print Group

This paper meets the requirements of

ANSI/NISO Z39.48-1992

(Permanence of Paper).

Binding materials have been chosen for durability.

♾ ♻

Library of Congress Cataloging-in-Publication Data

Ammerman, Loren K.

    Bats of Texas / Loren K. Ammerman, Christine L.
Hice, and David J. Schmidly ; illustrated by Carson
Brown ; photographs by J. Scott Altenbach.—1st ed.

        p. cm.—(W.L. Moody Jr. natural history series ;
v. 43)

    Revision and enlargement of: Bats of Texas / David J.
Schmidly. 1991.

    Includes bibliographical references and index.

    ISBN 978-1-60344-476-7 (book/pb-flexibound : alk.
paper) 1. Bats—Texas. I. Hice, Christine L. II. Schmidly,
David J., 1943– III. Title. IV. Series: W.L. Moody, Jr.,
natural history series ; no. 43.

QL737.C5A77 2011

599.4—dc22

                                2011010695

Contents

In loving memory of my father,
Charles H. Ammerman, who showed me
the meaning of a strong work ethic.
—LKA

In loving memory of my grandfather,
Harold Hice, for his unwavering belief
in me.—CLH

In memory of Terry L. Yates, one of my
graduate students, for his contributions
to mammalogy.
—DJS

Preface

Bats are among nature's most misunderstood animals. No other group of mammals has been so shrouded in mystery, myth, and misinformation as bats. They abound in mythology and folklore, where they are normally portrayed as sinister, demoniac, and generally undesirable creatures. The German word for bat, *fledermaus*, translates as "flying mouse."

Historically, bats have suffered from a bad public image. Many people regard them as common carriers of rabies and as blind, dirty, and ugly creatures that try to become entangled in human hair. In fact, however, bats are not blind (they see quite well, especially in the dark), they rarely carry rabies, they don't fly into people's hair or make them go crazy, and they are not even remotely related to mice. Moreover, Chinese culture regards bats as symbols of good fortune and happiness.

In reality, bats are intelligent and extremely interesting, and they possess fascinating abilities. Not only can they fly, which is unique among mammals, but they also can do so in complete darkness—and negotiate obstacles and catch insects at the same time. Their varied diet includes insects, pollen, nectar, fruits, flesh, and blood. Their social organization ranges from solitary through small family groups and harems, to immense colonies of several million individuals. Some species are able to slow their body processes down to an absolute minimum in order to survive cold temperatures. Others are known to migrate hundreds or even thousands of kilometers in search of suitable living conditions.

Bats have been the subject of increased interest and study in the United States in recent years primarily because of their intriguing and in many ways unique biological properties. The increased attention is also due to the fact that some species have been associated with diseases that affect people (rabies and histoplasmosis); in addition, some bats are highly beneficial to us. Nectarivorous and frugivorous bats are important pollinators and seed dispersers for an array of useful plants. Insectivorous bats are practically the only predators of night-flying insects and are responsible for destroying tons of such pests annually. Bat guano is collected for fertilizer, and bats also are valuable as study animals for scientific research.

The present treatment is a synopsis of current knowledge of Texas bats with a very extensive review of information about their natural history throughout North America and especially in Texas and surrounding states and regions. There are 4 families and 33 species of bats in Texas. Although 4 of these species are known on the basis of a single specimen captured in the state and may be regarded as vagrants, no other state has a bat fauna as rich as that of Texas. Our bat fauna includes all of the families and all but 14 of the species that occur in the United States.

The ultimate aim of this book is to stimulate interest in and provide a better understanding of these often maligned mammals, as well as to promote a greater appreciation of their role in our biological communities and of the need to conserve them. *Bats of Texas* (Schmidly 1991) was the first book to integrate available information about the distribution, systematics, and biology of Texas bats in a single reference. Subsequently, Merlin Tuttle (2003) published a popular account about Texas bats written primarily for amateur naturalists as opposed to professional ones. This latest edition of *Bats of Texas* incorporates new data about the natural history of bats, including updated and revised distribution maps, an improved key to the identification of bats, the most recent application of taxonomic names, and updated species accounts. Also included are a synopsis of bat-detection techniques and a discussion of the effects of climate change and anthropomorphic impacts on bats in Texas.

The first chapter, which is about bats in general, provides basic information about their appearance, distribution, classification, evolution, biology, and life history, as well as about public health, detection techniques, climate change, and bat conservation. However, it is not a comprehensive treatise on bat biology (for this see Kunz 1988; Kunz and Fenton 2003; Neuweiler 2000; Zubaid et al. 2006). The second chapter contains a dichotomous key with illustrations to aid in the identification of Texas bats. The third and most extensive chapter presents a synopsis of current knowledge of the 33 species of bats known to occur in our state, plus a short treatment of species that could possibly be found in the state in the future. This is followed by a detailed bibliography and list of references for bats in Texas, other parts of the United States, and northern Mexico. All of the literature cited in the text has been listed, as well as several other general references

and technical reports for the interested reader seeking greater detail.

Accounts for each species have been arranged so that they contain the following information in sequence: (1) remarks about the source of the scientific Latin name (known as the etymology); (2) the name and appropriate scientific authority for all of the subspecies of a particular bat in Texas, as well as the most recent taxonomic treatment of the species; (3) a brief description of the bat, accompanied by a photograph, with special emphasis on distinguishing features and comparisons with similar species; (4) a description of the distribution of the species in Texas, with reference to a map; (5) a discussion of the animal's life history, including habitat/roosting preferences, food habits, and reproduction; (6) conservation status of the species according to the International Union for Conservation of Nature and Natural Resources (IUCN) and federal and state governments; (7) interesting facts and more detailed information about the species not presented elsewhere; (8) the total number of specimens examined by the authors that are deposited in natural history collections, as well as other known literature records; and (9) a comprehensive listing of published material about each species, indicated by numbers that correspond to the numbered citations provided in the literature and references section. The life histories include observations recorded by other researchers and reported in the literature, as well as our personal experiences based on more than 60 years of combined field work in Texas.

Texas exhibits a wide range of climate, landforms, and vegetation that combine to

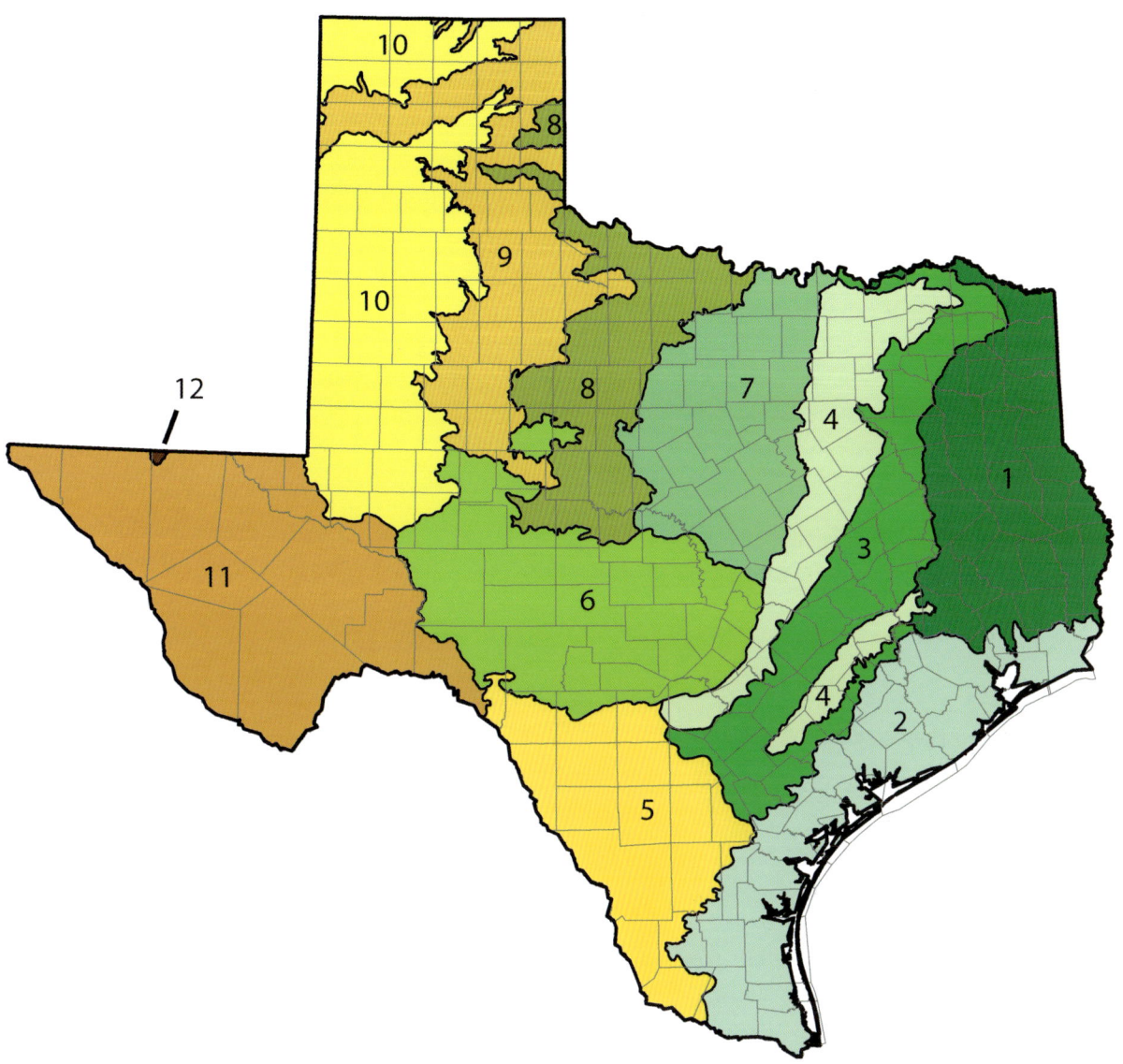

FIGURE 1. Ecological regions of Texas: (1) South Central Plains, (2) Western Gulf Coastal Plains, (3) East Central Texas Plains, (4) Texas Blackland Prairies, (5) Southern Texas Plains, (6) Edwards Plateau, (7) Cross Timbers, (8) Central Great Plains, (9) Southwestern Tablelands, (10) Western High Plains, (11) Chihuahuan Deserts, and (12) Arizona/New Mexico Mountains (modified by Jerry Dragoo from Griffith, G.E., Bryce, S.A., Omernik, J.M., Comstock, J.A., Rogers, A.C., Harrison, B., Hatch, S.L., and Bezanson, D., 2004, Ecoregions of Texas (color poster with map, descriptive text, and photographs): Reston, Virginia, U.S. Geological Survey [map scale 1:2,500,000] http://www .epa.gov/wed/pages/ecoregions/tx_eco.htm.

TABLE 1. *Museum collections containing Texas bat specimens and acronyms used to identify the various collections.*

Acronym	Collection
ACUNHC	Abilene Christian University, Natural History Collection, Abilene, TX
AMNH	American Museum of Natural History, New York City, NY
ANSP	Academy of Natural Sciences of Philadelphia, Philadelphia, PA
ASNHC	Angelo State Natural History Collection, San Angelo, TX
BBNHA	Big Bend Natural History Association, Big Bend National Park, TX
CCSU	Texas A&M University-Corpus Christi, Vertebrate Collection, Corpus Christi, TX
CM	Carnegie Museum of Natural History, Pittsburgh, PA
DMNH	Dallas Museum of Natural History, Dallas, TX
FMNH	Field Museum of Natural History, Chicago, IL
FWMSH	Forth Worth Museum of Science & History, Forth Worth, TX
KU	University of Kansas, Museum of Natural History, Lawrence, KS
LACM	Natural History Museum of Los Angeles County, Los Angeles, CA
LSUMZ	Louisiana State University Museum of Natural Science, Baton Rouge, LA
MMNH	University of Minnesota, James Ford Bell Museum of Natural History, Minneapolis, MN
MSB	University of New Mexico, Museum of Southwestern Biology, Albuquerque, NM
MSU	Michigan State University Museum, East Lansing, MI
MVZ	University of California at Berkeley, Museum of Vertebrate Zoology, Berkeley, CA
MWSU	Midwestern State University, Wichita Falls, TX
NTSU	University of North Texas, Denton, TX
NWMSU	Northwest Missouri State University, Maryville, MO
OMNH	University of Oklahoma, Oklahoma Museum of Natural History, Norman, OK
SFASU	Stephen F. Austin State University, Nacogdoches, TX
SHSU	Sam Houston State University, Vertebrate Natural History Collection, Houston, TX
SM	Baylor University, Strecker Museum, Waco, TX
SRSU	Sul Ross State University, Alpine, TX
TAIU	Texas A&M University-Kingsville, Kingsville, TX
TCWC	Texas A&M University, Texas Cooperative Wildlife Collection, College Station, TX
TNHC	University of Texas at Austin, Texas Natural History Collection, Austin, TX — now housed at TTU
TTU	Texas Tech University, Museum of Texas Tech University, Lubbock, TX
TWC	Texas Wesleyan University, Museum of Zoology, Forth Worth, TX — now housed at ASNHC
UIMNH	University of Illinois, Museum of Natural History, Urbana, IL — now housed at MSB
UMNH	University of Utah, Utah Museum of Natural History, Salt Lake City, UT
UMMZ	University of Michigan, Museum of Zoology, Ann Arbor, MI
USNM	United States National Museum of Natural History, Washington, D.C.
UTA	University of Texas at Arlington, Arlington, TX — now housed at FWMSH
UTEP	University of Texas at El Paso, Centennial Museum, El Paso, TX
WMM	Witte Memorial Museum, San Antonio, TX

produce a diverse state geography. Twelve ecological regions have been identified and described in the state (fig. 1), and these are used to describe the ranges of bats in the state. The interested reader is referred to Griffith et al. (2004) for a detailed description of each of the major ecological regions.

Altogether we have examined approximately 9,000 specimens of bats from Texas that have been collected by professional mammalogists and deposited in various collections throughout the United States. In preparing distribution maps we have taken into account reports on bats from states adjacent to Texas, including Lowery (1974) for Louisiana, Harvey (1986) and Sealander (1979) for Arkansas, Caire et al. (1989) for Oklahoma, Findley et al. (1975) and Findley (1987) for New Mexico, and Hoffmeister (1986) for Arizona, as well as reports from those Mexican states that border Texas, namely Alvarez (1963) for Tamaulipas, Baker (1956) for Coahuila, Anderson (1972) for Chihuahua, and Ceballos and Oliva (2005) for all of Mexico.

On distribution maps, the shaded area is our estimate of the probable distribution in Texas of the species concerned; if the entire map is shaded, the species is to be expected (in suitable habitats) in any part of the state. On these maps, localities represented by specimens that we have examined are indicated by black dots, whereas those based on literature records are indicated by black squares. Black triangles represent bats reported to the Texas Department of State Health Services (DSHS), which we have not examined or identified.

A list of specimens examined, where the specimens are housed, and locality data can be found at www.batsoftexas .com. Although measures given in the book are in metric, measures used to describe collection localities are in imperial measure on the website, as they were originally recorded. Acronyms for the various collections (see table 1) follow Hafner et al. (1997). Collecting localities recorded in the literature for which we did not examine the specimens are listed separately as additional records, along with the appropriate literature citation.

Common and scientific names of Texas bats follow those suggested by Manning et al. (2008), except where recent taxonomic study has required changes. As is customary, the authority that first described a species and the date of description are named following the Latin name of the bat. If, subsequent to the original description, the bat has been placed in a different genus, the describer's name and the date are placed in parentheses. This style of reporting names should not be confused with literature citations. Because of recent taxonomic revisions based on study of new material or the use of new technologies, only 7 Texas bats have retained their original generic names, with the describer listed without parentheses.

Acknowledgments

This project could not have been completed without the assistance of many people at numerous institutions. More than 2,400 new specimens were examined for the updated version of *Bats of Texas*; we thank the following individuals and their respective institutions for providing access to collections and specimen data bases and for shipping specimens: Robert Baker (Texas Tech University, TTU), Lois Balin (Texas Parks and Wildlife Department), Amy Bishop (Angelo State Natural History Collection, ASNHC), Michael Bogan (United States Geological Survey, Albuquerque, New Mexico, USGS), Lisa Bradley (TTU), Robert Bradley (TTU), Andy Bradstreet (Stephen F. Austin State University, SFASU), Wes Brashear (ASNHC), Brent Burt (SFASU), Joseph Cook (University of New Mexico, UNM), Paul Cryan (USGS), Michael Dixon (University of Minnesota), Beverly Dole (University of Michigan Museum of Zoology), Robert Dowler (ASNHC), Jon Dunnum (UNM), Cody Edwards (George Mason University), Heath Garner (TTU),Keith Geluso (University of Nebraska at Kearney), Mark Hafner (Louisiana State University), Tom Lee (Abilene Christian University), Jessica Light (Texas A&M University, TAMU), Ben Marks (TAMU), Ray Matlack (West Texas A&M University), Bonny Mayes (Texas Department of State Health Services), Molly McDonough (TTU), Jim Mueller (Tarleton State University), Julie Parlos (TTU), Steven Platt (Sul Ross State University, SRSU), Heather Prestridge (TAMU), Cindy Ramotnik (USGS), Marcy Revelez (Oklahoma Museum of Natural History), Eric Rickart (Utah Museum of Natural History), Chris Ritzi (SRSU), Frederick Stangl (Midwestern State University), Monte Thies (Sam Houston State University), Robert Timm (University of Kansas), Marie Tipps (ASNHC), and Don Wilson (United States National Museum of Natural History).

The illustrations for the keys and the general anatomy of bats were prepared by Carson Brown. His graceful and detailed drawings greatly enhance the clarity of the keys. Photographs of each bat species were included to aid in identification and to provide readers with a rare view of these elusive creatures. The photographs were graciously provided by Scott Altenbach. His exceptional work brings each bat species to life and is a major improvement over the first edition. His willingness to share his photographs is gratefully acknowledged and warmly appreciated. Several people willingly shared their knowledge of bats with us, including Louise Allen, Robert Baker, Robert Barclay, Gerald Carter, Tom Kunz, Ray Matlack, DeeAnn Reeder, and Ernest Valdez. We are thankful for their insight and enthusiasm. Thanks to Bill Gannon for providing the spectrographs. Tom Kunz and Jennifer Miller graciously provided photographs. Distributional maps were modified from *The Mammals of Texas: Revised Edition* by David J. Schmidly (2004), courtesy of the University of Texas Press. Website development was provided by Jay and Amy Packer, to whom we are indebted. The maps in the book would not exist if it were not for the efforts of

Jerry Dragoo in the design of the map of Texas ecoregions and Brian Beck for aid in creating the distribution maps. We sincerely appreciate their efforts. Michael Bogan critiqued the keys and added helpful insight into how to best distinguish some of the more cryptic species. His experience with bat identification was inimitable. The Interlibrary Loan Departments at UNM and ASU rapidly provided dozens of papers which would have taken weeks to track down otherwise. They did a tremendous job. We acknowledge financial support in the form of an Angelo State University Faculty Development Grant for Loren Ammerman to visit museum collections and examine the specimens used in this revision.

Special thanks to those individuals who have aided in field work on Texas bats with Loren Ammerman over the last 15 years: Faisal Anwarali, Jana Higginbotham Baldwin, Sarah Bartlett, Amy Bishop, Wes Brashear, Scott Burt, Carson Brown, James Dixon, Leanne Dixon, Michael Dixon, Richard Dolman, Robert Dowler, Carla Ebeling, Adam Ferguson, Travis Fisher, Candace Frerich, Gema Guerra, Nick Hristov, Tom Kunz, Bob Lee, Dana Lee, David Long, Amanda Matthews, Steve McAllister, Molly McDonough, Michael Moreno, Amy Vestal Nalls, Steve Oertling, Austin Osmanski, Jay Packer, Fiona Reid, Aimee Roberson, Rogelio Rodriguez, Michael Ryan, Raymond Skiles, Jason Strickland, Rustin Tabor, Marie Tipps, and Suzanne Tomlinson. We look forward to future research on the bats of Texas.

Bats of Texas

Bats are mammals and, as such, possess all of the features characteristic of this vertebrate class, including a body covering of hair (pelage or fur) and mammary glands for the production of milk to suckle their newborn young. Bats differ from other mammals, however, in possessing wings that flap—a character that, coupled with their nocturnal habits, has played a prominent role in their portrayal in folklore and superstition. Other so-called flying mammals, such as flying squirrels and flying lemurs (or colugos), which possess expanded flaps of skin, are not able to undertake powered flight—they merely glide.

Bats are so highly specialized for flight (fig. 2) that locomotion by other means is accomplished with difficulty. The wing pattern is essentially similar in all species, but differences in shape reflect the variety of ecological niches and feeding behaviors bats exhibit. The upper arm, or humerus, is shorter than the forearm, which is composed of a thin, threadlike ulna fused to a longer, somewhat curved radius. All bats have a short, clawed thumb, used mainly for moving around roosts, and 4 greatly elongated fingers that serve to spread and manipulate the wing. The thumb also serves as an attachment for the propatagium, which is that part of the wing membrane in front of the forearm.

The hind limbs of bats are relatively small and attached at the hip in a reverse manner from those of other mammals. Thus, the knee is directed outward due to a 90° rotation of the hind leg, and the foot has rotated 180° from the usual mam-

FIGURE 2.
Anatomy of a typical bat labeled to show names of parts, as used in text.

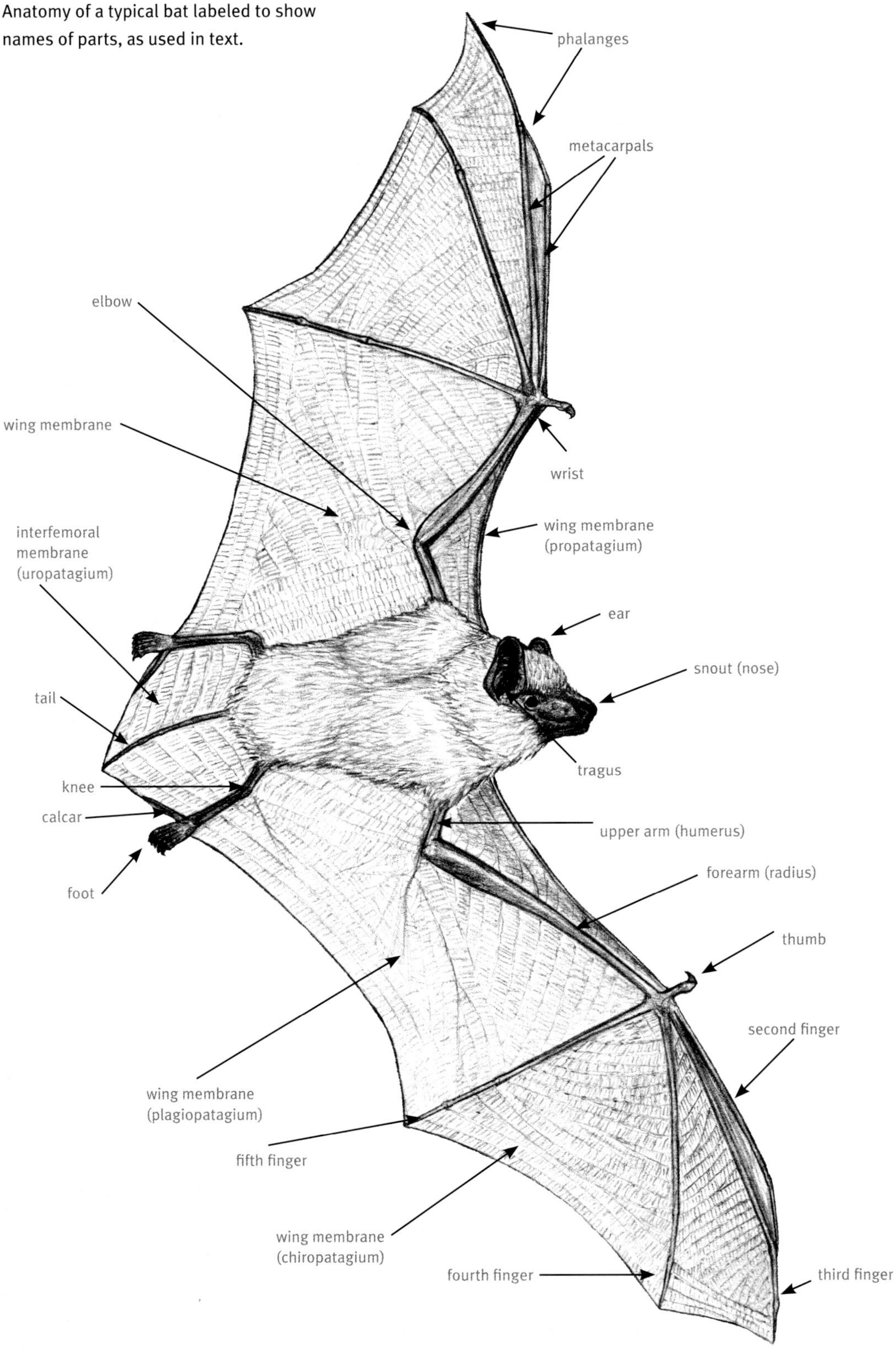

phalanges

metacarpals

elbow

wing membrane

wrist

interfemoral
membrane
(uropatagium)

wing membrane
(propatagium)

ear

snout (nose)

tail

tragus

knee

calcar

upper arm (humerus)

forearm (radius)

foot

thumb

second finger

wing membrane
(plagiopatagium)

fifth finger

wing membrane
(chiropatagium)

fourth finger

third finger

2

malian position so that it points backward and facilitates a head-down suspension of the animal. The hind foot has 5 toes of approximately equal length, all bearing well-developed curved claws with which bats use to hang upside down.

The wing membrane, called the patagium, consists of an upper and lower skin layer, between which are sandwiched bundles of elastic tissue and muscle fiber. This structure is attached along the sides of the body and hind legs and is braced by the elongated finger bones, arms, and legs. Four distinct flight membranes are recognized. The propatagium extends from the shoulder to the wrist anterior to the upper arm and forearm. The chiropatagium is that portion of the wing membrane that forms an elastic webbing between the fingers of the hand. The plagiopatagium extends from digit 5 to the side of the body and the hind leg. Finally, the area of the wing membrane between the hind limbs and the tail is called the uropatagium, or interfemoral membrane. The tail membrane is further supported in many bats by a long cartilaginous spur, called the calcar, which articulates with the heel of the foot. Tail membranes help increase aerial maneuverability and serve as scoops to assist insect-eating bats in catching prey in midair, as pouches to receive baby bats during delivery, and as blankets to help foliage-roosting bats conserve body heat.

Bats exhibit a wide variety of head shapes, which is primarily a reflection of their wide variation in diet and method of food capture (fig. 3a–i). The faces of many species look peculiar because of their oddly shaped and often enlarged ears; leaflike projections and other elaborate structures

on or around their noses; and wrinkles, lumps, and bumps on their lips. Among the most unusual of head shapes in Texas bats are those found in the nectar and pollen feeders, in which the snout is greatly elongated and the back of the head is low and rounded. Although not present in all species of bats, an interesting structure called the tragus exists in the ears of all Texas bats. The tragus is a thin, erect, fleshy projection that rises from the inner base of the ear. This structure varies considerably in size and shape and is occasionally used for identification purposes. It is used in echolocation and is particularly important in the vertical location of sound (Lawrence and Simmons 1982; Neuweiler 2000; Wotton et al. 1995).

Worldwide, bats vary substantially in size, ranging from the world's smallest mammal, Kitti's hog-nosed bat (wingspan, 130–145 mm; weight, 1.5 g), which lives in Thailand and is about the size of a large bumblebee, to the large "flying foxes" of Africa, Asia, Australia, and many Pacific islands, which have a reported weight of 1.5 kg and a wingspan of nearly 2 meters. Such size variations are less evident in Texas bats, with the smallest species (American parastrelle, *Parastrellus hesperus*) having a wingspan of 190 mm and the largest (western mastiff bat, *Eumops perotis*) 550 mm (Table 2).

Nearly all Texas bats have a tail, although as with many of the structures discussed earlier, there is considerable variation (fig. 4a–i). The bats of the family Molossidae are all characterized by having at least half of the tail projecting beyond the rear margin of the uropatagium, which is where they get their common

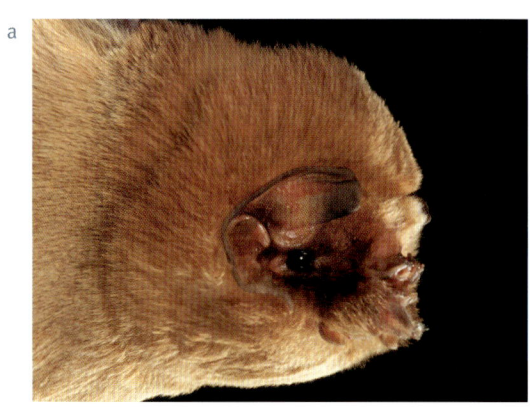

FIGURE 3. Faces of several Texas bats:
(a) *Mormoops megalophylla*; (b) *Leptonycteris nivalis*;
(c) *Euderma maculatum*; (d) *Corynorhinus townsendii*;
(e) *Antrozous pallidus*; (f) *Parastrellus hesperus*;
(g) *Myotis velifer*; (h) *Eumops perotis*; and
(i) *Eptesicus fuscus*.

name, "free-tailed" bats. Bats of the family Vespertilionidae have long tails that are completely bound within the uropatagium. In the family Mormoopidae, the tail protrudes for about 10–15 mm from the top surface of the interfemoral membrane at about the level of the knee. The 3 species of the family Phyllostomidae that occur in Texas either lack a tail altogether or have a very short one (less than 10 mm in length).

In common with other nocturnal or crepuscular mammals, bats have, as a rule, much more somber coloring than diurnal mammals. Texas bats are mostly drab shades of brown, black, and gray, with the ventral surface a lighter tint of the same color as the back. Several species in our state, however, diverge in color from the quiet hues of their relatives. Among these are the spotted bat (*Euderma maculatum*; see photo in species account, p. 178), which is black on the dorsal surface, with 3 large white patches and enormous pink ears; the red bat (*Lasiurus borealis*; see photo in species account, p. 126); the hoary bat (*Lasiurus cinereus*; see photo in species account, p. 133), which has variegated, white-tipped fur; and the silver-haired bat (*Lasionycteris noctivagans*; see photo in species account, p. 154), whose white-frosted fur gives it a grizzled appearance.

The permanent teeth of bats, as in most other mammals, consist of 4 kinds: incisors, canines, premolars, and molars. All except the molars are deciduous, meaning they are replaced once in the life span of the individual. The milk teeth differ in both number and form from the permanent set. They are tiny, sharp pointed, and hooked to enable the young bat to cling more effectively to its mother's teat while she carries

FIGURE 4. External appearance of several Texas bats: (a) *Mormoops megalophylla*; (b) *Leptonycteris nivalis*; (c) *Diphylla ecaudata*; (d) *Euderma maculatum*; (e) *Lasiurus cinereus*; (f) *Tadarida brasiliensis*; (g) *Myotis thysanodes*; (h) *Lasionycteris noctivagans*; and (i) *Eptesicus fuscus*. Not to scale.

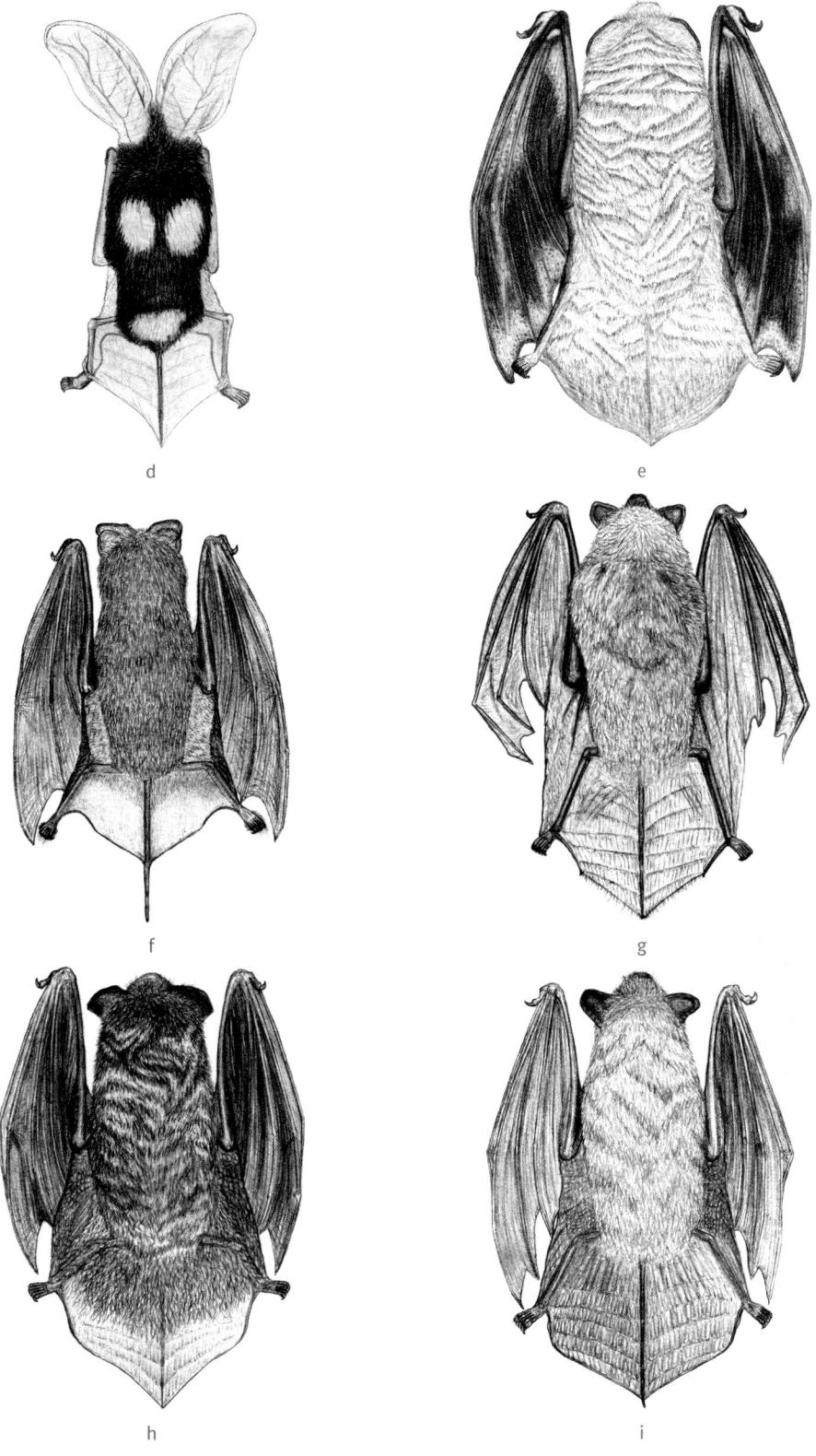

d

e

f

g

h

i

TABLE 2. *Ranges of body measurements and average wing measurements for Texas bats.*

Species	Weight (grams)	Forearm length (mm)	Wingspan (mm)	Wing area (cm²)
M. megalophylla	13–19	51–59	370	188
C. mexicana	10–20	43–45	345	166
L. nivalis	24–32	56–59	410	223
D. ecaudata	30–40	50–56	326	110
M. austroriparius	5–7	33–40	254	104
M. californicus	3–5	29–36	220	77
M. ciliolabrum	4–5	30–34	242	83
M. occultus	4–9	34–41	265	105
M. septentrionalis	5–9	32–39	241	91
M. thysanodes	6–11	40–45	285	125
M. velifer	8–16	37–47	296	131
M. volans	5–9	35–41	267	108
M. yumanensis	4–6	32–36	225	77
L. blossevillii	10–15	38–42	295	110
L. borealis	10–15	35–45	312	134
L. cinereus	20–35	46–58	400	180
L. ega	10–15	42–48	345	148
L. intermedius	14–31	45–56	370	203
L. seminolus	9–14	35–45	300	104
L. xanthinus	12–20	43–47	350	na
L. noctivagans	8–12	37–44	289	104
P. hesperus	3–6	29–33	190	60
P. subflavus	4–6	31–35	237	81
E. fuscus	13–20	42–51	325	162
N. humeralis	5–10	33–39	263	107
E. maculatum	13–18	44–55	365	181
C. rafinesquii	7–13	40–46	270	143
C. townsendii	7–12	39–46	293	136
A. pallidus	12–20	47–57	353	228
T. brasiliensis	11–15	36–46	301	110
N. femorosaccus	10–18	44–50	345	132
N. macrotis	24–30	58–64	426	178
E. perotis	61–84	72–83	550	322

her offspring in flight. The permanent teeth differ widely among genera and species, again in both number and form (Table 3). The greatest number of teeth present in a Texas bat is 38, which is the number in all but 1 species in the genus *Myotis*. The hairy-legged vampire bat (*Diphylla ecaudata*) has only 26 teeth, the fewest teeth of any bat in our state. Reduction in the number of teeth and simplification of the cusps of premolars and molars are adaptations for a liquid diet.

Distribution

Bats live nearly everywhere on the earth with the exception of the polar regions, the highest mountains, and some remote islands. However, their diversity and abundance are greatest in tropical regions, declining steadily north and south of the equator. Bats are common in the United States and can easily be found in most regions, although they are most abundant in the Southwest.

Bats occur in all of the major ecological zones of Texas (Table 4). Of the 4 major environmental components that determine their distribution and abundance (climate, roosts, food, and other animals—predators and competitors), the first and second of these factors are probably most important. Locations characterized by a high degree of topographic relief, indicative of a highly variable or patchy environment, typically support a high density and an abundance of bats primarily because these areas have more roosting sites and a greater diversity of food resources for bats to exploit.

The Chihuahuan Desert ecoregion (Trans-Pecos) in far West Texas, with its topographic pattern of high mountains and desert lowlands, supports more kinds of bats (27 species) than any other part of the state. Several extremely rare or unusual bats occur in this region in greater abundance than anywhere else in the country. The Edwards Plateau also maintains a high diversity of bats, chiefly cavern-dwelling species that inhabit the numerous caves of this region, often in staggering numbers. Because caves are excellent roosting and hibernation sites for bats, they play a prominent role in the distribution of many bats in our state. The diversity and abundance of bats are lower in the northern, eastern, and southern areas of the state, where topographic heterogeneity is low and caves are uncommon.

In many parts of North America bats either migrate or hibernate in winter. If they migrate, their distributions in winter and summer are often very different (Table 5). In parts of Texas where the climate is mild, bats may not migrate or hibernate if weather conditions permit a sufficient and suitable food supply year-round. In general, however, most species of bats in our state tend either to move from one region to another or to move out of the state altogether from November through March.

Classification

Bats belong to the mammalian order Chiroptera, which means "hand wing." Traditionally, the order is divided into 2 suborders, the Megachiroptera and the Microchiroptera. However, recent interpretations of higher relationships of bats based on molecular data suggest a different classification that includes the suborders Pteropodiformes (mostly Megachiroptera) and Vespertilioniformes (mostly Microchiroptera; Hutcheon

TABLE 3. *Dental formulas for Texas bats.*

Species	Upper Teeth[1]				Lower Teeth[1]				Total
	I	C	Pm	M[2]	I	C	Pm	M[2]	($\times$ 2)
M. megalophylla	2	1	2	3	2	1	3	3	34
C. mexicana	2	1	2	3	0	1	3	3	30
L. nivalis	2	1	2	2	2	1	3	2	30
D. ecaudata	2	1	1	2	2	1	2	2	26
M. austroriparius	2	1	3	3	3	1	3	3	38
M. californicus	2	1	3	3	3	1	3	3	38
M. ciliolabrum	2	1	3	3	3	1	3	3	38
M. occultus	2	1	2	3	3	1	3	3	36
M. septentrionalis	2	1	3	3	3	1	3	3	38
M. thysanodes	2	1	3	3	3	1	3	3	38
M. velifer	2	1	3	3	3	1	3	3	38
M. volans	2	1	3	3	3	1	3	3	38
M. yumanensis	2	1	3	3	3	1	3	3	38
L. blossevillii	1	1	2	3	3	1	2	3	32
L. borealis	1	1	2	3	3	1	2	3	32
L. cinereus	1	1	2	3	3	1	2	3	32
L. ega	1	1	1	3	3	1	2	3	30
L. intermedius	1	1	1	3	3	1	2	3	30
L. seminolus	1	1	2	3	3	1	2	3	32
L. xanthinus	1	1	1	3	3	1	2	3	30
L. noctivagans	2	1	2	3	3	1	3	3	36
P. hesperus	2	1	2	3	3	1	2	3	34
P. subflavus	2	1	2	3	3	1	2	3	34
E. fuscus	2	1	1	3	3	1	2	3	32
N. humeralis	1	1	1	3	3	1	2	3	30
E. maculatum	2	1	2	3	3	1	2	3	34
C. rafinesquii	2	1	2	3	3	1	3	3	36
C. townsendii	2	1	2	3	3	1	3	3	36
A. pallidus	1	1	1	3	2	1	2	3	28
T. brasiliensis	1	1	2	3	2/3	1	2	3	30/32
N. femorosaccus	1	1	2	3	2	1	2	3	30
N. macrotis	1	1	2	3	2	1	2	3	30
E. perotis	1	1	2	3	2	1	2	3	30

[1]Number of teeth in *each side* of jaw.

[2]I = Incisors; C= Canines; Pm = Premolars; M = Molars

TABLE 4. *Distribution of Texas bats among the twelve ecological regions of Texas.*

Species	South Central Plains	Western Gulf Coastal Plains	East Central Texas Plains	Texas Blackland Prairies	Cross Timbers	Southern Texas Plains	Edwards Plateau	Central Great Plains	Southwestern Tablelands	High Plains	Chihuahuan Desert	Arizona/New Mexico Mountains
M. megalophylla		X				X	X				X	
C. mexicana		X					X			X	X	
L. nivalis											X	
D. ecaudata											X	
M. austroriparius	X	X	X		X				X			X
M. californicus		X							X	X	X	X
M. ciliolabrum										X	X	X
M. occultus									X		X	
M. septentrionalis						X						
M. thysanodes		X	X	X	X	X		X	X	X	X	X
M. velifer							X		X	X	X	X
M. volans						X					X	X
M. yumanensis											X	
L. blossevillii	X	X				X					X	
L. borealis	X	X	X	X	X	X	X	X	X	X	X	
L. cinereus	X	X	X	X	X	X	X	X	X	X	X	X
L. ega	X	X										
L. intermedius	X		X	X				X	X	X	X	
L. seminolus			X	X				X	X	X		
L. xanthinus					X		X	X	X	X	X	
L. noctivagans	X	X					X	X		X	X	X
P. hesperus	X	X		X	X	X	X			X	X	X
P. subflavus	X	X		X	X						X	
E. fuscus	X	X		X	X	X					X	X
N. humeralis	X	X	X				X					
E. maculatum											X	
C. rafinesquii	X	X									X	
C. townsendii		X				X	X	X	X	X	X	X
A. pallidus		X					X	X	X	X	X	X
T. brasiliensis	X	X	X	X	X	X	X	X	X	X	X	X
N. femorosaccus									X		X	
N. macrotis										X	X	X
E. perotis		X	X				X				X	
Total	11	18	10	9	9	11	13	10	13	16	27	13

TABLE 5. *Seasonal occurrence (by month) of bats in Texas based on capture and literature records.*

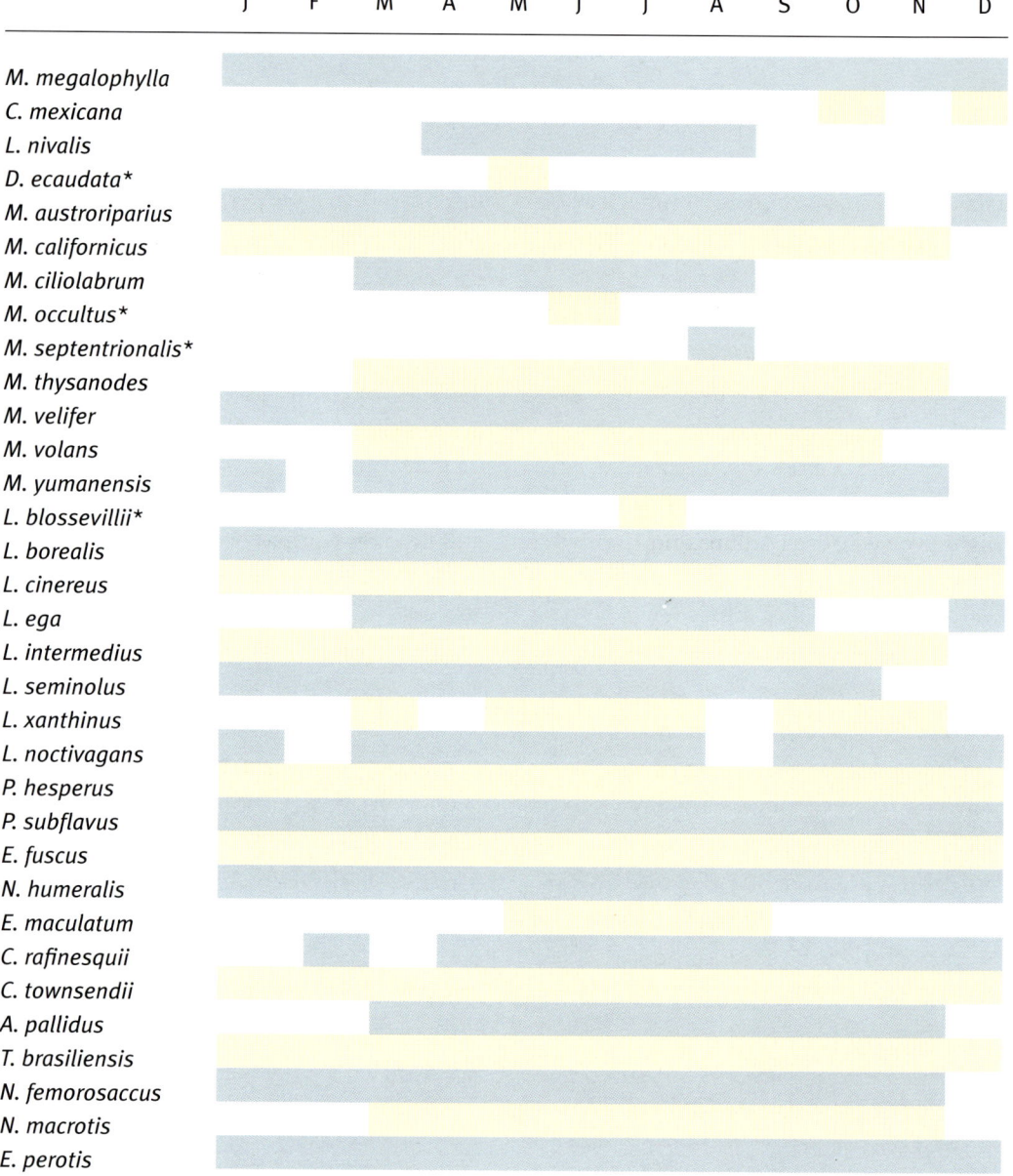

	J	F	M	A	M	J	J	A	S	O	N	D
M. megalophylla												
C. mexicana												
L. nivalis												
D. ecaudata*												
M. austroriparius												
M. californicus												
M. ciliolabrum												
M. occultus*												
M. septentrionalis*												
M. thysanodes												
M. velifer												
M. volans												
M. yumanensis												
L. blossevillii*												
L. borealis												
L. cinereus												
L. ega												
L. intermedius												
L. seminolus												
L. xanthinus												
L. noctivagans												
P. hesperus												
P. subflavus												
E. fuscus												
N. humeralis												
E. maculatum												
C. rafinesquii												
C. townsendii												
A. pallidus												
T. brasiliensis												
N. femorosaccus												
N. macrotis												
E. perotis												

*based on a single specimen

and Kirsch 2006). All families of Texas bats would be classified in the suborder Vespertilioniformes. For details, we direct the serious reader to Teeling et al. (2002, 2005), Hoofer et al. (2003), and Eick et al. (2005).

In this book we retain the traditional suborders—Megachiroptera and Microchiroptera. The former is distinguished from the latter by the fact that most of its genera have a claw on the second digit, whereas in the Microchiroptera only the thumb bears a claw. Moreover, all but one genus (*Rousettus*) of megachiropterans do not echolocate, whereas all microchiropterans do. The early 1990s witnessed some controversy about the evolutionary history of bats, specifically about whether flight developed once (Baker et al. 1991a, 1991b; Simmons et al. 1991) or twice (Pettigrew 1991a, 1991b; Pettigrew et al. 1989). These hypotheses were based on anatomical features, which can be problematic when constructing relationships among mammals. With the advance of molecular techniques, it has been determined that both types of bats belong to one order (Chiroptera) and that flight evolved once (Adkins and Honeycutt 1991; Ammerman and Hillis 1992; Gunnell and Simmons 2005; Simmons 1994).

The Chiroptera is the largest mammalian order after the rodents, with 18 families, 202 genera, and approximately 1,116 species. The Megachiroptera comprises only 1 family with 186 species distributed in the tropical and subtropical regions of the Old World. The suborder Microchiroptera contains 17 families and 1,030 species and is cosmopolitan (Simmons 2005).

Evolution

Bats are thought to have originated in the Paleocene or mid- to late Cretaceous period of prehistory (some 70 to 100 million years ago), but their fossil record is poor. Approximately 25 genera of bats (18 extant and 7 from fossils) are known (Simmons 2005; Simmons and Conway 2003), with the earliest fossils dating back to the early Eocene (approximately 50–52 million years ago; Simmons and Geisler 1998; Teeling et al. 2005). Many of these genera are represented only by fragmentary and incomplete fossils, but some of the earliest known bats are quite well preserved. Even these, however, are well developed and modern in appearance; therefore, transitional links between modern bats capable of flight and their nonflying progenitors have yet to be discovered. Recently, a new family and genus of fossil bat was described that is decidedly more primitive than other fossils in several features; in particular, the development of the inner ear would not have facilitated echolocation (Simmons et al. 2008). However, it still had well-developed wings and was capable of powered flight. Although fossils of more primitive bats are still unknown, it has been suggested that bats may have evolved from small, arboreal tree shrew-like mammals (like the genus *Tupaia*), which had gliding abilities (Neuweiler 2000). However, genetic data do not support this interpretation of morphological characteristics; instead, these data support an affinity among bats, carnivores, artiodactyls, and some insectivores (Murphy et al. 2001). Whatever their origin, the available evidence indicates that bats quickly evolved into their modern form and then rapidly specialized to become the large and diverse order of mammals seen today (Jones et al. 2005).

Fossil bats from Texas (Table 6) are primarily of living species that still occur in the state. However, 2 extinct bats have been recorded in Texas (*Desmodus stocki* and *Myotis rectidentis*), and 2 species (*Macrotus californicus* and *Myotis evotis*) for which there are fossil records no longer occur in the state. The California leaf-nosed bat (*Macrotus californicus*) is a big-eared phyllostomid found in the southwestern United

TABLE 6. *Fossil bats from Texas.*

Species	Reference(s)
Macrotus californicus*	Cookerell (1930), Ray and Wilson (1979)
Desmodus stocki[†]	Cockerell (1930), Ray et al. (1988)
Myotis lucifugus	Dalquest and Stangl (1984b), Roth (1972)
Myotis velifer	Choate and Hall (1967), Czaplewski (1993), Dalquest and Stangl (1984a), Dalquest and Stangl (1984b), Dalquest et al. (1969), Dalquest and Schultz (1992), Dorsey (1977), Graham (1987), Harris (1985), Kurtén and Anderson (1980), Logan (1983), Logan and Black (1979), Lundelius (1967), Patton (1963), Roth (1972), Valastro et al. (1977), Van Devender et al. (1977)
Myotis volans	Roth (1972)
Myotis evotis*	Dalquest et al. (1969)
Myotis californicus	Roth (1972)
Myotis rectidentis[†]	Choate and Hall (1967), Dorsey (1977)
Myotis sp.	Semken (1961)
Lasionycteris noctivagans	Dalquest (1978), Logan and Black (1979)
Parastrellus hesperus	Roth (1972)
Perimyotis subflavus	Dalquest et al. (1969)
near *Eptescius fuscus*	Dalquest (1978)
Eptesicus fuscus	Czaplewski et al. (2008), Dalquest and Carpenter (1988), Dalquest and Stangl (1984b), Dalquest el at. (1969), Logan and Black (1979), Patton (1963), Slaughter and McClure (1965)
Lasiurus borealis	Dalquest (1978)
Lasiurus cinereus	Dalquest et al. (1969)
Lasiurus sp.	Dalquest (1967)
Plecotus townsendii	Logan and Black (1979)
Antrozous pallidus	Czaplewski (1993), Dalquest (1978), Logan and Black (1979), Van Devender et al. (1987)
near *Tadarida*	Dalquest (1975)

* No longer occurs in Texas
[†] Extinct

States and Mexico, and the long-eared myotis (*Myotis evotis*) is a common bat of western and northwestern North America.

Flight

Obviously, the central feature in the evolution of bats and their most diagnostic feature is the development of wings and their ability to fly. The power of flight has allowed bats to exploit a large food base unavailable to terrestrial mammals (flying insects), reduces potential predation, and allows bats to have large geographic ranges. These abilities are responsible for the great diversity and abundance of bats worldwide.

Compared with the wings of birds, bat wings are simple in structure. The wing skeleton contains all of the bones evident in the forelimbs of most mammals (fig. 2) and, in fact, is merely an elongated hand. A flexible membrane, or patagium, is stretched between the body and the fingers of this hand to complete the wing structure. The bat can alter the curvature of the entire structure (camber) at will to increase or decrease lift.

Nine muscles or muscle groups are involved in bat flight. Four large muscles that attach to the upper arm (humerus) and scapula are responsible for the down stroke, or power stroke. Eight smaller muscles act on the upstroke of the bat wing. The power-stroke muscles are located in the chest, and the upstroke muscles on the back. This arrangement differs markedly from the flight muscles of birds, which have only 1 muscle responsible for the power stroke and 1 for the upstroke. In birds, both flight muscles are located in the chest.

The specialization of bats for different flight styles, feeding strategies, and habits is often evident in the shape and size of their wings. Three parameters characterize wing shape: aspect ratio (calculated as the wingspan squared, divided by the area of the wing), tip-length ratio (calculated as hand length divided by arm length), and wing loading (calculated as the mass of the bat divided by the area of the wing). High aspect ratios denote a narrow wing and low aspect ratios a broad one. Increasing aspect ratio decreases drag, thus permitting greater speed, but at the same time reduces lift. Wings with a low-aspect-ratio generate

considerable drag at higher speeds but also promote maximal lift at low speeds. A high tip-to-length ratio indicates a relatively long wing span, which maximizes speed at the expense of maneuverability. A shorter wingtip allows for maximum maneuverability in cluttered environments such as forests. Wing loading is directly proportional to body weight: Heavier bats have higher wing loading. Those bats with the lowest wing loading are able to hover. Based on these statements, one would expect that long, narrow wings (with high aspect and tip-to-length ratios) are indicative of rapid and/or sustained flight, whereas short, wide wings (with low aspect and tip-to-length ratios) indicate slow and maneuverable flight (for a more detailed treatment of bat flight see Altringham 1996; Holderied and Jones 2009; Neuweiler 2000; Norberg 1987, 1994, 1998).

Table 2 provides comparisons of wing measurements for Texas bats. The species in the family Molossidae have high aspect ratios with long, very narrow wings. These bats, such as the Brazilian free-tailed bat (*Tadarida brasiliensis*), are capable of swift flight and are known to forage over wide open areas and migrate long distances. Conversely, most vespertilionids and necta-rivores have low aspect ratios, indicative of broad wings useful for slow (fluttery) flight and hovering. These bats typically forage along vegetation, forest canopies, and narrow canyons and over streams and ponds, where slow and highly maneuverable flight is required. Exceptions to this generalization among the vespertilionids include the highly migratory species, such as the hoary bat (*Lasiurus cinereus*) and silver-haired bat (*Lasionycteris noctivagans*), which also have the highest aspect ratios in this family of bats in Texas.

Echolocation, Hearing, and Vocalization

Bats produce a wide variety of sounds designed for communication, some of which are well within our normal audible range, whereas others are ultrasonic and above the range of human hearing. Bats emit low-frequency "squeaks and squawks," which humans can usually hear, to facilitate social interaction, such as territorial spacing among individuals, mating, mother/infant communication, recognition and to serve as warning calls. Bats use high-frequency, ultrasonic sounds (generally above 20 kHz) outside our hearing range to navigate, avoid obstacles, and capture prey in the dark. This capability of producing ultrasonic pulses and interpreting the echoes rebounding from objects in their path is called echolocation. Not all bats echolocate, however. Most megachiropterans, which are crepuscular (active at dawn and dusk), cannot echolocate, although 1 genus (*Rousettus*) has developed a crude echolocation system using tongue clicks. Echolocation is highly developed in the microchiropterans inasmuch as it is the primary means of orientation for most nocturnal, insectivorous species.

Echolocation calls originate in the larynx and may be emitted orally, as in the vespertilionids, or nasally, as in the phyllostomids. Apparently a few species produce echolocation pulses that humans can hear. For example, the spotted bat (*Euderma maculatum*), which occurs in the Big Bend region of Texas, produces signals that sweep from 15 to 9 kHz, which is within the audible range of most people.

The complex ultrasonic pulses are species specific and may incorporate a combination of frequencies in differing sequences, intensities, and duration to suit flying conditions. A bat traveling from its roost to a feeding area may produce 5 pulses each second, but the same individual trying to catch an insect may increase its rate of pulse production to more than 200 per second as it closes in on its prey. Such bursts of calls, known as "feeding buzzes," provide subtle details on the location, size, and texture of the insect immediately before it is consumed.

Our understanding of the mechanisms of bat echolocation has increased greatly over the past 20 years, particularly how the brain processes echolocation signals. The maximum range of the echolocation "apparatus" is about 20 m, but within that distance bats are able to determine the direction, distance, velocity, shape, size, and even texture of their prey while tracking and closing in for the kill. To measure the distance to a target, bats compute the time elapsed between sound production and echo return, basing the distance on the speed of sound. This requires a highly accurate neural "stopwatch," which provides bats with the ability to resolve delay differences as small as 60 microseconds. The horizontal location of the target is calculated by using the differences in the sounds that reach the right and left ears. The details of the target (shape, texture, movement, etc.) are determined by comparing the frequencies of the original call with those of the echo. All of this information is processed in the auditory regions and associated structures of the bat's brain, undoubtedly involving an overwhelming array of neural operations.

The bat broadens the area from which it collects information by moving its head and ears. In fact, the external ears (pinnae) act as moveable external antennae that amplify the incoming signal. The size of the pinnae is related to the frequency of calls a bat produces. Bats that use higher frequencies (sounds with shorter wavelengths) have relatively small ears, as seen in the Brazilian free-tailed bat (*Tadarida brasiliensis*) and members of the genus *Myotis*. Bats that use lower frequencies have much larger ears, like those of the spotted bat (*Euderma maculatum*) and the 2 species of *Corynorhinus*. Among vespertilionids, bats that use high-frequency sounds detect and capture free-flying insects, whereas bats that glean insects from vegetation use lower-frequency sounds (Neuweiler 2000; Schnitzler and Kalko 2001). Moreover, high-frequency calls are used to detect smaller insects, and lower-frequency calls are used to detect larger insects. Members of the family Molossidae generally use lower frequencies than do vespertilionids even though they also capture free-flying insects. This is particularly true for the western mastiff bat (*Eumops perotis*) and the 2 species of *Nyctinomops*.

Bats are able to detect objects as small as 0.06 mm in diameter (approximately the diameter of a human hair) and can avoid wires of that diameter when flying. How they do this is not well understood because the width of the wires is 20–280 times smaller than the wavelengths the

bats use to detect them. In fact, the sound energy that such objects reflect has not been successfully measured in the laboratory (Neuweiler 2000). This high level of sensitivity allows bats to rapidly detect and capture flying insects. The little brown myotis (*Myotis lucifugus*) is capable of capturing fruit flies (*Drosophila* spp.) at the rate of one every 3 seconds in a closed room and one every 7 seconds in the wild.

The unusual facial structures present on many bats are thought to be important in echolocation, although the exact function of some of these appendages is not clear. One example of such a structure is the nose leaf, which is characteristic of phyllostomid bats. The shape of this structure coincides with the variation in diet of bats in this family but not with variation in echolocation calls (Bogdanowicz et al. 1997). However, this structure has been found to direct the nasally emitted calls in the vertical plane in some frugivorous species (Hartley and Suthers 1987). The tragus also serves to locate the source of the sound vertically by producing a secondary echo of sounds entering the external ear. The delay between the time the actual sound and the secondary echo enter the ear encodes the angle of the vertical direction of the sound source (Lawrence and Simmons 1982).

Several insects, most notably moths and lacewings, have ears that are sensitive to ultrasonic sounds; consequently, their chances of being caught by bats may be up to 40% lower than those of "deaf" insects (Pavey and Burwell 1998). The moth's ears provide information about the distance to the bat based on the bat's call. A faint call from a distant bat prompts the moth to turn and fly away. A more intense call from a closer bat causes the moth to fly in an erratic zigzag pattern or fold its wings and dive toward the ground as an evasive maneuver. A moth resting on a leaf will freeze and stay motionless when it hears an ultrasonic signal. Some moths have taken this antibat strategy one step further by emitting ultrasonic signals of their own, which warn the bat it is unpalatable (Barber and Conner 2007; Hristov and Conner 2005), or in some cases actually jam the bat's echolocation apparatus (Corcoran et al. 2009; Fullard et al. 1994).

Bats also can hear normal (not just ultrasonic) sounds produced by external sources, such as the rustle of leaves and sounds of flying and crawling insects. The ability of some species to hear is quite acute, approximately 10 times greater than that of humans (Neuweiler 2000). This is especially true of carnivorous bats and of bats that glean insects crawling on leaves or on the ground (such as the pallid bat, *Antrozous pallidus;* Fuzessery et al. 1993).

Vision and Olfaction

A common misconception about bats is that they cannot see and must rely solely upon echolocation to orient themselves. To be as "blind as a bat" is a poor analogy, however, as all bats have eyes, and many can see quite well. The vision of megachiropterans (which are crepuscular in their habits and do not echolocate) is better than that of humans, particularly at low levels of illumination. Even those microchiropterans with highly developed echolocation systems have retained their eyes, although they are often tiny. Visual acuity is poorest in the vampire bats and aerial insectivorous

species and best in frugivorous phyllostomids and carnivorous species. The vision of microchiropterans is adapted to low light levels, and some species can distinguish horizontal from vertical stripes when it is so dark that the bat cannot be seen. The big brown bat (*Eptesicus fuscus*) can do this until nearly absolute darkness (Neuweiler 2000). Pallid bats (*Antrozous pallidus*) can see well in light levels equivalent to a clear, moonless night (Bell and Fenton 1986), although vision is not important for locating prey in this species (Bell 1982; Fuzessery et al. 1993). Some species of bats, such as the California leaf-nosed bat (*Macrotus californicus*), use vision extensively to locate and capture prey (Bell 1985). Brazilian free-tailed bats (*Tadarida brasiliensis*) prefer visually (versus acoustically) guided escape routes with a positive response toward light (Mistry 1990). Vision is important to bats for general orientation when leaving and entering caves, looking for landmarks during long-distance flights, controlling flight altitude, and undertaking their extended seasonal migrations (Neuweiler 2000).

The role of olfaction, or the sense of smell, in bats is not entirely clear. Frugivorous bats use smell to locate fruit when foraging (Rieger and Jakob 1988), and insectivorous bats use it to locate and discriminate prey (Neuweiler 2000). Many bats have large and highly developed glands that secrete odoriferous substances, and others have distinctive body odors. Buchler (1980) found evidence that the little brown bat (*Myotis lucifugus*) established and maintained a scent post near a nursery colony. He suggested these bats used the post as an orientation cue for young bats during early foraging flights.

Pipistrelle bats (*Pipistrellus pipistrellus*) use odor to discriminate among conspecifics with regard to gender and reproductive condition (de Fanis and Jones 1995). Male sac-winged bats (*Saccopteryx bilineata*), a small South American bat, use odor extensively during courtship. The sacs store secretions that are "fanned" onto females during complex hovering displays (Voigt and von Helversen 1999). Olfaction also is thought to play an important role in the courtship ritual of big-eared bats (genus *Corynorhinus*), 2 of which occur in Texas. Both of these species possess pairs of large glands protruding from the rostrum. Male Townsend's big-eared bats (*Corynorhinus townsendii*) have been seen to rub the snout vigorously over the female's body prior to mating (Pearson et al. 1952).

The Brazilian free-tailed bat (*Tadarida brasiliensis*), which is one of Texas's most ubiquitous bats, has such a distinctive body odor that roosts can often be located by smell. Whether this scent is useful in guiding bats to the roost is not known, but the odor is unmistakable (similar to taco shells) to anyone familiar with it. The volatile organic component responsible for this odor has recently been isolated using gas chromatography techniques (Nielsen et al. 2006). Other volatile compounds found in the pelage (fur) of Brazilian free-tailed bats (*Tadarida brasiliensis*) have antimicrobial properties (Wood and Szewczak 2007).

Female Brazilian free-tailed bats (*Tadarida brasiliensis*) use olfaction to distinguish their pup in large maternity roosts, both by marking them with odors from the scent glands on their muzzle and by recognizing the scent the pup itself produces (Loughry and McCracken 1991).

However, pups do not distinguish between the scent of their mother and those of randomly selected lactating females from the same colony (Gustin and McCracken 1987).

Roosts

When not foraging, bats rest, groom, and interact with other bats at sites known as roosts. Roosting sites may vary greatly according to species, season, and even time of day. Bats roost in a variety of naturally occurring structures such as caves, rock crevices, and the interior of termite mounds or birds' nests; they also roost in most parts of trees (e.g., inside hollow logs, under loose bark, inside furled leaves) or just hang from branches. They also roost in many anthropogenic structures, including abandoned mines, buildings, bridges, and culverts (Kunz and Lumsden 2003; Pierson 1998). For example, 3 species of bats, the southeastern myotis (*Myotis austroriparius*), big brown bat (*Eptesicus fuscus*), and American perimyotis (*Perimyotis subflavus*), have been found roosting in the winter in concrete culverts under a major freeway (I-45) in Texas (Walker et al. 1996).

Another species in Texas, the Brazilian free-tailed bat (*Tadarida brasiliensis*), commonly uses bridges and culverts as roosting sites, perhaps the most famous of which is the Congress Avenue Bridge in Austin, Texas. This species also uses these anthropogenic structures as maternity roosts and stopover roosts during migration (Fraze and Wilkins 1990; Sgro and Wilkins 2003). Structures such as these are becoming increasingly important to bats as both habitat destruction and disturbance of caves reduce the number of natural roosts

(Keeley and Tuttle 1999; Sherwin et al. 2009).

Many species tolerate very narrow environmental gradients of factors such as temperature, relative humidity, or light intensity and will occupy sites only where such factors are stable within specific limits. Thus, certain bats may roost only in caves and rock crevices, others in hollow trees or foliage, while still others are more general in their living requirements and can be found in a variety of sites.

Roost fidelity also is highly variable among different species of bats and is related to roost permanency and availability (Lewis 1995). Moreover, roosts used during the daylight hours often are quite different from sites used at night. Night roosts are used for ingesting food transported from nearby feeding areas, resting, and escaping from inclement weather or predators. Day roosts are used for sleeping and raising young.

Bats (such as tree bats of the genus *Lasiurus*) may occupy roosts either singly or in small groups. Alternatively, tremendous numbers may inhabit a single site, as in the famous central Texas colonies of Brazilian free-tailed bats (*Tadarida brasiliensis*), which some scientists have speculated may contain up to 20 million individuals during summer population peaks. It is not unusual for several species to use the same roost site, but, when they do, different species use particular areas of the site.

Many bats raise their young in special roosts known as nursery or maternity roosts. The populous summer colonies of Brazilian free-tailed bats (*Tadarida brasiliensis*), which inhabit caves, are classic examples of nursery roosts. These bats arrive in spring and give birth to their offspring in the caves, where the young are subsequently placed in clusters on the ceiling. Clusters of more than a million baby bats have been observed in several of these caves. After the young have matured in the fall, the nursery roosts are abandoned. Adult males are rarely found in nursery roosts.

Bats that hibernate have very strict requirements for hibernacula. Seemingly small fluctuations in temperature or relative humidity may critically affect survival during the long winter months, and changes in these factors will cause torpid bats to arouse and move to more favorable locations. Generally, temperatures must not drop below freezing, and preferred sites seem to have high relative humidity. Although caves and rock crevices are the most common hibernacula, a few species hibernate in relatively unprotected sites like hollow trees and behind tree bark, notably the hardy eastern red bat (*Lasiurus borealis*). Buildings also are used as hibernacula. In far eastern Texas, a subspecies of the Brazilian free-tailed bat (*Tadarida brasiliensis cynocephala*) hibernates almost exclusively in buildings.

Bats are extremely sensitive to disturbance at roosts, and many species will quickly abandon a site if harassed. Obvious overt disturbance includes touching hibernating bats, making excessive noise, or throwing things at bats. However, even nontactile disturbance by humans, such as quietly walking in a cave, can have a significant impact on the arousal of hibernating bats (Thomas 1995). Thus, continued disturbances at natural roosting sites are discouraged. Bats may occasionally take up residence in buildings, but they can be evicted by locating and covering their entranceways. This method is generally simpler, cleaner, less costly, and more humane than killing the animals or using noxious chemicals.

Thermoregulation

Due to their small body size, bats have a relatively large surface area from which energy (heat) readily escapes; therefore, they are extremely sensitive to changes in their thermal environment. They typically respond to the changing seasons by moving to more favorable areas (migration) or becoming torpid to conserve energy (hibernation).

During hibernation, bats conserve energy by reducing their metabolism to the absolute minimum required to sustain life. The bat lives off body fat stored during the preceding autumn and arouses at periodic intervals to assess the condition of the hibernaculum. If conditions have become unfavorable, the bat will move either to a different location in the same roost or to a completely different site. Because such periods of arousal consume a great deal of energy, repeated arousals will endanger the bat's life as it quickly expends its stored fat reserves. Thus, an exceptionally prolonged winter or disturbances at hibernacula may cause starvation and contribute to high mortality at winter roosts.

Many species of Texas bats become torpid during winter. The cave myotis (*Myotis velifer*), Townsend's big-eared bat (*Corynorhinus townsendii*), and pallid bat (*Antrozous pallidus*) are examples of species that hibernate in tightly packed clusters in caves (Twente 1955a). Clustering is common in

bats but not well understood. This behavior probably helps them stabilize their body temperature when ambient conditions fluctuate. Bats near the center of the cluster are cooler than those at the periphery and thus are better able to conserve their precious supply of body fat. Although this may seem counterintuitive, clusters of bats function like a thermos of cold water on a warm day. Since hibernating bats produce negligible amounts of heat, any increase in temperature originates from the surrounding air. This is absorbed by bats on the periphery of a cluster, thus maintaining a lower temperature in the center of a cluster.

Not all bats are able to hibernate, and those that cannot often migrate to warmer latitudes at the onset of winter. Migration has not yet been well studied in bats, and little is known about their destinations, the distances they travel, and their navigational mechanisms. Many Texas bats are seasonal residents of the state, migrating to Mexico for winter. The best known of these is the Brazilian free-tailed bat (*Tadarida brasiliensis*), which arrives abruptly in spring, completes its cycle of raising young, and then departs equally suddenly in fall. These bats winter in Mexico, and migratory moves of up to 1,300 km have been recorded (Villa-R. and Cockrum 1962). The Mexican long-nosed bat (*Leptonycteris nivalis*) also migrates to Mexico for the winter. Many species of *Myotis* are either less active during the winter months or migrate out of the state during that time. Other migratory species that occur in Texas include tree-dwelling species such as the hoary bat (*Lasiurus cinereus*), the eastern red bat (*Lasiurus borealis*), the northern yellow bat (*Lasiurus intermedius*), and the silver-haired bat (*Lasionycteris noctivagans*), which move as far northward as Canada in the summer months and then southward to the southern United States and Mexico in winter (Fleming and Eby 2003; Hill and Smith 1984). These bats have been recorded throughout the year in Texas during both the northward and southward phases of their migratory cycles. They also are known to enter torpor (temporary hibernation) during particularly cold periods (Genoud 1993).

Reproduction and Life Expectancy

Most hibernating bats do not follow the basic mammalian reproductive pattern of fertilization of the eggs by sperm immediately after copulation. Instead, they exhibit a reproductive pattern characterized by 2 phenomena: delayed ovulation (egg releases) and overwinter storage of sperm in the female reproductive tract. For these species most mating occurs in the fall prior to hibernation, but fertilization does not take place then because eggs have not yet been released into the female's reproductive tract. Instead, sperm are stored in the female's uterus and/or oviduct, where they remain alive throughout her winter dormancy. Eggs are shed and fertilized soon after the females emerge from hibernation in the spring (Crichton 2000).

With delayed ovulation and fertilization, there is no waiting for males to produce sperm in early spring. This reproductive strategy allows maximum time for the young to mature, learn to fly, and store fat for the upcoming winter. In addition, it permits adults to attain breeding condition and to copulate during late summer and fall when insects—their energy supply—are

plentiful. Delayed ovulation and fertilization are widespread among Texas bats, including species of the genera *Antrozous, Corynorhinus, Eptesicus, Euderma, Lasiurus, Myotis,* and *Perimyotis* (Wilkinson and McCracken 2003).

In addition to sperm storage in the female genital tract and delayed fertilization, at least 2 Texas bats (big brown bat, *Eptesicus fuscus,* and southern yellow bat, *Lasiurus ega*) also store sperm in the epididymis of the male reproductive tract, which can be used during arousals from hibernation (Crichton 2000; Mendonça and Hopkins 1997). The pallid bat (*Antrozous pallidus*) may also store sperm based on the structure of the epididymis (Crichton et al. 1993). Contrary to Racey (1979), males of the Brazilian free-tailed bat (*Tadarida brasiliensis*) do not store sperm during hibernation but produce it in late winter/ early spring and copulate in spring. This is followed immediately by fertilization and then by birth in the summer (Krutzsch 2000; Krutzsch et al. 2002).

Bats give birth in a manner similar to that of most other mammals. The annual litter is usually produced in May or June, although parturition (birth) may occur from April through July, varying with species and climate. Litter sizes range from 1 to 5 young per pregnancy, but most species give birth to only 1 young per year. Among Texas bats, twinning is common in the big brown bat (*Eptesicus fuscus*), southeastern myotis (*Myotis austroriparius*), American parastrelle (*Parastrellus hesperus*), American perimyotis (*Perimyotis subflavus*), evening bat (*Nycticeius humeralis*), pallid bat (*Antrozous pallidus*), silver-haired bat (*Lasionycteris noctivagans*), and hoary bat

(*Lasiurus cinereus*; Barclay and Harder 2003; Tuttle and Stevenson 1982). The red bats, yellow bats, and Seminole bat of the genus *Lasiurus* usually have 3 or 4 young per litter (the average is 3) and sometimes as many as 5 (Tuttle and Stevenson 1982). All Texas bats produce only 1 litter annually, and a few species (for example, all 3 species of phyllostomid bats) are thought not to reproduce while living here.

To bear and raise their young, most bats assemble in nursery colonies that range in size from a dozen or so to several million bats. Pregnant females typically locate in buildings, caves, mines, tree hollows, or other dark, secluded places for this purpose. No nest is built. The mother holds her head upward as the young is born, and the baby is received in a pocket formed by the interfemoral membrane. In those species with a poorly developed interfemoral membrane, the female hangs her head downward when giving birth.

Baby bats are large and well developed at birth. As soon as they are born, they crawl to the mother's breast and attach to a nipple. Mother bats, like all mammals, produce milk in mammary glands to nourish the young during their initial growth period. Females of most bat species have only 1 pair of functional mammary glands. Members of the genus *Lasiurus* have 2 pairs of mammary glands.

Typically the young remain attached to the mother throughout the day but are left behind, usually in clusters, when the mother emerges to feed in the evening. Mothers return periodically throughout the night to nurse the small babies, but as the young mature, the mothers return less frequently. Most bats apparently can recognize their offspring and pick them out from the clusters to nurse. Young bats grow rapidly, and most are able to fly within a few weeks. As they become increasingly adept at flying and obtaining their own food, they depend less on their mothers. The maternity colonies begin to disperse as the young are weaned.

Bats partly compensate for their low reproductive rate by having a long life expectancy. A conservative estimate for the average life span of a bat after surviving its first year is about 16 years. Records of 20-year-old and older individuals living under natural conditions are not uncommon (Barclay and Harder 2003). One bat is known to have survived at least 34 years in the wild (R. M. R. Barclay, pers. comm.). Such a reproductive strategy that involves a low reproductive rate and long life expectancy is unusual for a small mammal. Similarly sized rodents, by contrast, typically have many young per litter and several litters each year, but they usually live no longer than a few months or years. Nevertheless, because of their much higher reproductive rates, rodents still produce more offspring per mother than do bats.

Populations

Due to the nocturnal and secretive habits of bats, it is difficult to study their population dynamics. Determining the age and sex structure of bat populations and even obtaining accurate estimates of numbers are complicated by many factors, including the location of roosts in unknown or inaccessible sites and the difficulty of catching and observing bats. Also, determining the age of bats is practically impossible beyond the simple classifications of "juvenile" and "adult." As a result, detailed information on population size, mortality, longevity, home range use, and migration pattern is sparse for most species.

Studies of banded bats have provided some information. Bat banding is similar to bird banding, although in bats the coded bands are usually placed on the forearm rather than around the leg. As marked individuals are eventually recovered, they provide clues to the movements of the population. The first bats (a group of 5 American perimyotis, *Perimyotis subflavus*) were banded in 1916 by an ornithologist, A. A. Allen. The first bands (used until about 1960) had sharp edges and, if applied improperly, caused serious injury to the bats by damaging the wing membrane. In 1958 the U.S. Fish and Wildlife Service developed a band with smooth, rounded edges that were issued thereafter. However, bat-banding projects were restricted after 1976 to limit the mortality allegedly caused by the disturbance of hibernating roosts by banding activity (Hill and Smith 1984). Demographic bat-banding studies conducted in the United States are reviewed in Tuttle and Stevenson (1982) and Ellison (2008).

Several banding studies have been carried out with Texas bats. Early studies demonstrated that Brazilian free-tailed bats (*Tadarida brasiliensis*) from Central Texas annually migrate to Mexico in winter and then back to Texas in spring (Davis et al. 1962; Short et al. 1960). A more recent study used banding to help differentiate how these bats use roosts during their seasonal stay in Texas; some roosts were used for mating, whereas others for were used for giving birth and raising young (Fraze

and Wilkins 1990). Twente (1955b) showed that cave myotises (*Myotis velifer*) that hibernate in north Texas caves periodically move between caves in winter and in summer do not always return to the same roost on succeeding nights. The most recent banding study on bats in Texas shows that the American perimyotis (*Perimyotis subflavus*) chooses hibernacula based on the extent of forest and agricultural areas surrounding the site, not on microclimatic parameters alone (Sandel et al. 2001).

More recent studies have examined the population structure of Brazilian free-tailed bats (*Tadarida brasiliensis*) using genetic techniques (McCracken and Gassel 1997; Russell and McCracken 2006; Russell et al. 2005). They have found that the entire population of these bats in North America is panmictic (all individuals can potentially breed with all other individuals), suggesting there is no subspecific differentiation among the different migrating and nonmigrating groups as Cockrum (1969) originally proposed. This makes the effective population size of this species in the hundreds of millions, something extremely rare in mammals. (However, in this book we continue to recognize 2 subspecies, *T. brasiliensis mexicana* and *T. brasiliensis cynocephala*, based on morphological and behavioral differences—as detailed in the species account.)

In other parts of North America, band recoveries and other studies have shown that significant mortality occurs during the first year of life, after which bats may be very long lived (Hill and Smith 1984). Reports of individuals recovered 20 to 30 years after they were initially banded are common (Barclay and Harder 2003; Tuttle and Stevenson 1982). Although bats do not have many natural predators, it is thought that the impact of predation on bats is higher than generally recognized (Speakman 1991; Tuttle and Stevenson 1982). No predator is known to prey exclusively on bats, although a variety of animals periodically attack and capture them. These include many species of snakes, which often frequent caves, as well as falcons, hawks, owls, raccoons, opossums, and feral cats. Owls in particular may depend on bats as a food source more than previously surmised (Hoetker and Gobalet 1999; Julian and Altringham 1994; Roberts et al. 1997b; Rupprecht 1979; Sealander 1979). Other raptors, including peregrine falcons and red-tailed hawks, also frequently depredate several species of bats, including solitary migrating bats (silver-haired bat, *Lasionycteris noctivagans*; big brown bat, *Eptesicus fuscus*; and eastern red bat, *Lasiurus borealis*) and large colonies of Brazilian free-tail bats (*Tadarida brasiliensis*) during their evening emergence (Byre 1990; Lee and Kuo 2001).

The only study that quantifies the impact of bird predation on bats was conducted in the British Isles, where it is estimated that more than 200,000 bats are taken by birds each year, 93% of which are taken by owls. This represents approximately 11% of the mortality of British bats (Speakman 1991), a substantial proportion that suggests predation by birds is a significant factor in bat behavior and population dynamics.

Young (less than 1 year old) bats face several daunting challenges before becoming adults. The first stage is simply to survive until first flight. As many as 75% of baby bats never make it past this stage (Hill and Smith 1984). Many simply fall out of the roost and perish. In cave situations, bats that fall to the floor often are eaten by dermestid beetle larvae. The next stage is to learn to fly and forage. Finally, yearling bats have to store sufficient body fat either to survive hibernation through the winter or to undertake a long migration. Many yearlings do not survive their first hibernation; the estimated mortality during this first critical year is 40–50% (Hill and Smith 1984; Stevenson and Tuttle 1981). Maternity roost selection, both for microclimatic factors (a higher temperature is usually better because it speeds development) and distance to suitable foraging sites and a water source, is perhaps the most important determinant of survival of juvenile bats (Tuttle and Stevenson 1982).

Estimating population numbers is extremely difficult. For most species, no accurate estimates exist. However, new thermal infrared imaging technology (which detects heat; see the New Technologies section) has considerable promise for enabling scientists to more directly and accurately measure bat populations, particularly in caves. In the past, crude population estimates have been attempted by manually counting bats in a roost or during the evening emergence from a roost (Kunz 2003). For example, Davis et al. (1962) estimated population numbers of Brazilian free-tail bats (*Tadarida brasiliensis*) for Central Texas caves by using a combination of counts inside the caves, trapping, and observing the density and duration of exodus flights. They estimated that between 95.8 and 103.8 million Brazilian free-tail bats (*Tadarida brasiliensis*) occupied Texas caves during summer, although they admitted their precision was low. Twente (1955a) estimated the numbers of bats in North Texas and Oklahoma caves by counting individuals in clusters and measuring the sizes of clusters.

Census of bat colonies in several caves in Texas have already been conducted using the new thermal imaging technology (Ammerman et al. 2009; Betke et al. 2008; Frank et al. 2003), which offers an objective technique for estimating the size of very large bat populations that in the past have been difficult to enumerate accurately. Historically, the census method involved watching a cave emergence and determining the number of bats emerging over time by approximation. The potential for this method to reflect observer bias and result in inaccurate estimates is high compared to thermal imaging technology according to Betke et al. (2008), who tested the method at several caves sites in Texas. Overall, their results suggest that methods used in the past tended to overestimate the number of Tadarida by several orders of magnitude. For example, the historically largest-known colony of Brazilian free-tailed bats in North America exists in Bracken Cave, Comal County, Texas. In 1957 the population was estimated to be 20 million using traditional methods. Using infrared thermal imaging, the population was estimated to be about 0.5 million in 2006 (Betke et al. 2008). The difference in estimates could reflect a difference in the accuracy of the techniques, a real decline in the population, or it might be because the census was taken early in the season before the colony had reached its peak. Nonetheless, determining the size of large colonies is challenging and the use of new

imaging technology provides a promising approach for future studies.

Unfortunately, research on changes in the status of bat populations in Texas and the Southwest is limited and typically has focused on a few cavernicolous species that aggregate in large colonies. O'Shea and Vaughan (1999) have reported observations made on populations of several species of bats, all of which also occur in West Texas, at a site in Central Arizona over a 25-year period from 1972–1997 to determine whether changes in populations of roosting bats had occurred during this period and to explore possible explanations for such changes. The major changes documented over the 25-year period included an absence of maternity colonies of cave myotis (*Myotis velifer*) and pallid bats (*Antrozous pallidus*), an absence of roosting groups of Yuma myotis (*Myotis yumanensis*) and Brazilian free-tailed bats (*Tadarida brasiliensis*), and the presence of small numbers of breeding Townsend's big-eared bats (*Corynorhinus townsendii*). American parastrelles (*Parastrellus hesperus*), big brown bats (*Eptesicus fuscus*), and California myotis (*Myotis californicus*) are continuing to use the site in small numbers as they had in 1972. Some of these species also were present in the area 66 years earlier, including far greater numbers of Townsend's big-eared bats (*Corynorhinus townsendii*), as documented in museum collections. The surrounding habitat did not appear substantially different between 1972 and 1997, and a reconstruction of possible impacts created by bat biologists did not suggest that researchers caused the local extinctions documented. The most obvious change during this quarter century was a dramatic increase in recreational use of the area, which led the authors to conclude that disturbances associated with recreationists resulted in the observed population changes, primarily through roost abandonment.

Parasites

Bats may be affected by a wide variety of parasites, most of which are specific to bats and not contagious to humans. These include both endoparasites, such as protozoans and helminthes (Duszynski 2002; Ubelaker et al. 1977), and ectoparasitic arthropods (Whitaker et al. 2009).

Protozoans are single-celled animals and are generally transmitted through the bites of intermediate hosts such as mosquitoes and flies. Protozoans responsible for diseases such as malaria, sleeping sickness, Chagas' disease, murrina, and mal de caderas have been found in bats from around the world; however, such pathogens isolated from bats do not appear to cause infection in humans (Hill and Smith 1984). Helminths, or parasitic worms, are internal parasites with complex life cycles that usually require mollusks (snails) as a host for the larval stages. Tapeworms may be found in the intestinal tract; flukes in the liver, lungs, or gall bladder; and roundworms may occupy almost any part of the body. Work on endoparasites in bats, particularly in Texas, is sparse in the recent literature (McAllister et al. 2007).

Ectoparasitic arthropods that subsist on bats include mites, batflies, ticks, kissing bugs, and fleas (Whitaker et al. 2009). Many of the mites and batflies are host specific, meaning they are associated only with particular bat species. Bat mites also may occupy only specific sites on their host,

such as the oral mites (Radfrodiella) of the Mexican long-nosed bat (*Leptonycteris nivalis*), which occurs in Trans-Pecos Texas. These mites are found only between the teeth and gums of this bat. Other locations for bat mites include in or at the bases of the ears, at the base of the tail and around the genitalia, on the wings, and around the eyes (Whitaker et al. 2007; Whitaker and Wilson 1974).

Batflies (families Nycteribiidae and Streblidae) are small flies that have become especially adapted for living on bats. Nycteribids have completely lost their wings and so are incapable of flight, whereas the streblids have retained wings (though these are much reduced) and are capable of short flights. Both of these families of flies feed upon the blood of bats, and their larvae are deposited directly on the bat rather than on an intermediate host.

Food

As a group, bats have the most diverse food habits of any group of mammals. Although there are no truly herbivorous (plant-eating) bats, the order Chiroptera includes members that specialize in almost every other diet conceivable. Some bats feed exclusively upon insects and other arthropods (insectivory), flesh (carnivory), fish (piscivory), fruit and flowers (frugivory), pollen and nectar (nectarivory), or blood (sanguivory), and some bats eat a variety of food items (omnivory). The bats of Texas are predominantly insectivores (30 species), but 2 nectarivores also occur here, and 1 sanguivore has been documented by a single specimen in the state.

Probably the most familiar of these specialized food habits is that of the true vampires, which feed exclusively on the blood of vertebrates. Worldwide, there are only 3 species of these unusual bats, and all are endemic to the New World tropics and subtropics. One of these species, the hairy-legged vampire (*Diphylla ecaudata*), has been recorded in Texas, although only once. It specializes in feeding on the blood of birds. The incisor teeth of vampire bats are enlarged and bladelike and are used to inflict a small wound from which blood is lapped with the tongue. Vampires also have dramatically modified stomachs, which vary from a baglike to a narrow, tubular structure with thin, elastic walls that easily stretch to accommodate large quantities of blood.

Equally interesting are the 2 nectarivorous bats (Mexican long-nosed bat, *Leptonycteris nivalis*; and Mexican long-tongued bat, *Choeronycteris mexicana*) known to occur in Texas. These bats, both of which reach the northern extent of their range in Texas and are rare in our state, exhibit structural modifications that allow them to feed upon the nectar and pollen of flowering plants. Nectarivores show an overall reduction in dentition, elongation of the rostrum, and an extremely long, protrusible tongue for extracting nectar, and they can hover in flight much like hummingbirds. Both species are important pollinators of flowering plants in their ranges. The Mexican long-nosed bat (*Leptonycteris nivalis*) is a particularly important pollinator of century plants or agave (*Agave* spp.) in the southwestern United States. Populations of this species are highly correlated with the abundance of agave near suitable roosting sites (Moreno-Valdez et al. 2004).

Insectivorous bats, which constitute the great majority of the Texas bat fauna, eat a wide variety of night-flying and ground-dwelling insects, including moths, beetles, grasshoppers, flies, mosquitoes, termites, lacewings, crickets, katydids, cicadas, true bugs, and night-flying ants, some of which are important agricultural pests. The size range of insects eaten by bats varies from small gnats, which weigh no more than about one-fifth of a milligram, to large moths that weigh up to 200 mg (Jones and Rydell 2003).

To capture insects, bats swoop low over the surface of bodies of water, snap prey out of the air, and even land on the ground to pursue them "on foot." Although some insect-eating bats grab their victims directly in their mouths, others scoop their prey out of the air with their wings or interfemoral membrane. The eastern red bat (*Lasiurus borealis*), for example, performs an aerobatic somersault while closing its wings and tail membrane around the captured insect. Before resuming flight, it tucks its head down into the pocket formed by the interfemoral membrane to collect the insect in its mouth (Hill and Smith 1984). Many other aerial insectivores capture flying insects in the wings or tail membrane and then transfer them to the mouth in flight (Jones and Rydell 2003; Webster and Griffin 1962).

Variety truly is the spice of life for insectivorous bats, whose menu may change from night to night and season to season. There is little evidence that any species of bat limits its diet to a particular type of insect, but some species seem to show distinct preferences for certain groups of insects. For example, Black (1974) investigated an insectivorous bat community in New Mexico and found that moths and beetles were the 2 insect groups most frequently used by the 16 species of bats at his study site. By using a percentage frequency of occurrence of 65% or greater as a criterion for differentiating food habits, Black was able to classify most species as either "moth strategists" or "beetle strategists."

In Texas, Rafinesque's big-eared bat (*Corynorhinus rafinesquii*) and the spotted bat (*Euderma maculatum*) are moth specialists, whereas the big brown bat (*Eptesicus fuscus*) is a beetle specialist. Several species specialize in flies, including species of the genus *Myotis*, the silver-haired bat (*Lasionycteris noctivagans*), American parastrelle (*Parastrellus hesperus*), and American perimyotis (*Perimyotis subflavus*). Several species of Texas bats have been documented eating unusual insects: The evening bat (*Nycticeius humeralis*) eats stinkbugs; the hoary bat (*Lasiurus cinereus*) has been known to eat dragonflies (Jones and Rydell 2003); and the pallid bat (*Antrozous pallidus*) has been documented consuming walking sticks (Lenhart et al. 2010).

The diets of 3 species of molossids have been examined in Texas. They consumed mostly moths and true bugs, although 1 species (big free-tailed bat, *Nyctinomops macrotis*) consumed a substantially higher proportion of moths than bugs, whereas the other 2 species (pocketed free-tailed bat, *Nyctinomops femorosaccus*, and Brazilian free-tailed bat, *Tadarida brasiliensis*) consumed relatively more bugs and beetles (Debelica et al. 2006). Although these latter 2 species have a similar diet for much of the year, in March their diets diverge, with the pocketed free-tailed bat (*Nycti-*

nomops femorosaccus) eating more bugs than the Brazilian free-tailed bat (*Tadarida brasiliensis*; Matthews et al. 2010). Whitaker et al. (1996) also found that Brazilian free-tailed bats (*Tadarida brasiliensis*) in Texas eat relatively more bugs and beetles in the evening and more moths during the morning hours.

Insect-eating bats use their long canines to seize and pierce their prey, which is then reduced to minute fragments by the sharp-edged premolars and bladelike crests of the molar teeth. The sharp cusps and ridges of the opposing teeth act as scissors to cut up the insect food into tiny pieces. Small, soft insects are probably chewed and swallowed directly. Larger insects, especially beetles with hard, tough exoskeletons, require some processing before they can be consumed. Some large bats will carry their victims to a convenient night roost to devour them.

Insectivorous bats, which are primarily nocturnal in their foraging habits, become active at varying intervals after sunset. Smaller bats, such as the American parastrelle (*Parastrellus hesperus*) and American perimyotis (*Perimyotis subflavus*), generally emerge early in the twilight period, while larger bats, such as western mastiffs (*Eumops perotis*), tend to be the last to commence foraging. This is due in part to the fact that smaller bats need to obtain a drink of water earlier because they dehydrate more rapidly than do larger bats. Smaller crepuscular bats also appear to fill their stomachs more quickly and return from foraging earlier than the larger species do.

Many insectivorous bats seem to be rhythmic in their feeding. They have 2 major feeding periods, the first of which begins with the onset of twilight or shortly thereafter. Apparently the more efficient feeders in a given species are satiated by 10:00 PM, while the feeding period appears to continue until about 12:00 AM for others less efficient. Then there is a period of relative inactivity until close to 3:00 AM, after which time bats may embark upon another brief foraging trip. However, some activity by stragglers may continue into the early daylight hours. At Frio Cave, in Uvalde County, Texas, Brazilian free-tailed bats (*Tadarida brasiliensis*) time their emergence and return times with sunrise and sunset, leaving approximately 25 minutes before sunset and returning within 90 minutes of sunrise (Lee and McCracken 2001).

The foraging habits of North American bats are highly variable (Kunz 2004). Three basic prey-capture methods have been described (Ross 1967). Hearing and vision are important for those species, such as the pallid bat (*Antrozous pallidus*), which rely on ground- or vegetation-dwelling prey. Others, such as the American parastrelle (*Parastrellus hesperus*) and the Brazilian free-tailed bat (*Tadarida brasiliensis*), are "filter feeders," which appear to seize prey at random within dense flights of insects. Finally, some bats, such as the hoary bat (*Lasiurus cinereus*) and Townsend's big-eared bat (*Corynorhinus townsendii*), use individual pursuit to catch prey. In this type of predation, the bat approaches the prey from the rear and engulfs its abdomen.

Bats consume large quantities of insects. It has been estimated that they eat from one-quarter to three-quarters of their body weight in insects nightly (Kunz et al. 1995). A 20-gram bat may thus eat 5–15 g of insects in one night. At this conservative

feeding rate such a bat might consume 1.8–5.4 kg in 1 year. In Texas, Brazilian free-tailed bats (*Tadarida brasiliensis*) are particularly important economically because they remove large quantities of agricultural pest insects, particularly the corn earworm, cotton bollworm, and fall armyworm, which destroy crops, including cotton and corn. One estimate is that the 100 million or so Brazilian free-tailed bats (*Tadarida brasiliensis*) in Texas alone consume on the order of 4 billion corn earworm moths (or similarly sized insects) each night!

Carbon isotope analysis of bat guano has shown that Brazilian free-tailed bats (*Tadarida brasiliensis*) preferentially feed on insects that utilize crop plants. In Arizona, two-thirds of the bat guano at Eagle Creek Cave comprised insects that fed on crop plants (Mizutani et al. 1992), and in Carlsbad, New Mexico, half of the insects eaten by bats exploited crop plants (des Marias et al. 1980). On the Edwards Plateau in Texas, dietary variation in Brazilian free-tailed bats (*Tadarida brasiliensis*) is closely correlated with the emergence, migration, and availability of adult moths of corn earworms and fall armyworms (Lee and McCracken 2005). In south-central Texas, it has been estimated that the economic value of the pest-control service provided by Brazilian free-tailed bats (*Tadarida brasiliensis*) is approximately $741,000 per year, or about 12% of the annual value of the cotton crop in this region (Cleveland et al. 2006).

New Technologies and Bat Natural History

Technology developed for other, unrelated uses has been adapted and used to further our understanding of natural history. This is perhaps truer for bats than for any other group of mammals because of their secretive nature, nocturnal habits, and ability to fly. The new technology available today allows us to capture bats on the wing, hear their echolocation calls, watch them feed in the dark, find their hidden roosts, and examine the genetic code in their DNA, the very thing that makes them bats (Kunz and Parsons 2009).

Technological advances that aid us in better understanding bat natural history can be divided into 3 major groups: those that enable us to capture bats, those that allow us to observe their behavior, and those that give us insight into their evolutionary history, population biology, and genetic diversity. Historically, the only methods available for capturing bats were shooting them with shotguns in the early evening or hand-capturing them in their roosts during the day. These methods were time consuming and uncertain in their return; only a fraction of the total number of species present in an area could be obtained in this manner. In 1937 A. E. Borell, while studying mammals in the Big Bend region, developed a method of capturing bats by stringing fine wire across a water tank frequented by bats coming to drink after dark (Borell 1937). The bats would hit the wires and drop into the water, and they could then be collected when they swam to the side of the tank. This method was more effective and economical than shooting and allowed bats to be banded and released unharmed if desired. Slow-flying species, such as pallid bats (*Antrozous pallidus*), American perimyotis (*Perimyotis subflavus*), and American parastrelle (*Parastrellus hesperus*), often avoided the wires, however, and the

technique could be used only on relatively small water tanks, not on ponds or streams from which bats also drink.

In 1954, the late Walter Dalquest, a mammalogist at Midwestern State University in Wichita Falls, Texas, published a technique for capturing bats using Japanese silk "mist nets" (Dalquest 1954). The use of this technique would soon revolutionize the study of bats, and today it is still the most effective method available for capturing them. The mist nets are commonly strung across ponds after dark but also can be set among bushes or trees or across openings of caves or other roost sites to capture bats as they leave their roosts in the evening. Mist nets offer the following advantages to the collector: Bats can be taken in large numbers; species otherwise difficult to collect are more easily obtained; and captured individuals can be released alive.

Another device for capturing bats alive is the harp trap, first developed by Constantine (1958c) and modified in various ways since then (Kunz and Anthony 1977; Palmeirim and Rodrigues 1993; Tuttle 1974). Harp traps consist of one or more frames with wires strung vertically. Bats hit the wires in flight and fall down into a collecting bag at the base of the trap. These traps have an advantage over mist nets in that bats are easier to remove once captured, thus facilitating the capture of large numbers of bats at a cave entrance, for instance. Their disadvantage is that they are bulky, heavy, and therefore difficult to transport to distant field sites. Nonetheless, they are useful and even preferred for capturing bats under some circumstances.

The most outstanding advances in our

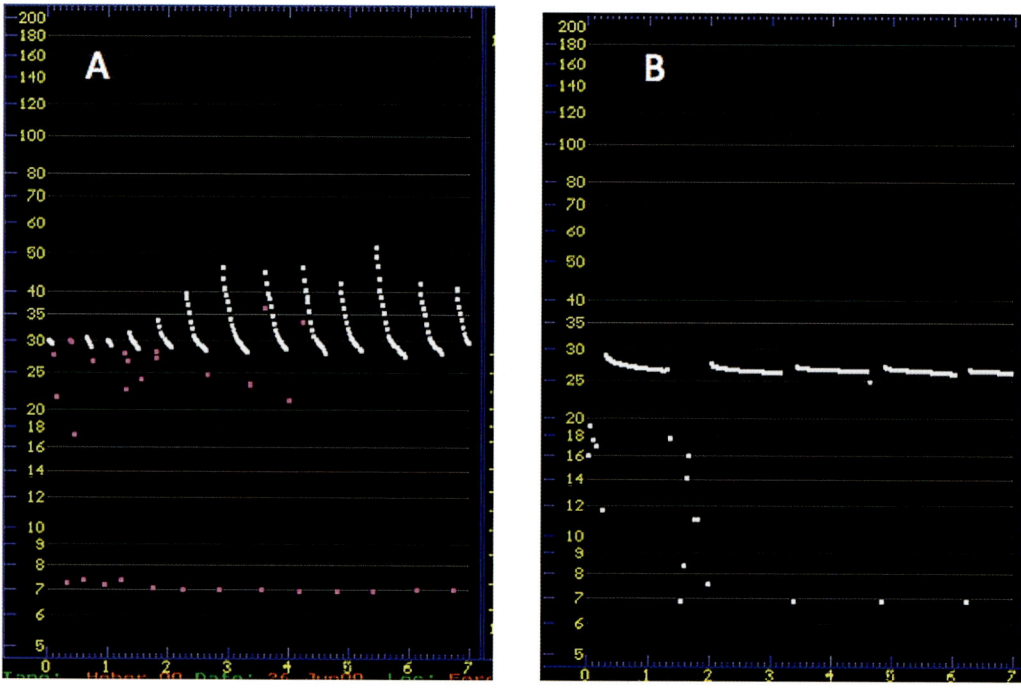

FIGURE 5. Spectrograph of the echolocation call of a gleaner (*A. Lasionycteris*) and an aerial hawker (*B. Tadarida*). Note the change in frequency (in kilohertz) of each call pulse over time (milliseconds). (*Courtesy William L. Gannon*)

understanding of bat natural history have come from our use of electronic devices, some of which were originally developed for wholly unrelated applications but have been adapted for use in the study of bat behavior. The first of these technologies to be widely used is the ultrasonic recorder, or "bat detector," which can detect, display, and record the echolocation calls of bats. The original detector was developed to study the sounds of insects (Noyes and Pierce 1937) and was used shortly thereafter to examine sounds produced by bats in the laboratory (Pierce and Griffin 1938). The first field studies were conducted in the 1950s and required a station wagon to carry all of the equipment (Griffin 2004). Today, bat detectors are handheld units the size of a large cell phone. Bat detectors

consist of an ultrasonic microphone and converter. The converter contains circuitry that transforms the ultrasonic call (18–80 kHz for Texas bats, which is beyond the range of human hearing) by lowering its frequency so it can be broadcast from a speaker in the range of human hearing (Limpens and McCracken 2004). The calls also can be displayed as spectrographs (fig. 5), which can be analyzed and compared to find differences in the calls that an individual bat makes, such as during feeding, socializing, and navigating, as well as differences among various species of bats (Parsons and Szewczak 2009).

The calls of many species have a characteristic frequency and shape. Because various species of bats have diverse feeding strategies, their spectrographs can be strikingly different. For example, those that capture insects in open space (such as the Brazilian free-tailed bat, *Tadarida brasiliensis*) have medium- to low-frequency calls of long duration and narrow bandwidth (fig. 5). Such calls carry well over long distances. Other species (such as members of the genera *Lasiurus*, *Lasionycteris*, and *Myotis*) that capture insects in more complex environments, such as forests, have calls that are multiharmonic and of short duration and have wide bandwidths (fig. 5). In reality, many species use a combination of call elements for detecting prey as they move through different habitats within a landscape.

Bat detectors are ideal for long-term monitoring of bats and are highly effective for detecting species that are difficult to catch by traditional methods (mist nets, roost searches, etc.) because of their life history characteristics. Although there has been some debate about the accuracy of species identification based solely on echolocation calls, it has been shown that, with experience, the number of misidentifications is negligible, and questionable calls should simply be judged as unidentifiable (Barclay 1999; O'Farrell et al. 1999a, 1999b).

The additional information collected by the use of bat detectors in bat inventories far outweighs any perceived problems with interpreting calls (Gannon et al. 2003). In fact, in one study that covered 57 sites in North America, only 63.5% of the species present were detected by traditional capture techniques, whereas 86.9% of species, on average, present at any individual site were detected by bat detectors (O'Farrell and Gannon 1999). This substantial differ-

ence reinforces the observation that the use of multiple detection techniques provides the most accurate inventory, particularly for bats because some species can be difficult to capture (Dixon et al. in press; Flaquer et al. 2007).

Radio transmitters have been used with great success in tracking large mammals across long distances. As the technology advanced, transmitters eventually became small enough to attach to a bat without adversely affecting its behavior and without long-term impacts on its survival, reproduction, body mass, or ability to forage (Aldridge and Brigham 1988; Hickey 1992; Neubaum et al. 2005). Today, transmitters weigh less than a gram, which is less than 10% of the mass of many adult bat species, and have a very short or no external antenna. Telemetry studies have provided important information about how far bats fly when they forage, when they are active, where they forage, how much area they cover when foraging, and where they roost (Amelon et al. 2009; Brigham et al. 1997; Henry et al. 2002; Kalko et al. 1999; Kurta and Murray 2002; Scales and Wilkins 2007; Weinbeer et al. 2006). However, modern radio transmitters can send information about much more than simply the location of an animal. They also can be equipped to measure surface body temperature, humidity, activity level, and other variables that provide insight into the conditions under which bats sleep, reproduce, and hibernate. Such transmitters have been important in increasing our understanding of the energetic demands of torpor and hibernation in bats (Barclay et al. 1996).

Flash photography, high-speed videography, and light-gathering technology used in night-vision devices have yielded tremendous insights into the feeding behavior and flight mechanics of bats (Altenbach and Dalton 2009). Since bats feed at night, it is difficult to directly observe them feeding and foraging in their natural habitat. The advent of flash photography and night-vision goggles has opened this arena to investigators. These tools have been splendidly combined with synchronized ultrasonic recordings of foraging bats to construct astonishingly complete natural histories for several species of bats (Kalko 1995; Kalko and Schnitzler 1989; Schnitzler et al. 1994). This includes information about their flight patterns, call structure, and prey-capture and handling mechanics at a level of detail not previously possible.

Infrared thermal imaging systems are perhaps the most promising technology for revolutionizing the study of bat ecology. Although its use in field ecology remains limited, reduced costs and increasingly compact sizes are making infrared thermal imaging systems a viable and affordable part of natural history studies of bats (Kunz et al. 2009a). This imaging, like all imaging approaches, is based on the detection of electromagnetic waves and their conversion to electric signals for visual display. Instead of measuring the radiation that is reflected, they measure the infrared radiation that objects produce and emit. The higher the temperature of an object, the brighter the image (fig. 6.; Kastberger and Stachl 2003).

The application potential for this technology is broad, from basic censusing of bat colonies (Ammerman et al. 2009; Betke et al. 2008; Frank et al. 2003; Sabol and

Hudson 1995) to behavioral studies (Kirkwood and Cartwright 1991, 1993) to studies of metabolic energetics of bats in their natural settings and not just in laboratories (Histrov et al. 2008). Metabolic studies of free-ranging animals have always been technically challenging and/or highly invasive. Infrared thermography provides an alternative, affordable way to measure the energy budgets of bats with little to no intervention. It has the potential to provide insight into questions about torpor, arousal from hibernation, and roosting behaviors that allow bats to regulate their temperature. Finally, and perhaps most obviously, infrared thermal imaging allows us to see bats in the dark and observe their behavior in a totally unobtrusive manner never before possible. Other imaging techniques generally disturb bats in some manner (because of the need for some ambient light, often a flash) and are limited in their spatial and temporal scope (Mistry and McCracken 1990). The number of behavioral questions that can be addressed with this technique is limited only by the investigator's creativity (Histrov et al. 2008; Kunz et al. 2008, 2009a).

Other new technologies that have increased our understanding of various aspects of bat natural history include helium balloons, Doppler radar, stable isotopes, force plates, and molecular techniques. Force plates have been used in conjunction with flash photography to help us understand the biomechanics of how bats land head down on a ceiling, as the vast majority of bats do (Riskin et al. 2009). Doppler radar has been used to follow the dispersal of Brazilian free-tailed bats (*Tadarida brasiliensis*) and their insect prey in Texas

FIGURE 6. Thermal infrared image of Brazilian free-tailed bats (*Tadarida brasiliensis*) in flight. Note the relatively dark (cool) wings of the bats versus the brighter (warmer) bodies of the bats punctuated with particularly bright (hot) spots, where blood flow and thus heat loss are higher. (*Courtesy Nickolay I. Hristov, Margrit Betke, and Thomas H. Kunz*)

(Horn and Kunz 2008; Williams et al. 1973; Wolf et al. 1995). Helium balloons were used to detect bats foraging at heights of 900 m (McCracken et al. 1997, 2008). Stable isotopes have been used to examine migration patterns in bats (Britzke et al. 2009), as well as diet (des Marias et al. 1980; Fleming 1995; Herrera M. et al. 1993; Mizutani et al. 1992).

Molecular techniques have provided insight into the taxonomic relationships of bats, their evolutionary history, and interactions among populations of bats (Jones and Teeling 2009). Deoxyribonucleic acid (DNA), found in the cell nucleus, and mitochondrial DNA (mtDNA), found in the cellular mitochondria, sequences have been used to elucidate relationships of bats at many taxonomic levels from order to species. Other genetic markers have been used to address additional questions about bat natural history. These include randomly amplified polymorphic DNA (RAPD), amplified fragment length polymorphism (AFLP), and restriction fragment length polymorphism (RFLP), which have been used to examine gene flow and genetic variability within and among bat species.

The tools of genetic analysis have allowed greater understanding of bat evolution, migration, gene flow, mating systems, diet, and disease. Compared to techniques used 20 years ago, only small amounts of DNA are necessary to conduct these studies. For instance, DNA has even been isolated from bat droppings (Puechmaille et al. 2007) and used to determine which species occupy particular redwood trees in California (Zielinski et al. 2007). As DNA sequence databases grow, it becomes possible to confirm species identification genetically and to fine-tune our understanding of species and subspecies boundaries. For example, genetic analyses of subspecies of Brazilian free-tailed bats (*Tadarida brasiliensis*) show that there is much gene flow even between migratory and nonmigratory populations, which brings into question the validity of these subspecies (Russell and McCracken 2006). Although bats typically are expected to exhibit high gene flow because of their highly mobile lifestyle, made possible by flight, Weyandt and Van Den Bussche (2007) have shown 3 geographically separate lineages of pallid bats (*Antrozous pallidus*) across the desert southwest, suggesting that interbreeding among populations is limited in this region. Also using genetic data, Weyandt et al. (2005) have found that females in some populations of Townsend's big-eared bat (*Corynorhinus townsendii*) in Oklahoma caves rarely disperse between caves that are separated by less than 20 km. These examples indicate that some species of bats have relatively sedentary lifestyles and exhibit high fidelity to particular roosting/foraging sites, which results in relatively small home ranges despite their ability to fly.

Several examples of discoveries pertaining to Texas bats have resulted from genetic analyses. First, taxonomic revision of the genus *Pipistrellus* (now *Perimyotis* and *Parastrellus* in North America) was largely supported by genetic data (Hoofer and Van Den Bussche 2003; Hoofer et al. 2006). Second, analysis of microsatellite loci provided evidence that both the big brown bat (*Eptesicus fuscus*) and the eastern red bat (*Lasiurus borealis*) have pups in a single litter that are sired by multiple fathers (Spradling et al. 2003; Vonhof et al. 2006). Third, the discovery of the western yellow bat (*Lasiurus xanthinus*) in Texas was confirmed using DNA sequence data (Higginbotham et al. 1999). Finally, levels of genetic variability in the endangered Mexican long-nosed bat (*Leptonycteris nivalis*) have been estimated by Brown (2008) and were not as alarmingly low as might be expected for a critically endangered species. The molecular tools that are being developed to examine whole genomes will certainly become integral to understanding conservation threats, predicting transmission of emerging diseases, and making management decisions that involve the bat populations in Texas.

Public Health

Although bats are highly secretive creatures, they occasionally come into contact with people. Usually this happens when a few bats take up residence in a building or a single one is found on the ground. Their unusual appearance and our fear of rabies inevitably lead to consternation in such events, but any danger bats present to people is greatly overstated.

Bats themselves can be afflicted with a variety of diseases and parasites, most of which are contagious only to other bats (see Parasites section). Worldwide, bats carry only a handful of diseases that are known to cause illness in humans (Messenger et al. 2003). Only 2 diseases, rabies and histoplasmosis, are known to have been transmitted by bats to people in Texas, and fears of acquiring even these from bats are often grossly exaggerated.

Rabies. Human cases of rabies caused by a bat in the United States were first reported in the 1950s (Brass 1994; Martin 1959; Scatterday and Galton 1954; Sulkin and Greve 1954). Since that time, 203 human cases of rabies have been reported in the United States (from 1950 to 2009), 22 (10.8%) of which were in Texas. Of the total number of cases, 48 (23.6%) were caused by bats (based on either the report of an encounter with a bat or by molecular sequencing of the virus), and 7 (31.8%) of the Texas cases could be linked to a bat. Bats as the causative agent of rabies has increased substantially since 1950, with 32 of 35 (91.4%) human rabies cases with an indigenous exposure caused by bats since 1990, but only 13 of 131 (9.9%) cases prior to 1990 (Mondul et al. 2003; Noah et al. 1998). This is due to a successful vaccination program in domestic dogs and cats in the United States.

Of the 33 species of bats that occur in Texas, 27 have been documented with rabies somewhere in their range, although not necessarily in Texas (Constantine 1979; Mondul et al. 2003). Of these, four are major reservoirs of rabies in the United States, including the Brazilian free-tailed bat (*Tadarida brasiliensis*), hoary bat (*Lasiurus cinereus*), eastern red bat (*Lasiurus borealis*), and big brown bat (*Eptesicus*

fuscus). Documentation of rabies in bats at the ordinal level (i.e., all bats without regard to species) is summarized annually for the United States by the Centers for Disease Control (CDC; Blanton et al. 2006, 2007, 2008, 2009; Krebs et al. 2000, 2001, 2002, 2003, 2004, 2005) and for Texas by the DSHS (Anonymous 1990, 1991, 1992, 1993, 1994). Although the number and proportion of bats submitted for testing has increased dramatically in the past 30 years, from less than 10% to nearly 25% of all animals submitted, the percentage of rabid bats has remained remarkably stable regionally (about 9–14% in Texas since 1990; Brass 1994; Yancey et al. 1997).

Documentation of rabies in bats at the species level has been less continuous (Mondul et al. 2003; Rohde et al. 2004; Schmidly 1991). From February 1984 to February 1987, mammalogists at Texas A&M University identified bats reported to the DSHS (Schmidly 1991). An additional survey of bats submitted to the DSHS for rabies testing from 1996 to 2000 also was conducted (Rohde et al. 2004). These data have contributed to knowledge of the distribution of many species and have furnished information on the prevalence of bat rabies in Texas. Of the 21 species examined for these surveys, only 11 included individuals that tested positive for the rabies virus. Infection rates were similar among species in both surveys (Table 7).

Nationwide, trends in rabies infection in bats have been summarized from 1993 to 2000 (Table 7; Mondul et al. 2003). Infection rates are similar to those found in Texas, with some notable exceptions. Several species have a much higher rate of infection nationwide than they do in

TABLE 7. *Summary of rabies prevalence in bats in Texas and the United States. Numbers are total number tested/number positive/percentage positive; blank cell indicates that no bats of that species were reported.*

Species	Texas, 1984–1987[1]	Texas, 1996–2000[2]	United States, 1993–2000[3]
M. megalophylla		1/0/0	
C. mexicana			14/0/0
L. nivalis			
D. ecaudata			
M. austroriparius	4/0/0	1/0/0	
M. californicus		1/0/0	338/12/3.6
M. ciliolabrum	1/0/0	1/0/0	
M. occultus			
M. septentrionalis			
M. thysanodes			4/0/0
M. velifer	82/2/2.4	172/4/2.3	2/0/0
M. volans			23/3/13
M. yumanensis			241/4/1.7
L. blossevillii			
L. borealis	626/46/7.3	714/48/6.7	520/47/9
L. cinereus	40/10/25	57/15/26.3	254/97/38.2
L. ega	1/0/0	80/2/2.5	32/7/21.9
L. intermedius	126/11/8.7	153/14/9.2	3/3/100
L. seminolus	186/17/9.1	14/2/14.3	1/0/0
L. xanthinus			
L. noctivagans	5/0/0	5/0/0	566/73/12.9
P. hesperus	1/0/0		193/41/21.2
P. subflavus	10/1/10	40/0/0	117/20/17.1
E. fuscus	19/1/5.3	14/1/7.1	20911/1216/5.8
N. humeralis	221/6/2.7	410/3/0.7	62/6/9.7
E. maculatum			1/1/100
C. rafinesquii	2/1/50		
C. townsendii			29/3/10.3
A. pallidus	2/0/0	3/0/0	100/21/21
T. brasiliensis	430/105/24.4	2062/338/16.4	673/214/31.8
N. femorosaccus			53/7/13.2
N. macrotis	2/0/0	5/0/0	14/3/21.4
E. perotis			5/0/0
Total	1758/200/11.4	3733/427/11.4	24156/1778/7.4

[1]Adapted from Schmidly (1991).
[2]Adapted from Rohde et al. (2004).
[3]Adapted from Krebs et al. (2003).

Texas. The difference seen in silver-haired bats (*Lasionycteris noctivagans*), American parastrelles (*Parastrellus hesperus*), American perimyotises (*Perimyotis subflavus*), and big free-tailed bats (*Nyctinomops macrotis*) could be attributed to sample size. However, the increased proportion of bats infected with rabies documented in hoary bats (*Lasiurus cinereus*), southern yellow bats (*Lasiurus ega*), evening bats (*Nycticeius humeralis*), and Brazilian free-tailed bats (*Tadarida brasiliensis*) probably reflects regional differences in infection rates in these species. Although data collected at a national level cannot be used to extrapolate rabies incidence in bats in Texas, they do suggest that additional Texas species that have not been reported to the DSHS could be infected with rabies.

It is important to note that in the Schmidly (1991) study, bats were identified to species by trained mammalogists, and voucher specimens were prepared and deposited in a natural history museum (the Texas Cooperative Wildlife Collection [TCWC] in this case). This is not the case for the Rohde et al. (2004) and the Mondul et al. (2003) studies even though the importance of involving a mammalogist in such research was recognized in the earliest surveys of rabies in bats in the United States, including those conducted in Florida (Scatterday and Galton 1954) and Texas (Burns et al. 1956). Even for experienced bat biologists, some species of bats are difficult to identify to species, and without voucher material, species identifications can never be verified (Brass 1994). The bats reported to the CDC and the DSHS represent important information about the distribution, seasonal occur-

rence, populations, and general natural history of bats, particularly in urban areas, in Texas and the United States. The DSHS worked in collaboration with mammalogists at Texas Tech University from 1993 to 1996 to ensure correct identification and preservation of the bats reported to the agency during this period. Unfortunately, no such collaboration existed from 1988 to 1992 or since 1996. We have lost valuable data about bats and understanding rabies infections in them during that time. Fortunately, mammalogists at Texas Tech University and Angelo State University are once again involved in the identification of bats reported to the DSHS as of 2009.

For reasons of sampling bias, it would be erroneous to conclude that 11.4% of the bat population in Texas or that 7.4% in the United States is infected with rabies. The DSHS and CDC test only bats that are submitted because they are rabies suspect; typically, these are bats found on the ground or in unusual places. Thus, the sample tested is not representative of the bat population as a whole. It has been found that less than half of 1% of free-flying, wild populations of bats are infected with rabies (Baer and Smith 1991; Brass 1994; Sheeler-Gordon and Smith 2001; Steece and Altenbach 1989; Yancey et al. 1997), a frequency no higher than that seen in many other animals (Tuttle 1988). Furthermore, it appears that rabid bats seldom transmit rabies to any animal except other bats. Carnivores, such as skunks, raccoons, and dogs, rarely are infected with the bat type of rabies virus. The bat rabies cycle seems to be entirely independent of carnivores (Brass 1994).

Bats are able to survive rabies infections

in controlled laboratory settings (Baer and Bales 1967), but whether this is true in natural settings is unknown (Messenger et al. 2003). A population of Brazilian free-tailed bats (*Tadarida brasiliensis*) at Lava Cave, New Mexico (Steece and Altenbach 1989), as well as populations from Texas caves (Burns et al. 1956; Constantine 1988), were found to have high levels (up to 80% of adults) of rabies-neutralizing antibodies, which suggests survival of bats after exposure to the rabies virus. This is further supported by the lack of rabies outbreak activity in dense populations of bats. However, it is obvious that some bats infected with rabies soon become paralyzed and die. These bats may represent immunodeficient or otherwise compromised individuals (Constantine 1988). The question of an acquired immunity to rabies in bat colonies is important to understanding how this disease impacts bat population dynamics and needs further examination (Messenger et al. 2003).

Logically, people are likely to encounter only injured or ill bats. For this reason, bats should not be handled unless one is wearing leather work gloves. Children should be warned never to pick bats up. If reasonable caution is taken, the danger of being bitten by a bat is very slim; consequently, the probability of contracting rabies is incalculably small. Although rabies is a serious zoonosis, bats should not be feared as vectors of this disease (Christiansen 2003). However, one should not take unnecessary chances with a fatal disease, and, in the event of any animal bite, medical advice should be sought immediately (Morimoto et al. 1996).

Histoplasmosis. This is a fungal disease that affects the lungs and presents symptoms similar to those of tuberculosis. The fungus and its spores are associated with bird and bat droppings. Human infection occurs through inhalation of airborne spores, and the severity of infection is normally proportional to the quantity of dust-laden spores one has inhaled. Symptoms normally include a cough and resemble those of influenza, although many infections in humans are asymptomatic and cause no illness. Bird droppings, frequently those of poultry or pigeons, are the primary source of infection for people. However, the fungus may occasionally be found in droppings associated with hot, dry attics where bats roost. One of us (David J. Schmidly) is personally familiar with a situation in which several biologists became infected with this disease after entering a cave that housed a large nursery colony of Brazilian free-tailed bats (*Tadarida brasiliensis*).

To avoid problems, one should be careful not to stir up and breathe dust in areas where bird or bat droppings have accumulated. Dry excrement should be dampened before removal to reduce the hazard of dust inhalation. When removing droppings or walking over a dry cave floor laden with droppings, one should use a properly fitted respirator that is capable of filtering particles as small as 2 microns in diameter; this precaution will greatly reduce the probability of exposure (Constantine 1988).

Conservation

Drastic reductions in bat populations have been reported in recent years not only in the United States but worldwide as well. Several species of bats already are extinct, and others are near extinction. As a group, bats are exceptionally vulnerable to extinction because they typically rear only 1 young per year and hence are slow to recover from major population declines. In addition, many species form large aggregations, which are vulnerable to mass destruction.

The causes of declining bat populations are not always evident, and many possible factors may be responsible. Global climate change is already indirectly impacting bat populations around the world and in Texas. In addition, major adverse effects, including the destruction of habitat, outright extermination, vandalism, excessive disturbance at roosts and maternity colonies, pesticide poisoning, and use of other chemical toxins, have been brought about by humans. Natural factors also can play a role in declining bat populations, the most serious of which is white-nose syndrome, a fungal infection that is devastating bat populations in the eastern United States and is rapidly moving west.

Climate change is generally defined as any long-term perturbation in the historical pattern of regional weather, particularly temperature and precipitation, over a substantial period of time. Today we usually think of climate change in terms of global warming, and in fact, global temperatures have increased an average of about 1°C during the past 150 years (IPCC 2007). Although many factors can contribute to global warming, including solar and orbital variation, volcanism, and various anthropogenic activities, the scientific consensus is that "Most of the observed increase in globally averaged temperatures since the mid-twentieth century is very likely due to the observed increase in anthropogenic greenhouse gas concentrations" (IPCC 2007, 10).

Regardless of the cause, global climate change has the potential to impact bat distributions worldwide by influencing local temperature and precipitation patterns. Although global warming could directly influence bat distributions by creating a local climate that is uninhabitable to bats, it could have a larger, indirect impact by changing the habitat distribution within landscapes. The type of habitat present at any given geographic location is dependent on a large number of variables, of which temperature and precipitation are 2 of the most important. This is because plant species that comprise a habitat type or an ecological region exist within fairly specific climatic parameters. That is why the plant communities of the southwestern deserts are different from those found in temperate deciduous forests. If temperature or precipitation changes significantly for a substantial period of time, the local flora will change to reflect the new climate. Whereas most bats in Texas are not directly dependent on plants for food or shelter, they are dependent on other animals (mostly insects) that are sensitive to climate and habitat (vegetational) change. Moreover, many roosts used by bats are immobile (caves, rock crevices, etc.), and bats are highly sensitive to changes in environmental variables around a roost site. If the climate were to change sufficiently, roosts could be abandoned in areas that presently harbor highly diverse bat communities that are dependent on such formations for shelter and hibernacula.

Climate-change models predict that climate change will produce a warmer climate in Texas that will be either wetter or drier than it is currently, depending on the predictive model (Cameron and Scheel 1993). A more recent study suggests the drier climate will prevail (Seager et al. 2007). Scheel et al. (1996) created a GIS model to predict the ecological effects of climate change in Texas. They used it to model distributions of vegetation and bat species throughout Texas in both wetter and drier climates. Under either scenario, no species currently occurring in the state is predicted to go extinct. In fact, bat diversity in Texas is likely to increase, mostly due to an influx of tropical species from the south. We are already seeing this in the expanded distribution of the southern yellow bat (*Lasiurus ega*), as well as the newly documented occurrence of the western yellow bat (*Lasiurus xanthinus*) in the Big Bend and El Paso areas. A new record of the Mexican long-tongued bat (*Choeronycteris mexicana*) in El Paso County (Balin 2009) suggests that individuals of this southern species are moving northward out of Mexico.

Climate-change scenarios predict that distributions of many cave- or cavity-dwelling species will contract and that those of tree-dwelling species will expand. This is because forested habitats are expected to expand spatially, particularly in the south. Several examples of tree-dwelling species that have increased their range from eastern Texas to the west have been documented, including the Seminole bat (*Lasiurus seminolus*), evening bat (*Nycticeius humeralis*), and American perimyotis (*Perimyotis subflavus*; Schmidly 2002). However, as suitable vegetation surrounding caves and in mountainous regions

is replaced by unsuitable vegetation and as climate change impacts the microclimates in caves and cavities, the number of species and individuals that these areas can continue to support could actually decrease. Unfortunately, not much is known about microscale changes in geographic distributions or species abundance of bats.

In summary, the evidence at hand strongly suggests that bat distributions and communities are already responding to climate change in Texas. Predicted climate change presently does not pose a grave threat to bat-species richness in Texas, although the composition of communities could change. It is important to protect large, complex caves that provide a variety of microclimates and to set aside areas of suitable habitat for mature forests to develop. This will enable us to continue to enjoy the most diverse bat fauna in the United States.

The use of pesticides has a direct and detrimental impact on many bat species. Bats are highly sensitive to some insecticides, especially the chlorinated hydrocarbons. Bats that feed on insects treated with these chemicals slowly accumulate pesticide residues in their fat during late summer and fall, which eventually results in their death as this fat is broken down and used during hibernation and migration (Clark 1981). A recent study of 2 populations of little brown bats (*Myotis lucifugus*), 1 from New York and 1 from Kentucky, found that these bats had extremely high concentrations of several pesticide residues (Kannan et al. 2010). To date, only outright mortality on a local population level has been identified as a threat to bats from pesticides, but subtle—and equally devastating—effects are possible on aspects such as reproduction, immune system function, acoustic behavior, and hibernation metabolism.

Owing to its geological complexity, Texas has many caves, and they are especially important for the conservation of bats because they support large populations throughout the year. Mining, human disturbance, and pollution have been and continue to be the biggest threats to bats that live in Texas caves. William Elliot (1993) has published a detailed account of caves and bat conservation.

Human disturbance of hibernating and maternity colonies is apparently a major factor in the decline of many species. Increased numbers of visitors to caves appear to have a strong deleterious effect since many bats are sensitive to intrusion and will leave a favored site after repeated disturbance. Frequently, well-meaning individuals, such as spelunkers (cave explorers) and biologists, are guilty of these disturbances. Unnecessary arousal of hibernating bats due to the presence of people may cause bats to arouse and use up the precious fat they need to survive the winter. Depleted energy supplies may then cause many bats to starve before spring, when their insect foods are again available.

Disturbance to summer maternity colonies also can be extremely harmful. Maternity colonies are very intolerant of disturbance, especially when flightless newborn young are present. Baby bats may be dropped to their deaths or abandoned by panicked mothers if disturbance occurs during this period.

Recently, a major disease known as white-nose syndrome, or WNS, has been detected, and it is causing major die-offs of hibernating bats in the northeastern United States (Kunz et al. 2009b). White-nose syndrome was first detected on February 16, 2006, in Howe's Cave, located 52 km west of Albany, New York (Blehert et al. 2009). Since that time, it has spread 1,600 km (1,000 miles) from its source to at least 115 hibernacula (Frick et al. 2010) in 17 additional states and 4 Canadian provinces (Castle and Cryan 2010; Lindner et al. 2010; Wibbelt et al. 2010). The fungus that causes WNS has been found on a bat (*Myotis velifer*) as far west as western Oklahoma, but no WNS has been detected in the bat population there as of 2010 (U.S. Fish and Wildlife Service 2010). The farthest west a bat confirmed with WNS has been reported is St. Louis, Missouri (Castle and Cryan 2010). Nonetheless, the rapid spread of the disease and the fungus that causes it could portend the spread of WNS into Texas in the near future (Holtcamp 2010).

White-nose syndrome is given its name due to its distinctive appearance as a white powder on the muzzle and wings of bats. It is caused by a previously undescribed fungus, *Geomyces destructans* (Gargas et al. 2009). Members of the genus *Geomyces* are terrestrial saprophytes that grow in cold temperatures and are known to infect the skin of animals in cold climates (Blehert et al. 2009). The source of *G. destructans* in the United States is unknown, but evidence is mounting that it could have been brought over from Europe by humans (Frick et al. 2010; Wibbelt et al. 2010). Anecdotal evidence of fungal growth similar in appearance to WNS on European bats is available from the early 1980s in Germany (Feldham 1984), and *G. destructans* has been isolated from infected bats in Europe (Stokstad 2010). However, *G. destructans* does not cause high levels of mortality in bats in Europe as it does in North America. It has been suggested that this is because European bats coevolved with the fungus, thereby providing them with behavioral or immunological resistance to WNS (Wibbelt et al. 2010).

In North America, WNS is highly lethal, killing 75–99% of the bats at infected hibernacula (Buchen 2010; Castle and Cryan 2010; Frick et al. 2010; Zimmerman 2009). Thus far, 6 species of bats have been documented with WNS, including the little brown bat (*Myotis lucifugus*), northern myotis (*Myotis septentrionalis*), eastern small-footed myotis (*Myotis leibii*), Indiana bat (*Myotis sodalis*), big brown bat (*Eptesicus fuscus*), and American perimyotis (*Perimyotis subflavus*; Castle and Cryan 2010; U.S. Fish and Wildlife Service 2010). An additional 3 species, including the gray myotis (*Myotis grisescens*), southeastern myotis (*Myotis austroriparius*), and cave myotis (*Myotis velifer*), have tested positive for the presence of *G. destructans*, but no mortality from WNS has yet been documented in them (U.S. Fish and Wildlife Service 2010). All but 3 of these species occur in Texas. The Brazilian free-tailed bat (*Tadarida brasiliensis*) is not thought to be as susceptible to the fungus because it does not overwinter in caves in North America.

It is unclear whether the fungus is the sole cause of death in bats or whether it is an opportunistic pathogen that is taking advantage of animals with immune systems weakened by other factors (Meteyer et al. 2009). It is generally thought that the presence of the fungus causes the bats to arouse from hibernation more frequently than normal, either to groom the fungus away or to mount an immunological response against the fungus, thus prematurely depleting the fat reserves, resulting in starvation (Reichard and Kunz 2009; U.S. Fish and Wildlife Service 2010; Zimmerman 2009). Bats with critically low fat reserves may arouse during midwinter to look for food during a time when no insects are present (Veilleux 2008). The fungus also causes long-term damage to the wing membrane, resulting in a lower body-mass index, attributed to reduced foraging success, and a lower reproduction rate (Reichard and Kunz 2009).

Since its discovery in 2006, WNS has killed more than 1 million bats (Frick et al. 2010; Turner and Reeder 2009). At the current mortality rate, regional extinction of the little brown bat (*Myotis lucifugus*), one of the commonest bats in North America, is estimated to occur within 16 years (Frick et al. 2010). A decline of 78% in the summer activity of little brown bats (*Myotis lucifugus*) has already been documented at sites along the Hudson River in New York (Dzal et al. 2010). This may result in unpredictable changes in ecosystem structure and function. For example, the reduced predation pressure on insects could result in increased numbers of insect pests, thereby increasing the amount of damage done to forests and crops (U.S. Fish and Wildlife Service 2010).

The disease is most likely transmitted from bat to bat (Zimmerman 2009), although there may be some evidence that humans transport the fungus from infected to uninfected caves on clothing and equipment (Crawley 2009; Turner and Reeder

2009). The presence of *G. destructans* in soil samples from caves within the known range of WNS has been confirmed, thus establishing the environment as a potential reservoir for the fungus (Lindner et al. 2010).

Scientists are working rapidly to understand how the fungus causes bat deaths, how it is spread, and how the spread can be stopped or mitigated. Several stop-gap measures have been proposed: These include the placement of small populations of endangered species potentially threatened by the disease in captivity to protect them (Zimmerman 2009); heating areas of hibernacula to provide warm refuges for hibernating bats (Boyles and Willis 2010); and closure of all caves and mines on U.S. Forest Service land to human intrusion (Leggett 2010). The efficacy of any of these methods is unknown, and the future remains bleak in the wake of this emergent epizootic syndrome.

Resources of renewable energy such as wind offer great promise of energy independence but could, if not properly implemented, have negative consequences for bats (Kunz and Arnett 2009). Wind power offers a renewable, "green" source of alternative energy and, as such, is one of the fastest-growing sectors of the energy industry. Since demand for electricity is only going to increase in the future, wind-generated power is an important alternative to traditional energy sources (i.e., fossil fuels). Unexpectedly, the large turbines used to harvest wind have a detrimental effect on volant wildlife and cause bird and bat fatalities.

Although there is a general paucity of data on the number of bats killed at wind facilities, Arnett et al. (2008) have summarized fatalities at 21 sites throughout North America. Unfortunately, no data exist for Texas even though the state has the largest installed capacity of wind energy in the continental United States. Of the examined sites, the closest to Texas was in Woodard, Oklahoma (Piorkowski 2006), which had a relatively low mean mortality rate per turbine when compared to other sites. Most of the bats killed at this site were Brazilian free-tailed bats (*Tadarida brasiliensis*), which are uncommon in or absent from the locations of the other 20 sites summarized by Arnett et al. (2008); they are common and abundant in Texas. Robert J. Baker, a well-known mammalogist in Texas with decades of experience with bats, told us that he has walked under the turbines of a wind facility near Lubbock, Texas, and in a year neither he nor his dogs ever found a dead bat on the ground. Research about the impact of wind farms on bat populations in Texas needs to be a priority for energy companies and wildlife-management programs.

Based on information currently available, there are several things that energy-producing companies can do to minimize the wind facilities' impact on bat populations. Location is an extremely important consideration when building a wind facility. With careful planning and by working with local biologists, energy companies can locate such facilities where energy production can be maximized and the impact on volant wildlife can be kept to a minimum. For example, it would not be a good idea to build a wind facility just outside of Bracken Cave! Turbines can be turned off or "feathered" (i.e., pitching the turbine blades parallel to the wind) during the times of the year when bats are migrating or times of the night when bats are most active. It is estimated that this could decrease mortality by 85% at some sites, with a minimal loss of energy production (Kunz and Arnett 2009). These simple steps will enable us to wisely produce energy in a renewable fashion and continue to enjoy the diverse bat fauna that Texas offers.

In addition to the unintentional killing of bats, large numbers are deliberately killed, and thousands of roosts are destroyed each year. Public health is the most common reason given for this destruction even though bats rarely transmit diseases to humans. The removal of bats from the wild for scientific and educational purposes, as well as for food (in some parts of the world), also has been a contributing factor in the decline of certain populations.

In recognition of the seriousness of bat population reductions, many countries now offer bats some legal protection. In the United States 6 species are protected under the federal Endangered Species Act, and for 1 of these species, the Mexican long-nosed bat (*Leptonycteris nivalis*), the only known colonies in the United States occur in Big Bend National Park in Brewster County, Texas, and in Hidalgo County, New Mexico (Hoyt et al. 1994, Paul Cryan, USGS, pers. comm.). In addition to the federal listing, the Texas Parks and Wildlife Department's list of protected nongame wildlife includes 3 species of bats (southern yellow bat, *Lasiurus ega*; spotted bat, *Euderma maculatum*; and Rafinesque's big-eared bat, *Corynorhinus rafinesquii*).

Some of the earliest efforts to give bats legal protection in the United States started in Texas. In 1917, the 35th Texas Legislature passed a general law (H.B. no. 40) that made it a misdemeanor to kill or injure bats because of their perceived value in controlling malarial mosquitoes. The entire contents of that law, as adopted, were as follows (*General Laws of the State of Texas*, 35th Legislature, 1917, p. 124):

PROTECTION OF BATS

H.B. No. 40   Chapter 65

An Act making it a misdemeanor to kill or in any manner injure the winged quadruped known as the common bat; repealing all laws in conflict therewith, and declaring an emergency.

Be it enacted by the Legislature of the State of Texas:

Section 1. Article 887a. If any person shall willfully kill or in any manner injure any winged quadruped known as the common bat, he shall be deemed guilty of a misdemeanor and upon conviction shall be fined a sum of not less than five ($5.00) dollars nor more than fifteen ($15.00) dollars.

Sec. 2. All laws and parts of laws in conflict with the above provision shall be and the same are hereby repealed.

Sec. 3. The fact that there is now no law in this State protecting bats creates an emergency and an imperative public necessity that the constitutional rule requiring bills to be read on three several days be suspended, and the same is hereby suspended, and that this Act take effect from and after its passage, and it is so enacted.

Approved March 9, 1917.

Takes effect 90 days after adjournment.

Effective conservation of bats will require the protection of hibernacula and maternity roosts and the prevention of the general degradation of summer foraging habitats. A number of conservation efforts have been designed specifically to reduce disturbances to roosting bats and to increase available roosting sites. Special gates have been installed in front of cave entrances to allow easy passage for bats while excluding people. Interpretive signs have been placed at caves to warn against human disturbance to bat colonies. Bat houses and towers are being erected in forests and backyards to attract insect-eating bats, and caves located on private lands have been purchased to protect important bat colonies.

Efforts to construct artificial structures for attracting bats also were pioneered in Texas. In the 1910s and 1920s, Dr. Charles A. R. Campbell of San Antonio promoted the construction of numerous "bat towers"—structures that looked like church belfries on stilts (Elliot 1993). As the chief public health physician for the city of San Antonio, Campbell was charged with maintaining quarantine camps. He theorized that bats would control mosquitoes, which would reduce malaria, still an important disease in Texas at that time. He also promoted guano as fertilizer; in a single year (1917) he recovered 2,996 pounds of guano from a tower at Mitchell's Lake, a sewage lake near San Antonio. Furthermore, it was Campbell who influenced the State Board of Health and the state legislature to pass the 1917 legislation that protected bats.

In 1914 Dr. Campbell received a patent for the design of his bat towers (U.S. Patent no. 1,083,318, issued January 6, 1914; *Patent Office Gazette*, vol. 198, 1914, pp. 20–21), and in 1919 the state legislature passed a resolution nominating him for the Nobel Prize. In 1925 he published a book, *Bats, Mosquitoes, and Dollars,* in which he claimed that the bat guano from his towers contained mosquito fragments, and he had many people attest in writing to a dramatic reduction in mosquitoes at Mitchell's Lake after the installation of a bat tower in 1911. He also urged communities to construct bat towers (estimated to cost from $2,500 to $3,500) for controlling mosquitoes and harvesting guano.

Professional mammalogists (Goldman 1926) were skeptical of the accuracy of Campbell's statements and his conclusions. In 1919 Tracy Storer visited the most successful of the bat roosts near Mitchell Lake, 16 km southwest of San Antonio. Storer (1926) published a paper describing the tower, which had become inhabited by a colony of Brazilian free-tailed bats (*Tadarida brasiliensis*). The tower was tall and pyramid shaped, measuring about 12 feet square at the base, about 6 feet square at the top, and about 20 feet in height (fig. 7). Storer obtained samples of guano from the tower and had them examined by several experts, who found no evidence of any mosquito fragments. Some of the bat towers are still in existence today, and apparently they continue to attract bats (Merlin Tuttle, pers. comm.). However, Campbell's claims about their utility in malaria control and profits generated by the harvest of guano have never been substantiated.

Probably the most useful outcome of Campbell's work was the development of a

FIGURE 7. Bat roost at Mitchell Lake, circa 1915–1920, part of a malaria eradication program directed by Dr. Charles A. R. Campbell, who stands at the roost's foundation; a flatbed truck loaded with bags of guano is in the foreground. (*Photo from* San Antonio Express-News *Collection, UTSA's Institute of Texan Cultures, #069-8391, Courtesy Hearst Corporation*)

generally more benign attitude about bats among the public. That changed, however, in the 1950s as a result of a series of rabies scares around the state (Elliot 1993). Consequently, the legislature unanimously rescinded the bat protection law (1957), and the public attitude toward bats hardened. Almost all of Campbell's bat towers were eventually torn down.

Recently, a Texas landowner and conservationist by the name of J. David Bamberger, who owns a 5,500-acre ranch in the Hill Country of Central Texas near Johnson City, constructed an artificial bat cave in an attempt to provide habitat for bats. His "chiroptorium" was built in

1997–1998 at a cost of about $170,000. The unique structure consists of 2 domes and a connecting arch that forms a giant, toadstool-like enclosure between them. The cave covers 3,000 square feet, and its walls offer roughly 8,000 square feet of roosting space—enough, he figures, for at least a million bats. The cave walls were formed from 20 tons of steel bars bent to shape and covered with gunite, the concrete used for swimming pools. In 1998 it looked like a modernist sculpture nestled into a scenic canyon. Now, covered with soil and native grasses except for the sculpted-concrete entrance, it seems like part of the natural landscape of the canyon (fig. 8).

FIGURE 8. Brazilian free-tailed bats (*Tadarida brasiliensis*) emerge on August 3, 2006, from an artificial cave, or "chiroptorium," built by J. David Bamberger near Johnson City. (*Photo by Jennifer J. Miller*)

For the first few years after construction, only very few bats frequented the artificial cave. Most seemed to be travelers that stayed for a few days or sometimes a few months. Then, in the summer of 2002, several hundred bats came to the cave, and on August 5, 2003, Mr. Bamberger witnessed a huge emergence at the cave entrance lasting many minutes. In June 2004, Dr. Gary McCracken, a noted bat biologist from the University of Tennessee, visited the chiroptorium. After an evening emergence of Brazilian free-tailed bats (*Tadarida brasiliensis*), the scientist and several staff went into the interior of the cave and discovered that this colony had become a maternity colony—harboring females and their young. Moreover, Dr. McCracken estimated that about a 1-meter-square area on the ceiling of the back dome was covered in bat pups, which would be about 5,000 babies. In May 2008, bat biologists from Boston University, using infrared video photography developed by Dr. Tom Kunz, determined that the cave contained about 116,000 bats. So, it can now be said with certainty that Bamberger's "chiroptorium" is officially a success, and it is predicted that more Brazilian free-tailed bats (*Tadarida brasiliensis*) will occupy the cave in the future. The cave myotis (*Myotis velifer*) has also been documented in the cave but in very small numbers, and it has not yet been documented to breed there.

Bats also use bridges and culverts as day and night roosts. As vulnerable natural roosting sites are disturbed or destroyed, these structures act as havens of last resort. Keeley and Tuttle (1999) conducted an extensive survey of bridges and culverts throughout North America. Of 1,060 such structures surveyed in Texas, 62 (5.8%) were used as day roosts (where bats sleep and rear their young), and 156 (14.7%) as night roosts (where bats eat and digest their meals). To attract bats, it is easy and cost effective to build new bridges and culverts or to retrofit existing ones. As mentioned earlier, one of the most famous examples of such a bridge is the Congress Avenue Bridge in Austin, Texas, where tourists come to watch the mass exodus of 1.5 million Brazilian free-tailed bats (*Tadarida brasiliensis*) each night. These bats consume an estimated 10–15 tons of insects every night.

Although conservation efforts such as those described earlier do provide some protection, they will never be totally effective unless people's attitudes toward bats also change. Negative attitudes must be replaced by an understanding of the ecological, economic, and scientific value of bats. This goal can be achieved only through education.

Concern about the worldwide decline of bats has led to the formation of Bat Conservation International (BCI), an organization whose purposes are to preserve bat populations around the world and to improve public attitudes toward them. Having originally started BCI in Milwaukee, Wisconsin, Dr. Merlin Tuttle, its founder and science director, moved the organization to Austin, Texas, in 1986. According to Dr.

Tuttle, the relocation of BCI was made, in part, because Texas harbors more species of bats than any other state and because the world's largest remaining bat cave (Bracken Cave) is located here. Information about this organization, including membership and subscription to its newsletter, may be obtained by writing to Bat Conservation International, P.O. Box 162603, Austin, Texas 78716, or by visiting www.batcon.org.

Several state and federal agencies and organizations are now actively involved in the bat conservation movement. These include the Texas Parks and Wildlife Department, Texas Natural Heritage Program, U.S. Fish and Wildlife Service, National Park Service, and Texas Nature Conservancy. Scientific societies concerned with the conservation of Texas bats include the Texas Society of Mammalogists, Southwestern Association of Naturalists, American Society of Mammalogists, and the Texas Chapter of the Wildlife Society. All of these groups are working to reverse the bad image of bats and to promote these animals as an important part of our wildlife heritage.

Finally, people can help preserve our bat resources by following a few common-sense guidelines. Maternity colonies and hibernating bats should be avoided because even a slight disturbance can be harmful. If exploring a cave, try to leave everything as you found it and avoid disturbing or harming the bats. Never shoot, poison, or otherwise harm bats. Bats are extremely beneficial insect predators, and nuisance bats generally can be encouraged to move elsewhere; this is highly preferable to killing them. More and more people, when they discover that bats are beneficial and not dangerous, are going out of their way to attract them by constructing backyard bat houses to take advantage of bats' insect-eating food habits. *The Bat House Builder's Handbook* is available for free from Bat Conservation International if you join.

It is true that, given the opportunity, large numbers of bats may take up residence in attics or other parts of buildings and become a nuisance. In most cases, eviction and exclusion are the only safe, permanent remedies. When control is necessary, those measures that are the most effective and the least harmful to the bats should be selected. The most effective way to keep bats out of a building is to locate and seal the entrance(s) through which they come and go. This should be done during the time of year when the bats are absent (usually October through March) or at night after the bats are out foraging. If blocking is not possible, lighting an attic can deter bats, but the light must be left on all the time. Poisons should rarely be used to solve bat problems in buildings, as they are usually unnecessary and may create far worse problems. A U.S. Fish and Wildlife Service publication titled "House Bat Management" is available and highly recommended as an aid in solving house bat problems (Greenhall 1982).

Given the diversity and large population size of bats in Texas, and especially in consideration of their overall ecological and economic importance in the state, it is crucial that we develop a far greater understanding of the ecological requirements of the various species in order to develop appropriate conservation policies and management plans. In particular, for effective conservation we need informa-

tion about the following: (1) current status—the distribution and abundance of species, including trends in population size or structure; (2) potential threats or causes of decline—as a result of natural or human-mediated changes; (3) ecological requirements—ecological factors essential to the continuation of species, including interrelationships with other organisms; and (4) conservation applications—research on approaches to avoid or mitigate predicted or actual threats (see Racey and Entwistle 2003).

Issues that greatly impede focused and comprehensive recommendations for management include insufficient knowledge of factors that influence population dynamics of bats and a poor understanding of their population status and trends (O'Shea et al. 2003), habitat requirements (Hayes 2003; Miller et al. 2003), and their role within ecosystems. Land-management practices have been implemented for more than a century on both public and private lands with little or no understanding or consideration of their effects on bats. In addition, the lack of available information about most species of bats means that conservation planning is difficult to undertake and that decision makers are difficult to convince regarding the most appropriate mechanisms to ensure their continuation.

Populations of bats are difficult to monitor. However, current recognition of the importance of bats to biodiversity, their ecological and economic value as ecosystem components, and their vulnerability to declines makes monitoring trends in their populations a much-needed cornerstone of their future management. Unfortunately, the complex and variable natural history of bats (small body size, long life span, nocturnality, and volancy) poses many challenges to monitoring, although significant progress is being made as noted earlier in this chapter.

Effective bat conservation in the future will depend on research, particularly in ecology and natural history, to elucidate fully the relationships among individual bat species, their environment, and humans. For most Texas bats, it rarely is clear just what factors limit their populations, although the protection of roosts (including summer roosts, nursery colonies, and hibernacula), food, habitat (mosaics or edges), and populations certainly seem to be the main focus for action to promote their conservation. However, for a few other species, such as the hoary bat (*Lasiurus cinereus*), it is not clear how we can ensure their protection because they roost alone in foliage and forage in a wide range of situations. Thus, bat biologists have a multitude of opportunities to use their scientific knowledge to advantage in conservation (Fenton 1997).

In an effort to address these shortcomings, the North American Bat Conservation Partnership (NABCP) was established in 1999 to provide organized support for continent-wide conservation efforts for bats and their habitats by advancing research, education, and management goals (Keeley et al. 2003). Similar partnerships, such as the North American Waterfowl Management Plan, have demonstrated the success of collaborating at local, regional, national, and international levels to promote the conservation of wildlife.

Most bat conservation programs must involve efforts to change public attitudes (Fenton 1997, 2003). Fortunately, the expanding information base about bats has made it easier to apply new information both to conservation and to the improvement of public awareness of bats. Furthermore, we now have more details about ways that human activities have benefited bats, providing "good news" stories so often lacking in discussions of biodiversity and conservation.

Illustrated Keys to the Bats of Texas

Two keys have been prepared for the 33 species of bats known to occur in Texas. The first is based on external characters of adult animals. Subadult or juvenile bats usually show the diagnostic characters of the adult, although their pelage color may be slightly different (usually lighter). The best way to distinguish a juvenile from an adult bat is to examine the cartilaginous area in the finger (metacarpal-phalangeal) joints (fig. 9); the larger the area, signifying that the joints have not fused, the younger the bat.

After an identification is made using the external key, refer to the account of the species, which provides a more extended description, including additional characters. The distribution maps also provide clues to help verify or question an identification. If a skull is available, it should be examined as a check on the identification made by using external characters. The second key, based on cranial and dental structures, follows the key based on external characters.

Both keys are arranged so that there is always a choice between 2 statements (a couplet) about some characteristic of a bat. To identify a bat, use a millimeter ruler or similar instrument to make the required measurements, and select the appropriate alternative from each couplet (1, 2, 3, etc.) in the key. At the end of the statement chosen is a number that indicates the location of the next choice to be made. The process should be repeated until a name instead of a number is given at the end of a line.

FIGURE 9. Wing of *Eptesicus fuscus*, labeled to show names of external parts and measurements used in the key to Texas bats. The insert drawing is an enlargement of the metacarpal-phalangeal joint in a juvenile (a) and an adult (b) bat.

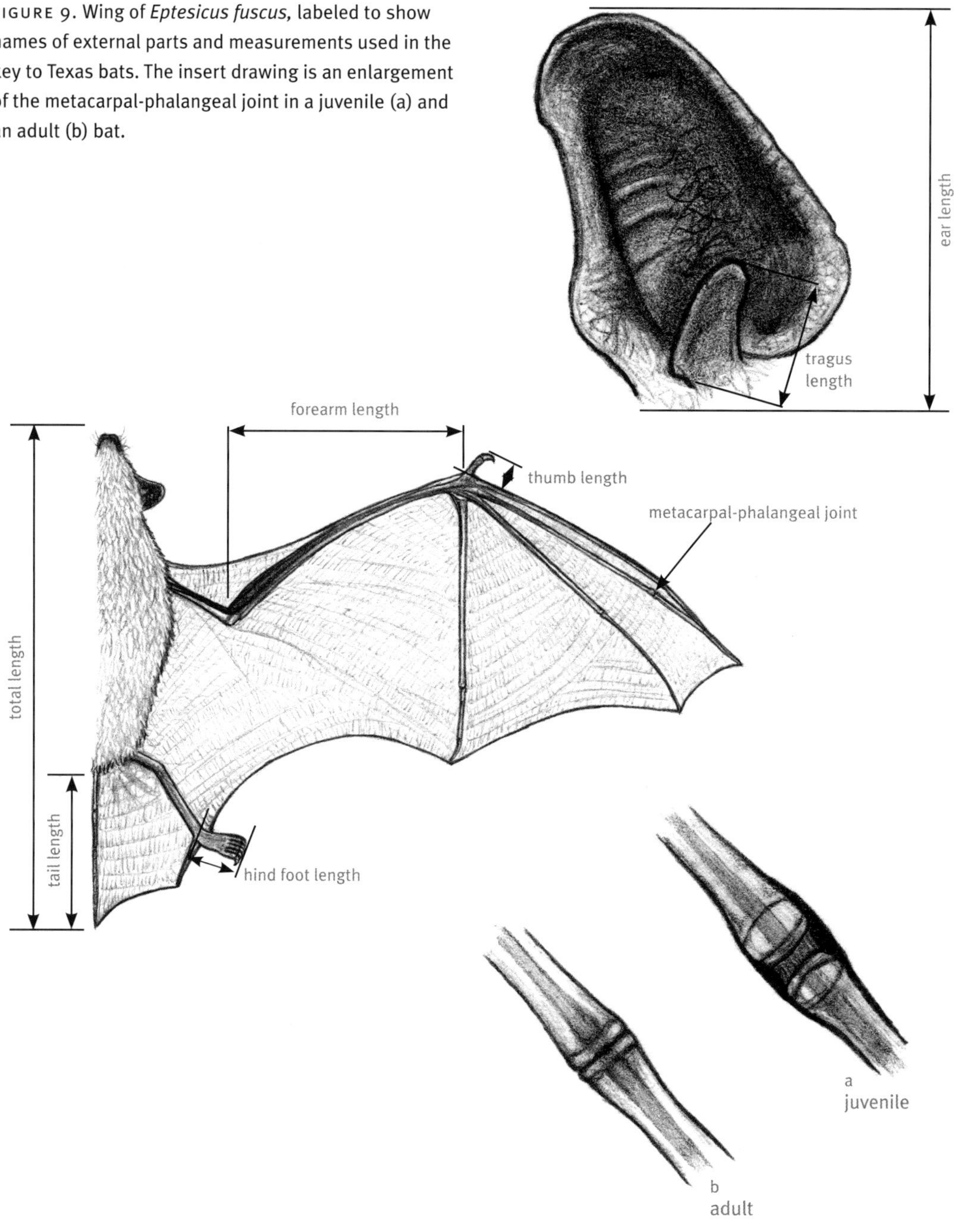

Key to External Characters

A millimeter ruler and, in some cases, a hand lens are required. When possible, the characters used in this key have been chosen so that the animal need not be sacrificed and also can be used with scientifically prepared voucher specimens (see Barbour and Davis 1969 for an explanation of this procedure). Some measurements and features are difficult to verify in voucher specimens because features shrink or deform when they are dry. This is particularly true of the ears, the tragus, (fig. 9) and any elaborate facial features.

All measurements (as defined and shown in fig 9) and weights are in millimeters and grams respectively unless stated otherwise. One of the external measurements taken on bats, the length of the forearm, differs from the usual standard measurements taken on small mammals. Forearm length is the best indicator of size in bats as it can be measured with precision.

1. Distinct, upwardly and freely projecting, triangular-shaped nose leaf at end of elongated snout (fig. 10a). . . . . Family Phyllostomidae (in part) 2
   Nose leaf absent, indistinct, or modified as lateral ridges or low moundlike structure; snout normal (fig.10b) 3
2. Tail evident, protruding about 10 mm from dorsal side of interfemoral membrane (fig. 11a); distance from eye to nose about twice the distance from eye to ear (fig. 11c); forearm less than 48 mm. . . . .*Choeronycteris mexicana*
   Tail not evident (fig. 11b); eye about midway between nose and ear (fig. 11d); forearm more than 48 mm. . . . . *Leptonycteris nivalis*
3. Thumb longer than 10 mm; hair straight, lying smoothly, glossy tipped . . . Family Phyllostomidae (subfamily Desmodontinae, vampire bats) . . . .
   . . . . *Diphylla ecaudata*
   Thumb less than 10 mm; hair slightly woolly, pelage lax, not usually lying smoothly, not glossy tipped 4
4. Prominent grooves and flaps on chin (fig. 12a); tail protruding from dorsal surface of interfemoral membrane (fig. 12b) . . . Family Mormoopidae. . . . .
   . . . . *Mormoops megalophylla*
   No notable grooves or flaps on chin; lumps above nose or wrinkled lips possible; most faces lack even these characteristics; tail extends to or beyond edge of interfemoral membrane. 5
5. Tail extends conspicuously beyond free edge of interfemoral membrane (fig. 13a)
   . . . . .Family Molossidae 6
   Tail extends to or just beyond the free edge of interfemoral membrane . . . .
   (fig. 13b) . . . . .Family Vespertilionidae 9
6. Forearm more than 70 mm; upper lips without deep vertical grooves . . . .
   (fig. 14a) . . . . *Eumops perotis*
   Forearm less than 70 mm; upper lips with deep vertical grooves (fig. 14b) 7

7. Forearm less than 52 mm 8

    Forearm more than 52 mm (58–64) . . . . *Nyctinomops macrotis*

8. Ears not united at base (fig. 15a); second phalanx of fourth finger more than 5 mm (fig. 15c) . . . . *Tadarida brasiliensis*

    Ears joined at base (fig. 15b); second phalanx of fourth finger less than 5 mm (fig. 15d) . . . . *Nyctinomops femorosaccus*

9. Ears proportionally large, more than 25 mm from notch to tip 10

    Ears of normal size, less than 25 mm from notch to tip 13

10. Color black with 3 large white spots on back, 1 just behind each shoulder and the other at the base of tail (fig. 16). . . . . *Euderma maculatum*

    Color variable but not black; no white spots on back 11

11. Dorsal color pale yellow; no distinctive glands evident on either side of the nose . . . . *Antrozous pallidus*

    Dorsal color light brown to gray; distinctive glands (large bumps) evident on either side of the nose (fig. 17) 12

12. Hairs on belly with white tips; strong contrast in color between the basal portions and tips of hairs on both back and belly (fig. 18a); presence of long hairs projecting beyond the toes. . . . . *Corynorhinus rafinesquii*

    Hairs on belly with pinkish buff tips; little contrast in color between basal portions and tips of hairs on back or belly (fig. 18b); absence of long hairs projecting beyond the toes. . . . . *Corynorhinus townsendii*

13. At least the anterior half of the dorsal surface of the interfemoral membrane is well furred (fig. 19). 14

    Dorsal surface of interfemoral membrane is naked, scantily haired, or at most lightly furred on the anterior third 20

14. Color of hair black, many of the hairs distinctly silver tipped (fig. 4h) . . . . *Lasionycteris noctivagans*

    Color various but never uniformly black 15

15. Color yellowish 16

    Color reddish, brownish, or grayish (not yellowish). 17

16. Total length more than 120 mm. . . . . *Lasiurus intermedius*

    Total length less than 120 mm . . . . *Lasiurus ega, Lasiurus xanthinus**

17. Forearm more than 45 mm; color wood brown and heavily frosted with white . . . . *Lasiurus cinereus*

    Forearm less than 45 mm; upper parts reddish or mahogany. 18

18. Upper parts brick red to rusty red, frequently washed with white 19

    Upper parts mahogany brown and washed with white . . . . *Lasiurus seminolus*

19. Color reddish with frosted appearance resulting from white-tipped hairs; interfemoral membrane fully haired . . . . *Lasiurus borealis*

    Color rusty red to brownish without frosted appearance; posterior third of interfemoral membrane bare or only scantily haired . . . . . . *Lasiurus blossevillii*

20. Tragus (projection within ear) short, blunt, and curved (fig. 20a–b)   21

    Tragus long, pointed, and straight (fig. 20c)   23

21. Forearm more than 40 mm . . . . *Eptesicus fuscus*

    Forearm less than 40 mm.   22

22. Forearm more than 32 mm; interfemoral membrane naked; color brown . . . . . .

    . . . . *Nycticeius humeralis*

    Forearm less than 32 mm; interfemoral membrane sparsely furred on anterior

        third of dorsal surface; color drab yellow to smoke gray . . *Parastrellus hesperus*

23. Dorsal fur tricolored when parted: black at base, wide band of light yellowish

        brown in middle, tipped with slightly darker contrasting color (fig. 18c); lead-

        ing edge of wing membrane noticeably paler than rest of membrane. . . . .

    . . . . *Perimyotis subflavus*

    Dorsal fur bicolored or unicolored with no light band in the middle (fig. 18d);

        leading edge of wing same color as other parts of membrane   24

24. Calcar with well-marked keel (fig. 21a)   25

    Calcar without well-marked keel (fig. 21b)   27

25. Forearm more than 36 mm; foot more than 8 mm long; underside of wing furred

        to elbow (fig. 22); pelage dark brown . . . . *Myotis volans*

    Forearm less than 36 mm; foot less than 8 mm long; underside of wing not

        furred to elbow; pelage light brown to buff brown.   26

26. Hairs on back with dull reddish-brown tips; black mask not as noticeable;

        thumb less than 4 mm long (6–7.5 mm, including wrist); naked part of snout

        about as long as the width of the nostrils when viewed from above (fig. 23a);

        absence of free tail beyond the border of the interfemoral membrane . . . .

    . . . . *Myotis californicus*

    Fur on back with long, glossy, brownish tips; black mask usually noticeable;

        thumb more than 4 mm long (8–8.5 mm, including wrist); naked part of

        snout approximately 1.5 times the width of the nostrils (fig. 23b); tail extends

        up to 4 mm beyond the border of the interfemoral membrane . . . .

    . . . . *Myotis ciliolabrum*

27. Forearm more than 40 mm   28

    Forearm usually less than 40 mm.   29

28. Conspicuous fringe of stiff hairs on free edge of interfemoral membrane (fig. 24)

    . . . . *Myotis thysanodes*

    No conspicuous fringe of stiff hairs on free edge of interfemoral membrane

    . . . . *Myotis velifer*

29. Fur thick and wooly, hair not bicolored, instead monocolored lead gray or bright

        buffy brown, sometimes bright orange overwash on part or all of body; pres-

        ence of toe hairs that extend beyond tips of claws . . . . . . *Myotis austroriparius*

    Fur thinner and not wooly, hair distinctly bicolored, some shade of brown over

        blackish.   30

30. Forearm length less than or equal to 36 mm; total length usually less than 80 mm
.... *Myotis yumanensis*

Forearm greater than 36 mm; total length usually greater than 80 mm 31

31. Ears longer, 15–18 mm from notch, and, when laid forward, extend conspicuously beyond tip of nose.......................... *Myotis septentrionalis*

Ears shorter, 15 mm or less from notch, and, when laid forward, extend only
near tip of nose.................................... *Myotis occultus*

* *Lasiurus ega* and *Lasiurus xanthinus* are excellent examples of cryptic species (i.e., genetically distinct but morphologically virtually identical species that only specialists can differentiate). The 2 species differ in their karyotypes and gene sequences (Baker et al. 1988). In appearance, *xanthinus* is typically brighter yellow in color, whereas *ega* is more yellow washed with black, particularly on the rump and extending onto the uropatagium. Ecologically, *xanthinus* is more frequently found in yucca plants of the Big Bend and Davis mountains of West Texas, whereas *ega* is more prominent in the palm trees of South Texas.

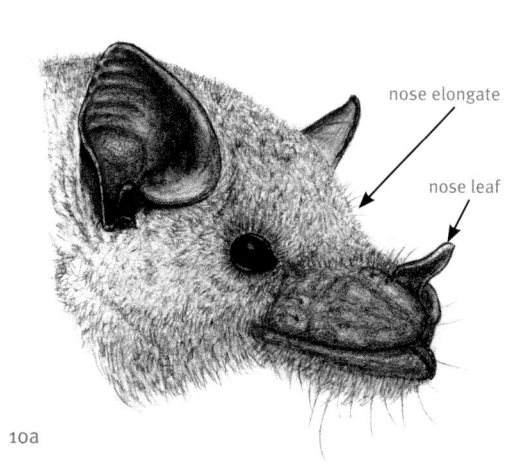

nose elongate

nose leaf

10a

10b

FIGURE 10.

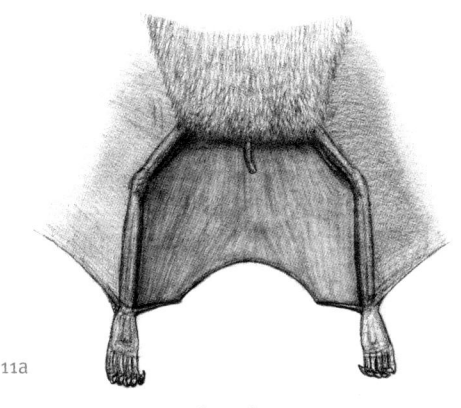

11a

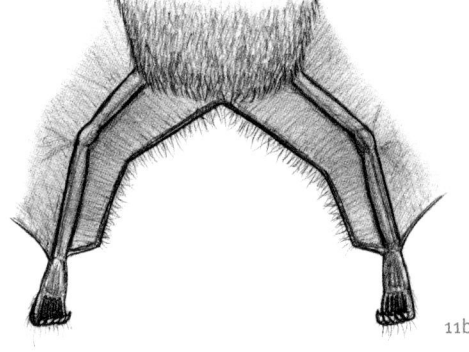

11b

C. mexicana

L. nivalis

11c

11d

FIGURE 11.

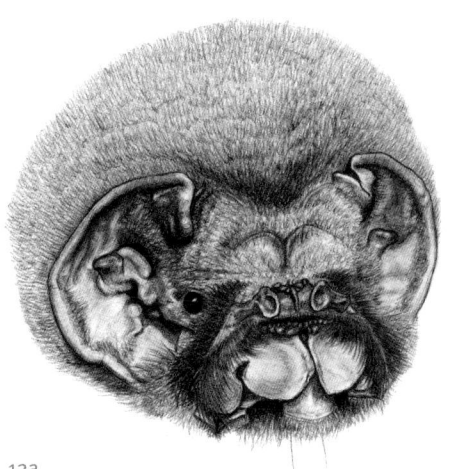

12a

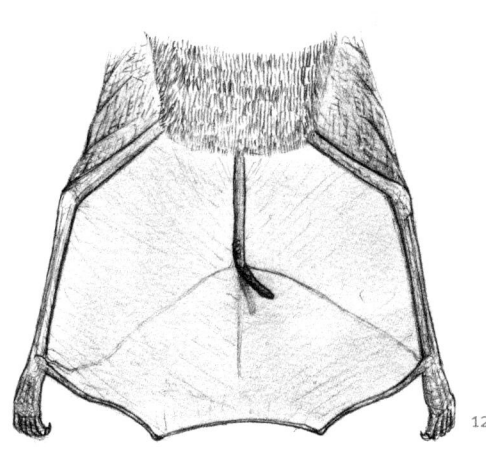

12b

FIGURE 12.

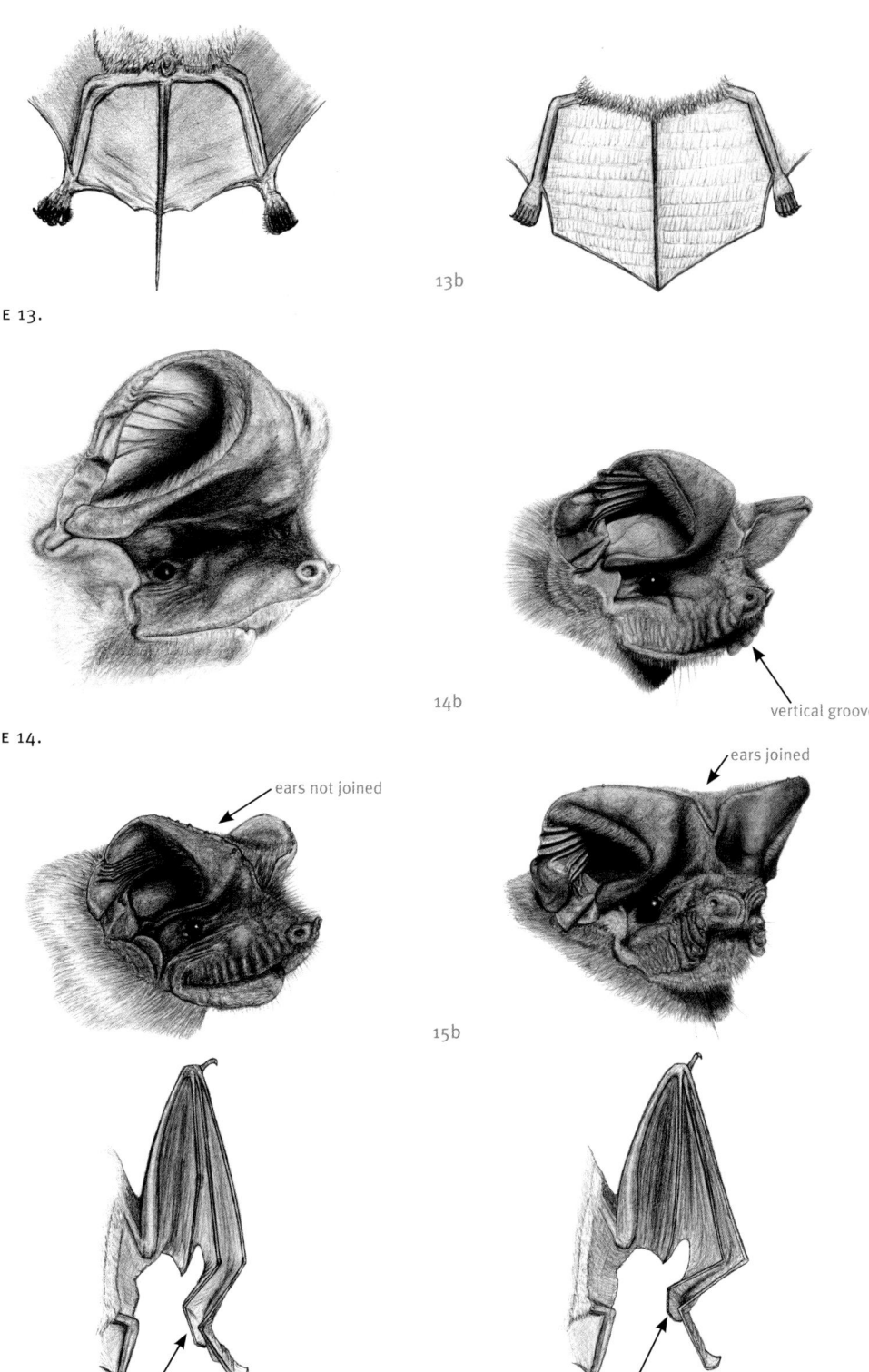

13a

13b

FIGURE 13.

14a

14b

vertical grooves

FIGURE 14.

ears not joined

15a

ears joined

15b

length >5mm

15c

length <5mm

15d

FIGURE 15.

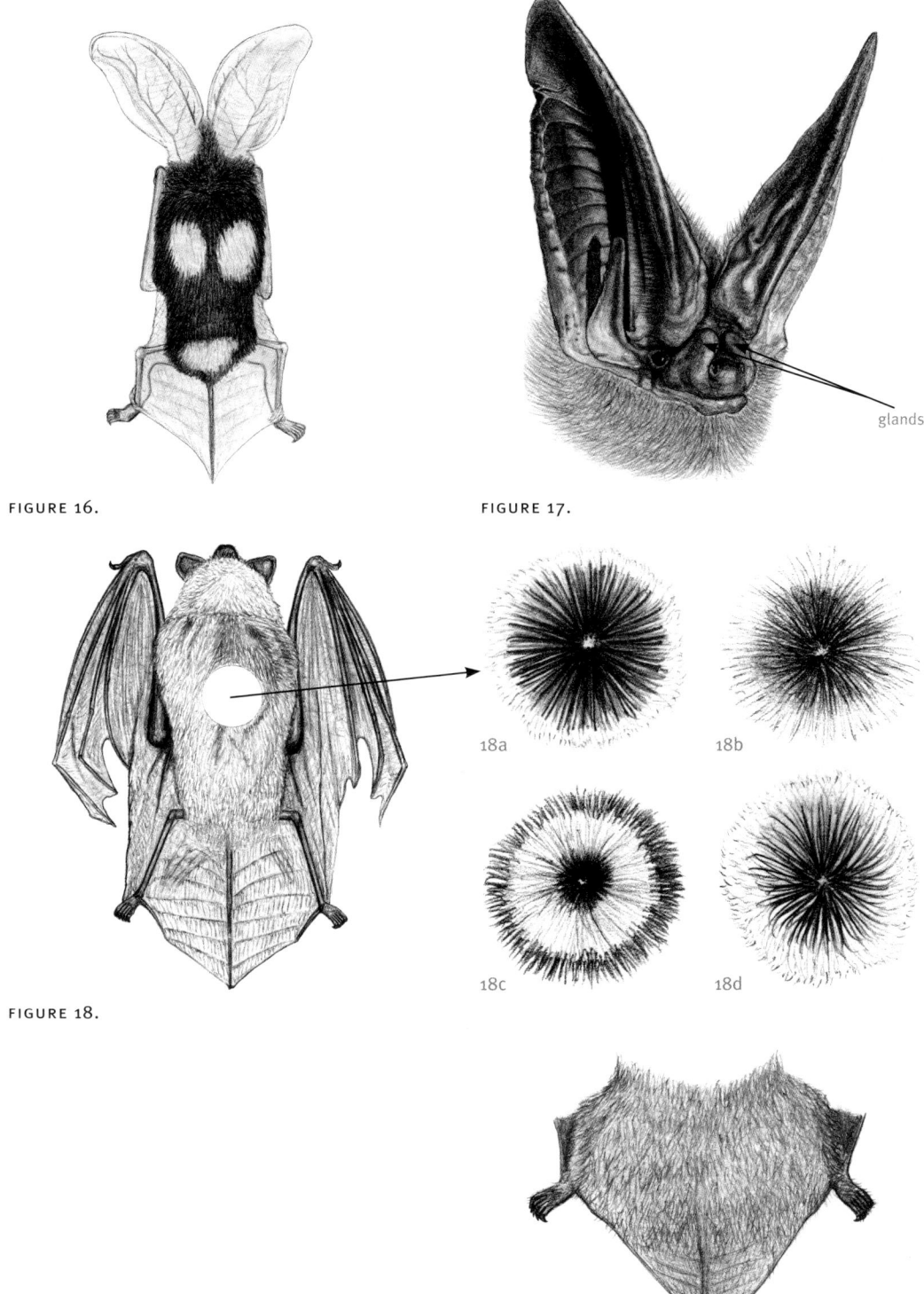

FIGURE 16.

FIGURE 17.

glands

FIGURE 18.

18a

18b

18c

18d

FIGURE 19.

20a

broad, rounded tragus

20b

curved, blunt tragus

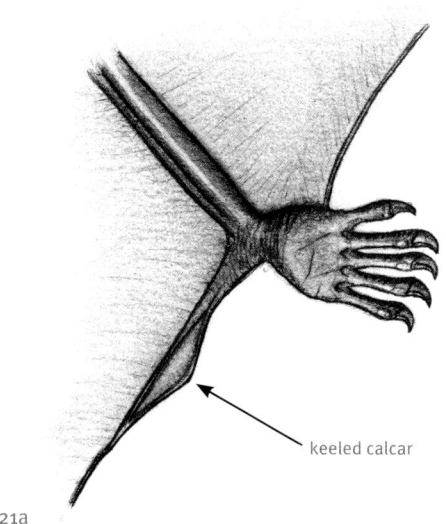

FIGURE 20. (a) broad, rounded tragus; (b) curved, blunt tragus; (c) straight, pointed tragus

20c

straight, pointed tragus

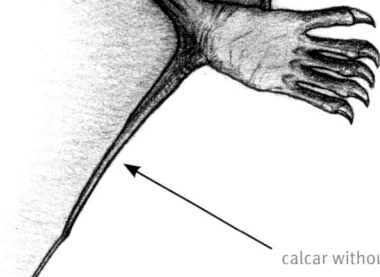

keeled calcar

calcar without keel

21a

21b

FIGURE 21.

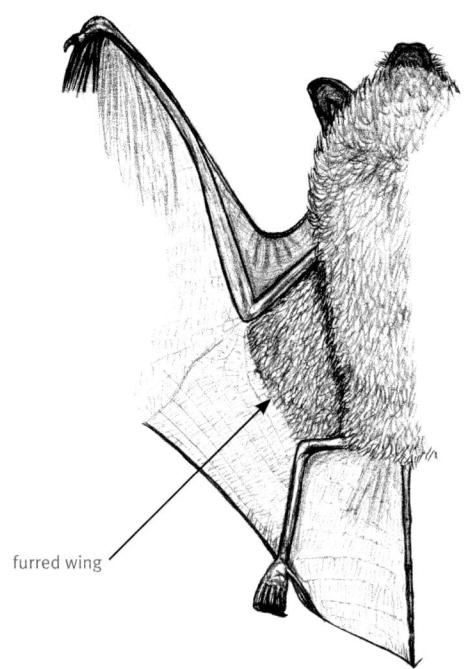

furred wing

FIGURE 22.

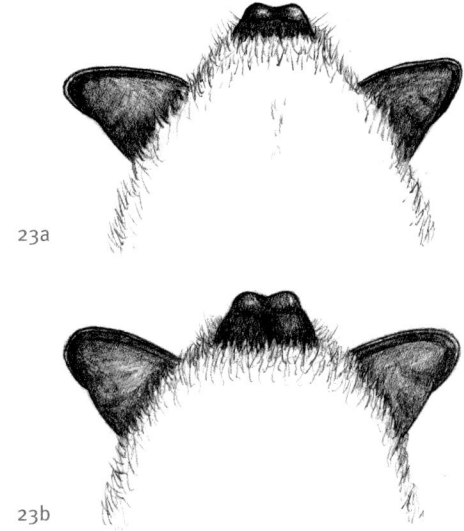

23a

23b

FIGURE 23.

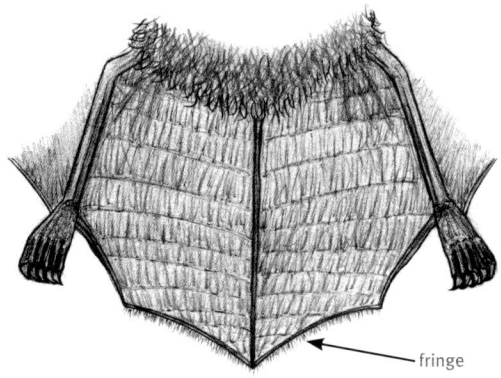

 fringe

FIGURE 24.

Key to the Skulls

This key, which is based on cranial and dental features of adults, includes the same species that appear in the preceding key, which is based on external features. The user will need a cleaned skull, a binocular dissecting microscope with ocular micrometer, and dial calipers capable of measuring to the nearest 0.1 mm. Measurements used in the key should be taken as described in fig. 25.

1. Upper incisors enlarged and bladelike; only 3 pairs of small teeth behind upper canines (fig. 26) . . . . *Diphylla ecaudata*
   Upper incisors small, much smaller than canines; more than 3 pairs of teeth behind upper canines (fig. 25)   2
2. Skull long and narrow; rostrum elongated, at least as long as braincase (fig. 27) .   3
   Skull not especially long and narrow; rostrum not greatly elongated   4
3. Rostrum exceptionally long; zygomatic arches of skull incomplete (fig. 27a); no lower incisors . . . . *Choeronycteris mexicana*
   Rostrum less elongated; zygomatic arches of skull complete (fig. 27b); 2 lower incisors . . . . *Leptonycteris nivalis*
4. Rostrum abruptly upturned; dorsal profile of skull is concave in lateral view (fig. 28a) . . . . *Mormoops megalophylla*
   Rostrum not abruptly upturned; dorsal profile of skull usually flat or not concave (fig. 28b)   5
5. Six postcanine teeth in upper jaw (fig. 29a)   6
   Fewer than 6 postcanine teeth in upper jaw (fig. 29b-l)   14
6. Greatest length of skull more than 15.5 mm   7
   Greatest length of skull less than 15.5 mm.   8
7. Rostrum relatively massive; sagittal crest well defined (fig. 30a); breadth across canines more than 4.3 mm. . . . . *Myotis velifer*
   Rostrum relatively narrower and less massive; sagittal crest not well defined (fig. 30b); breadth across canines less than 4.3 mm . . . . *Myotis thysanodes*
8. Occurring east of 100th meridian (fig. 31).   9
   Occurring west of 100th meridian (fig. 31)   10
9. Interorbital breadth usually more than 4 mm . . . . *Myotis austroriparius*
   Interorbital breadth usually less than 4 mm . . . . *Myotis septentrionalis*
10. Braincase (cranial breadth) broad (usually 7.1 mm or more)   11
    Braincase (cranial breadth) narrow (usually 7.0 mm or less)   13
11. Braincase flattened, resulting in a gradual slope from rostrum to top of braincase (fig. 32a) . . . . *Myotis occultus*
    Braincase rises abruptly from rostrum (fig. 32b, 33b)   12
12. Skull broader (cranial breadth usually 7.2 mm or more). . . . .*Myotis volans*
    Skull narrower (cranial breadth usually 7.2 mm or less). . . . . . *Myotis yumanensis*[1]

13. Braincase rises abruptly from rostrum; skull usually less than 13.6 mm in total length (fig. 33a). . . . . *Myotis californicus*

Braincase flattened, resulting in a gradual slope from rostrum to top of braincase; skull usually more than 13.6 mm in total length (fig. 33c) . . .*Myotis ciliolabrum*

14. One pair of upper incisors (fig. 29f, h–l) 15

Two pairs of upper incisors (fig. 29b–e, g) 26

15. One pair of upper premolars (fig. 29f, h, i) 16

Two pairs of upper premolars (fig. 29f, j–l). 19

16. Two pairs of lower incisors (fig. 34h). . . . .*Antrozous pallidus*

Three pairs of lower incisors (fig. 34f, i) 17

17. Upper incisors in contact with canine (fig. 29f); third lower incisor crowded when viewed from front and smaller than first and second incisors (fig. 34f) . 18

Upper incisors separated from canine (fig. 29i); third lower incisor not crowded and equal in size to the first and second (fig. 34i). . . . . *Nycticeius humeralis*

18. Length of maxillary toothrow 5.5 mm or more . . . .*Lasiurus intermedius*

Length of maxillary toothrow less than 5.5 mm . *Lasiurus ega, Lasiurus xanthinus*[2]

19. Breadth across canines greater than length of maxillary toothrow 20

Breadth across canines less than length of maxillary toothrow 23

20. Greatest length of skull more than 15 mm. . . . .*Lasiurus cinereus*

Greatest length of skull less than 15 mm 21

21. Lacrimal ridge present (fig. 35a). 22

Lacrimal ridge not developed (fig. 35b) . . . . *Lasiurus seminolus*

22. Greatest length of skull usually more than 12.8 mm in males and 13.0 in females; length of maxillary toothrow usually more than 4.2 mm in males and 4.6 in females . . . . *Lasiurus borealis*[3]

Greatest length of skull usually less than 12.8 mm in males and 13.0 in females; length of maxillary toothrow usually less than 4.2 mm in males and 4.6 in females . . . . *Lasiurus blossevillii*[3]

23. Greatest length of skull less than 30 mm; premaxillary gap present on front part of bony palate (fig. 36a) 24

Greatest length of skull more than 30 mm; no premaxillary gap on front of bony palate (fig. 36b). . . . . *Eumops perotis*

24. Greatest length of skull more than 21 mm. . . . . *Nyctinomops macrotis*

Greatest length of skull less than 21 mm 25

25. Greatest length of skull more than 18 mm. . . . . *Nyctinomops femorosaccus*

Greatest length of skull less than 18 mm . . . . *Tadarida brasiliensis*

26. One pair of upper premolars (fig. 29g) . . . .*Eptesicus fuscus*

Two pairs of upper premolars (fig. 29b–e) 27

27. Two lower premolars in each side of jaw (fig. 34c, d) 28

Three lower premolars in each side of jaw (fig. 34b, e). 30

28. Auditory bullae very large and elongate; canines small and weak, the lower one with a distinct accessory cusp behind; zygomatic arches abruptly widened in the middle (fig. 37a) . . . . *Euderma maculatum*

Auditory bullae neither greatly enlarged nor elongate; canines large and strong, unicuspidate; zygomatic arches of uniform width throughout (fig. 37b) 29

29. Skull nearly straight in dorsal profile; palate extends far behind molars (fig. 37b–c) . . . . *Parastrellus hesperus*

Skull concave in dorsal profile; palate extends little behind molars (fig. 37d) . . . . *Perimyotis subflavus*

30. Auditory bullae much enlarged; rostrum narrow, evenly convex above or slightly concave medially; forehead conspicuously elevated (fig. 38a–b) 31

Auditory bullae not especially enlarged; rostrum broad, concave on each side; forehead nearly flat (fig. 39a–b) . . . . *Lasionycteris noctivagans*

31. First upper incisor with 1 cusp (fig. 40a) . . . . *Corynorhinus townsendii*

First upper incisor with 2 cusps (fig. 40b) . . . . *Corynorhinus rafinesquii*

[1] One specimen of this species has been reported east of the 100th meridian in Starr County along the Rio Grande.

[2] There is no way to distinguish *Lasiurus ega* and *Lasiurus xanthinus* using characteristics of the skull (see key to external features).

[3] *Lasiurus blossevillii* and *Lasiurus borealis* are another example of cryptic species that are genetically distinct but morphologically similar. They are very difficult to identify morphologically by anyone other than a trained expert.

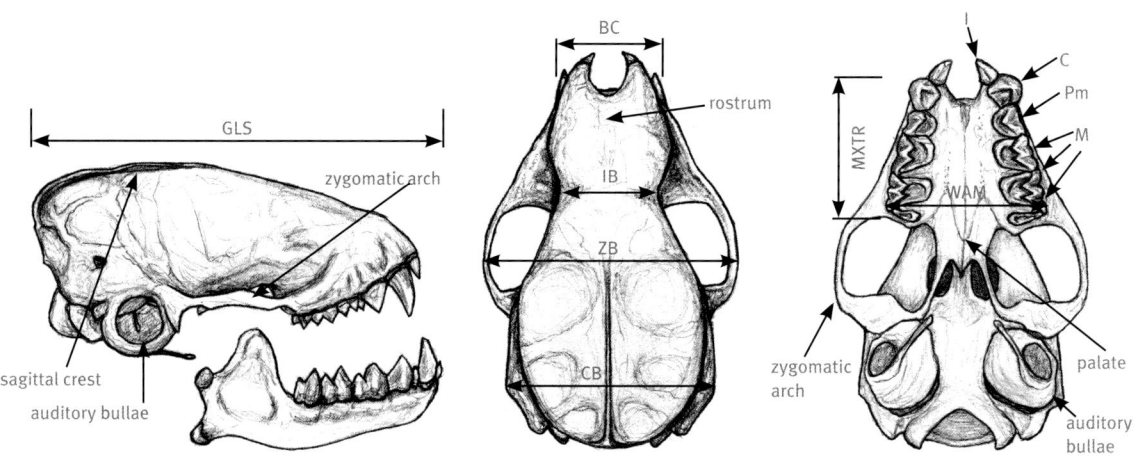

FIGURE 25. Skull of *Antrozous pallidus,* labeled to show the various parts and measurements used in the key to Texas bats. Skull measurement abbreviations are as follows: GLS, greatest length of skull; BC, breadth across canines; IB, interorbital breadth; ZB, zygomatic breadth; CB, cranial breadth; MXTR, length of maxillary toothrow; WAM, width across molars. Teeth are labeled as follows: incisors, I; canines, C; premolars, Pm; and molars, M.

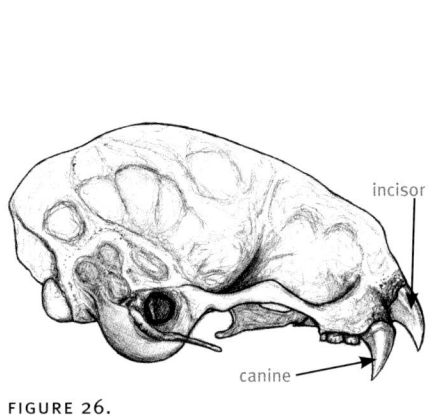

FIGURE 26.

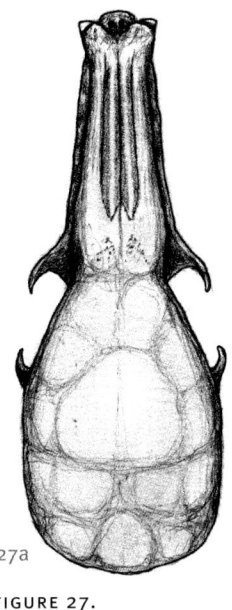

27a

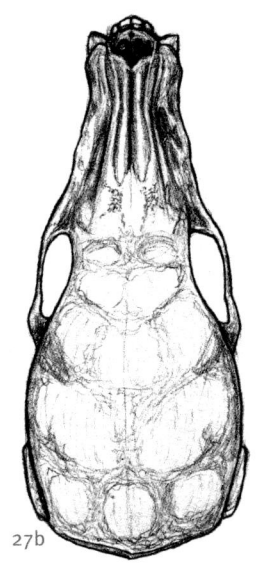

27b

FIGURE 27.

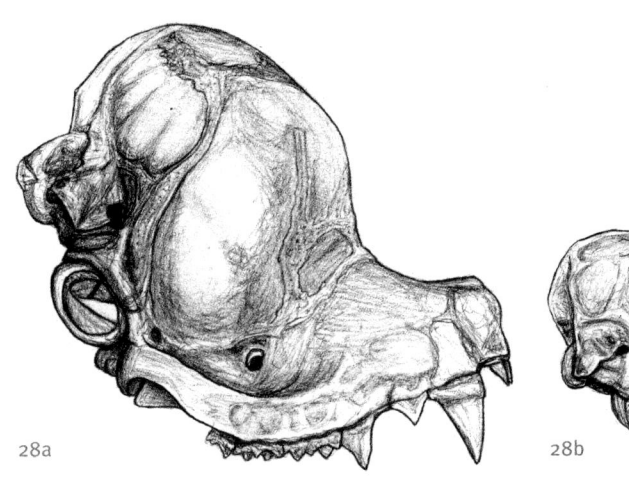

28a

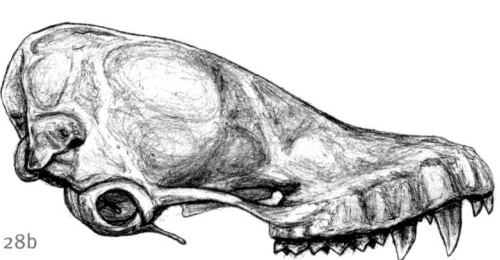

28b

FIGURE 28.

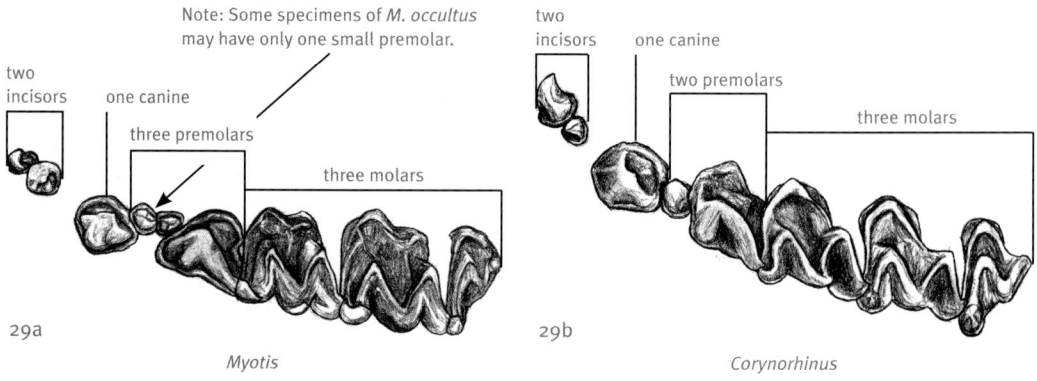

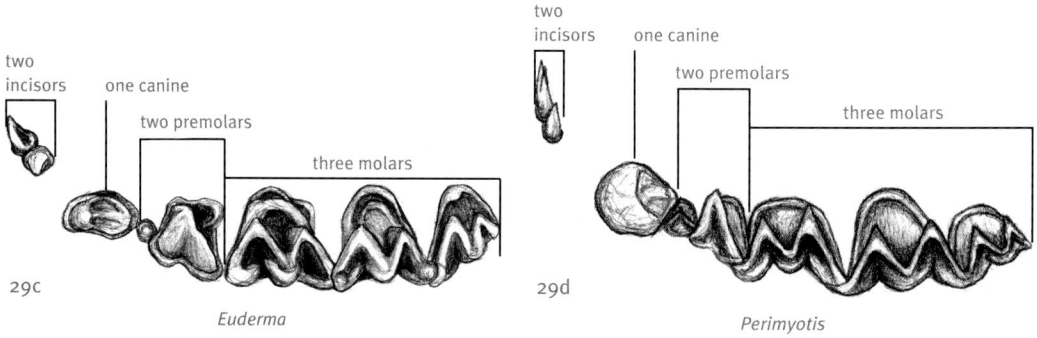

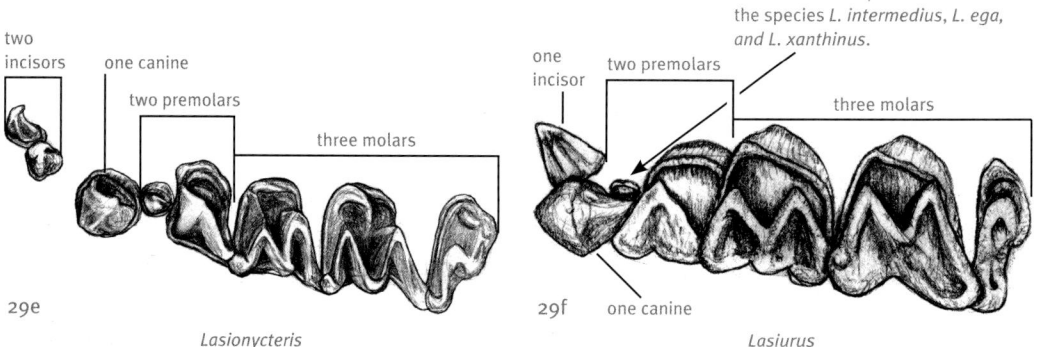

FIGURE 29. Upper teeth of Texas bats: (a) *Myotis*; (b) *Corynorhinus*; (c) *Euderma*; (d) *Perimyotis*; (e) *Lasionycteris*; (f) *Lasiurus*; (g) *Eptesicus*; (h) *Antrozous*; (i) *Nycticeius*; (j) *Eumops*; (k) *Nyctinomops*; and (l) *Tadarida*. Refer to table 3 for the dental formula of each genus.

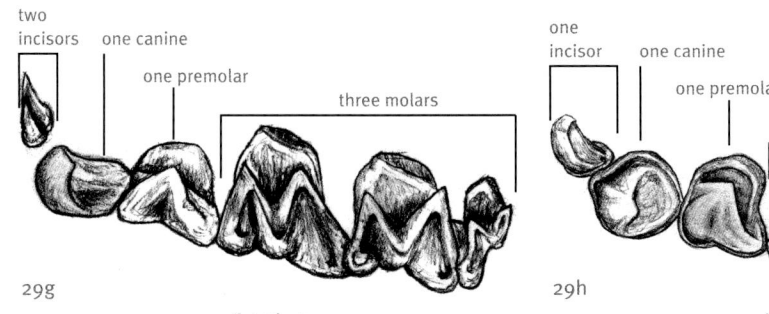

two incisors
one canine
one premolar
three molars

29g

Eptesicus

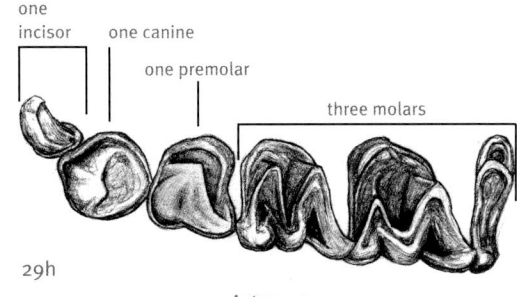

one incisor
one canine
one premolar
three molars

29h

Antrozous

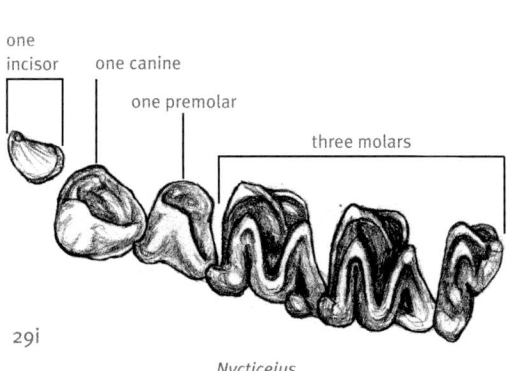

one incisor
one canine
one premolar
three molars

29i

Nycticeius

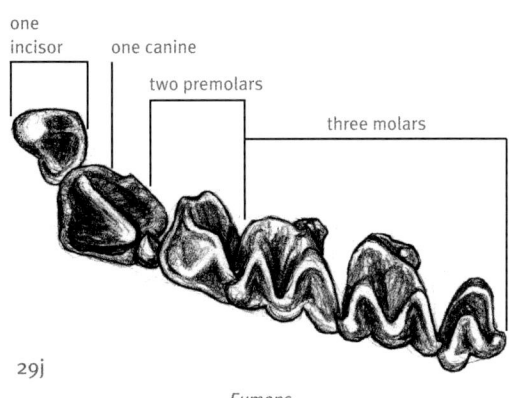

one incisor
one canine
two premolars
three molars

29j

Eumops

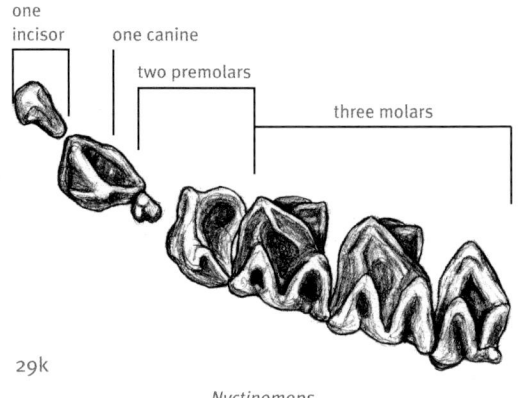

one incisor
one canine
two premolars
three molars

29k

Nyctinomops

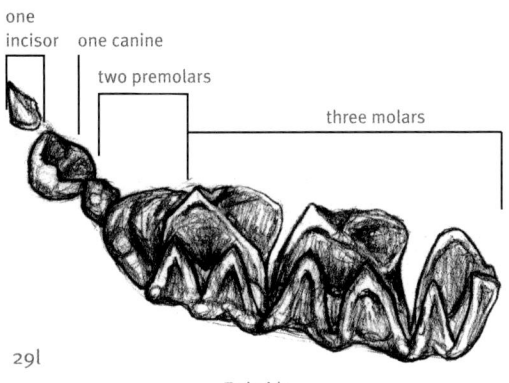

one incisor
one canine
two premolars
three molars

29l

Tadarida

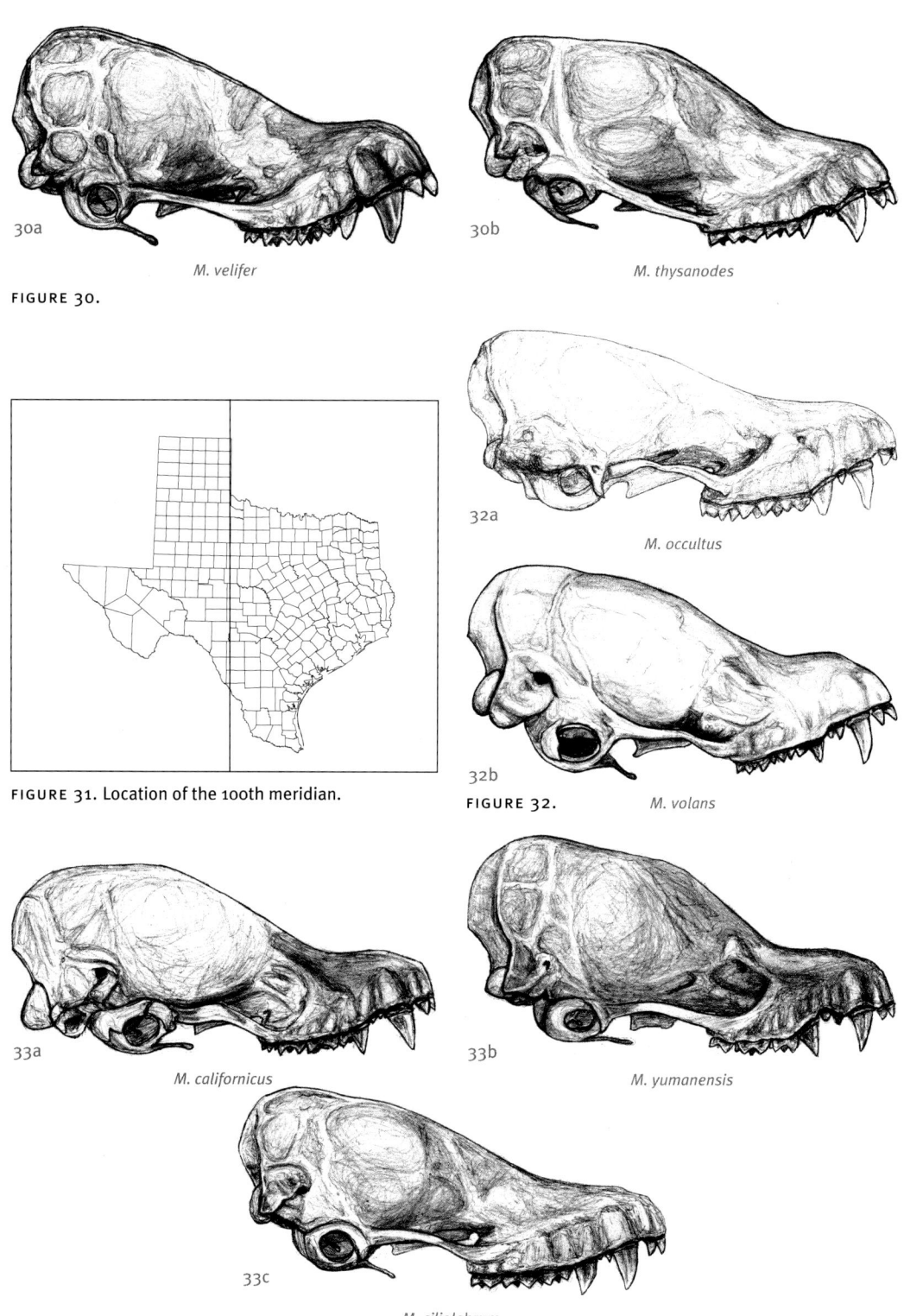

30a *M. velifer*

30b *M. thysanodes*

FIGURE 30.

FIGURE 31. Location of the 100th meridian.

32a *M. occultus*

32b *M. volans*

FIGURE 32.

33a *M. californicus*

33b *M. yumanensis*

33c *M. ciliolabrum*

FIGURE 33. (a) *M. californicus*; (b) *M. yumanensis*; and (c) *M. ciliolabrum*.

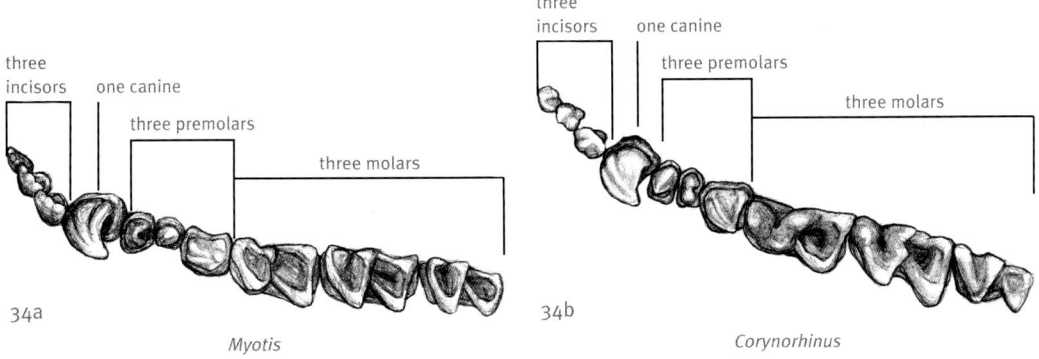

34a

Myotis

three
incisors

one canine

three premolars

three molars

34b

Corynorhinus

three
incisors

one canine

three premolars

three molars

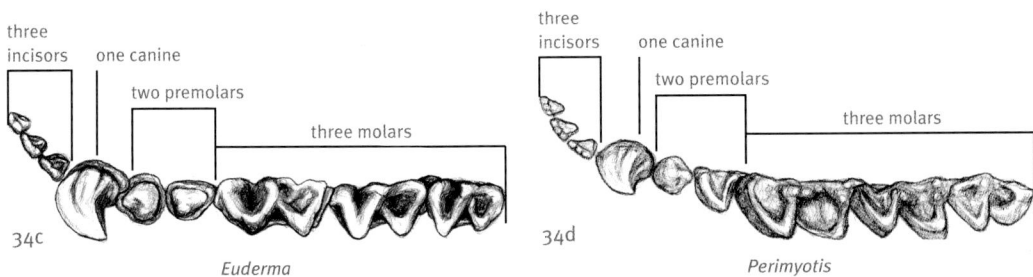

34c

Euderma

three
incisors

one canine

two premolars

three molars

34d

Perimyotis

three
incisors

one canine

two premolars

three molars

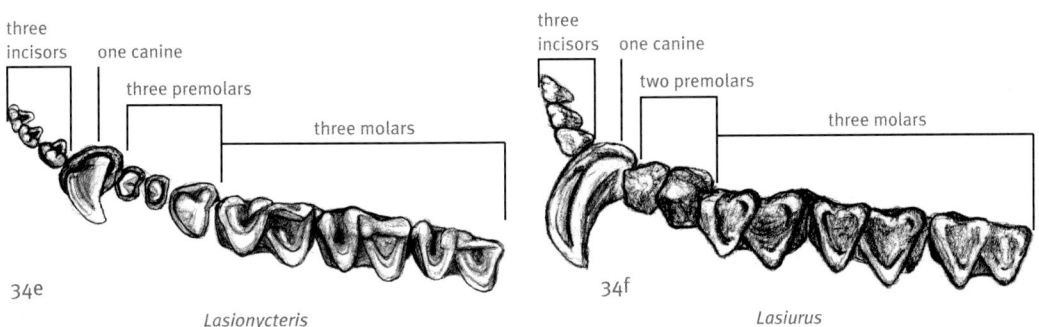

34e

Lasionycteris

three
incisors

one canine

three premolars

three molars

34f

Lasiurus

three
incisors

one canine

two premolars

three molars

FIGURE 34. Lower teeth of Texas bats: (a) *Myotis*; (b) *Corynorhinus*; c) *Euderma*; (d) *Perimyotis*; (e) *Lasionycteris*; (f) *Lasiurus*; (g) *Eptesicus*; (h) *Antrozous*; (i) *Nycticeius*; (j) *Eumops*; (k) *Nyctinomops*; and (1) *Tadarida*.

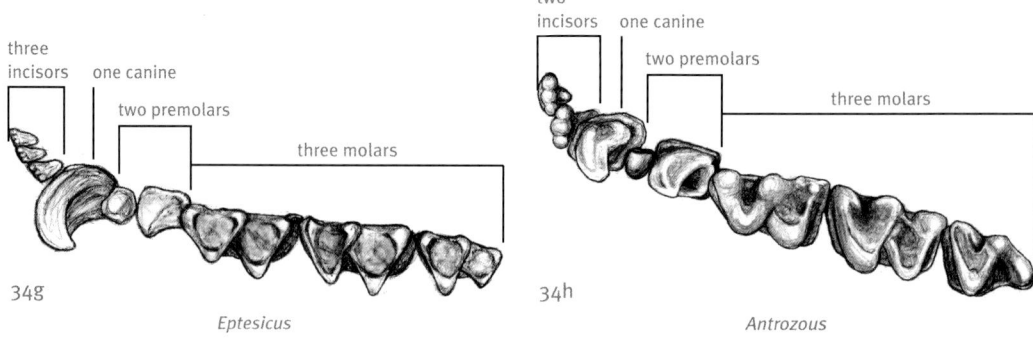

three
incisors

one canine

two premolars

three molars

34g

Eptesicus

two
incisors

one canine

two premolars

three molars

34h

Antrozous

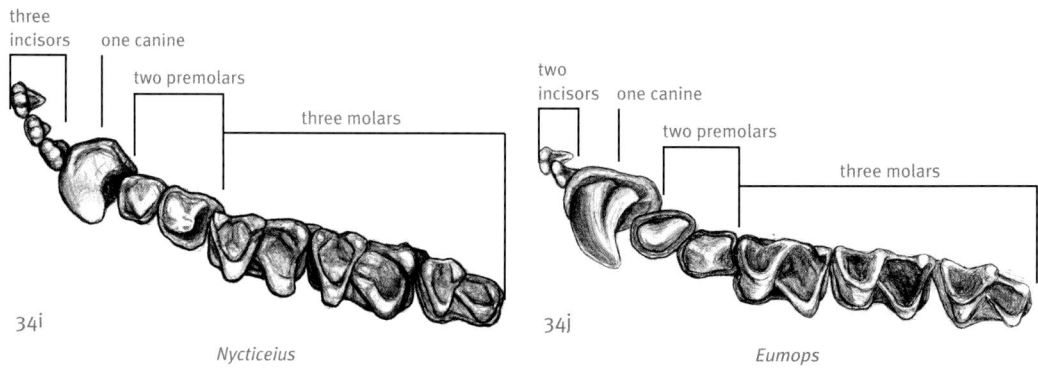

three
incisors

one canine

two premolars

three molars

34i

Nycticeius

two
incisors

one canine

two premolars

three molars

34j

Eumops

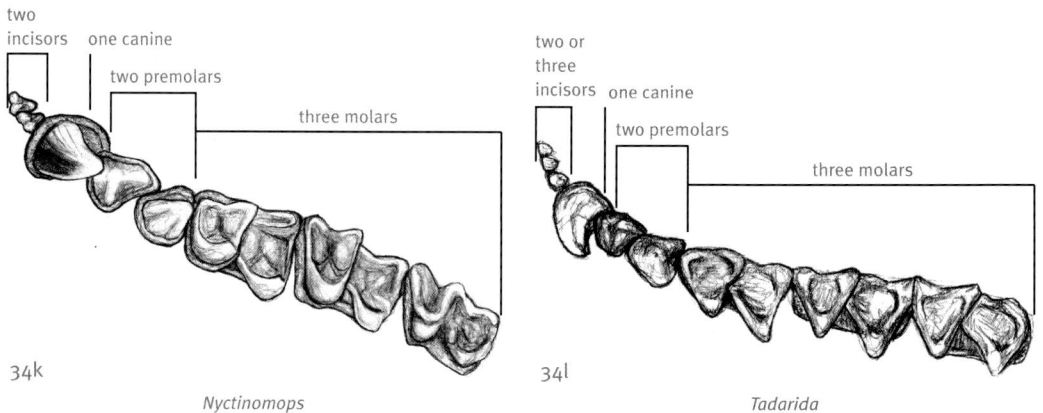

two
incisors

one canine

two premolars

three molars

34k

Nyctinomops

two or
three
incisors

one canine

two premolars

three molars

34l

Tadarida

FIGURE 34. continued

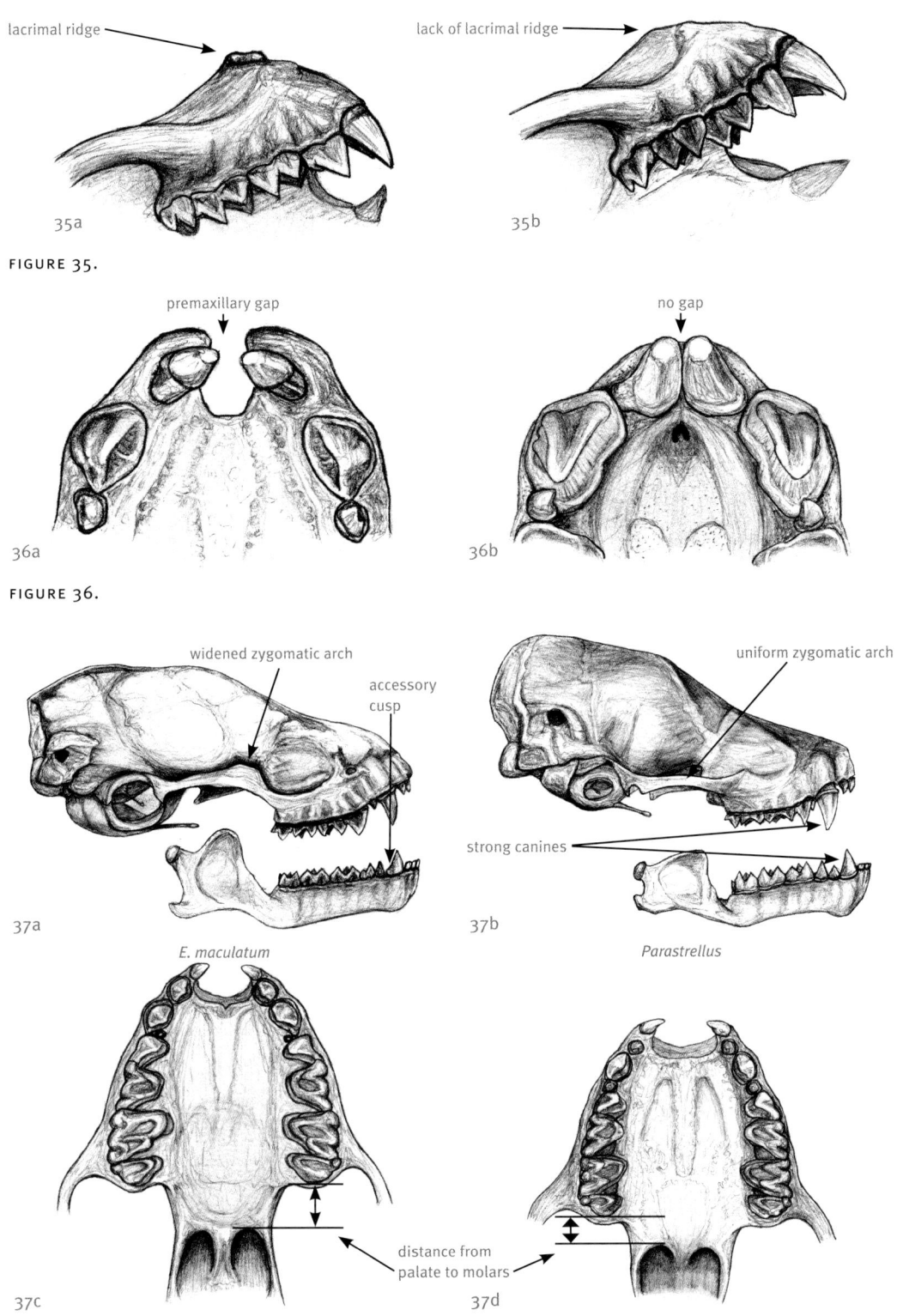

lacrimal ridge

lack of lacrimal ridge

35a

35b

FIGURE 35.

premaxillary gap

no gap

36a

36b

FIGURE 36.

widened zygomatic arch

accessory cusp

uniform zygomatic arch

strong canines

37a

E. maculatum

37b

Parastrellus

37c

Parastrellus

37d

Perimyotis

distance from palate to molars

FIGURE 37. (a) *Euderma*; (b) *Parastrellus*; (c) palate of *Parastrellus*; and (d) palate of *Perimyotis*.

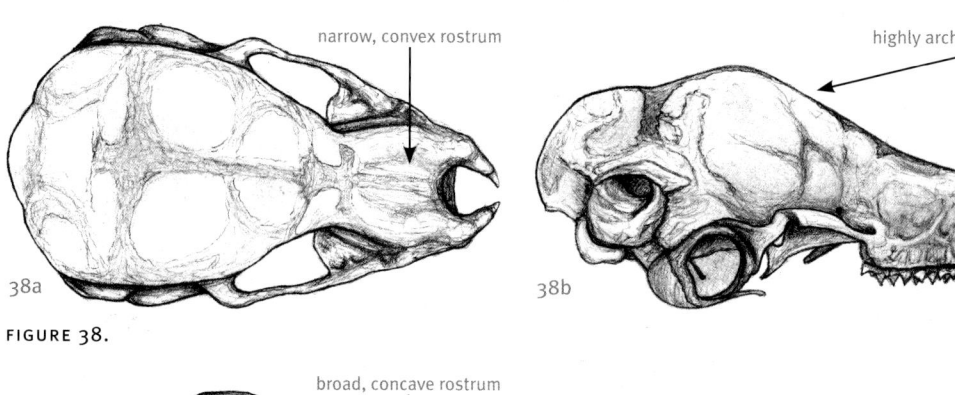

narrow, convex rostrum

highly arched profile

38a

38b

FIGURE 38.

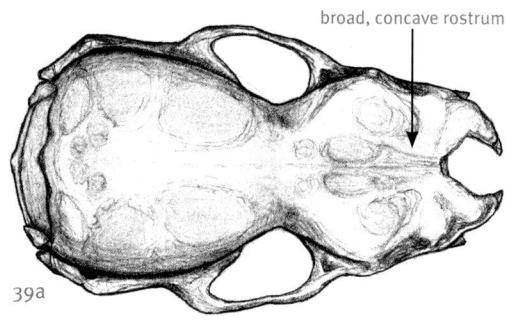

broad, concave rostrum

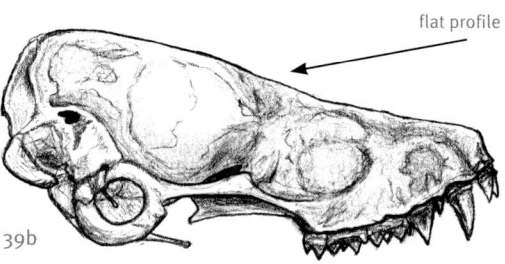

flat profile

39a

39b

FIGURE 39.

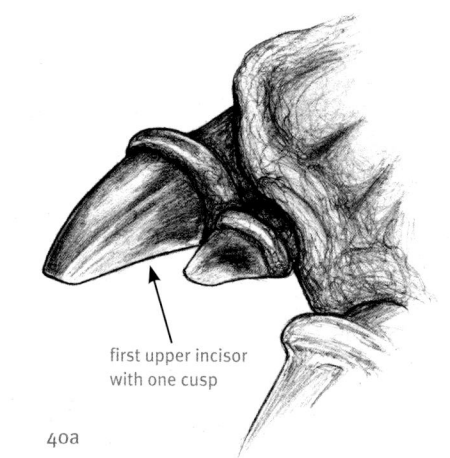

first upper incisor
with one cusp

40a

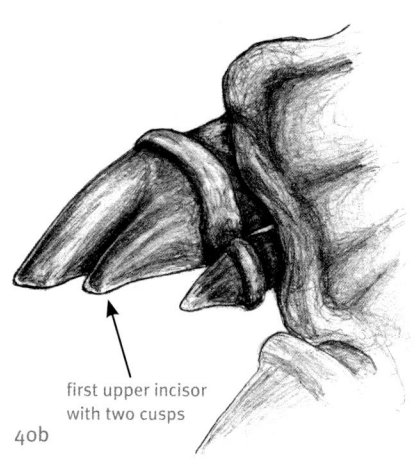

first upper incisor
with two cusps

40b

FIGURE 40.

Accounts of Species

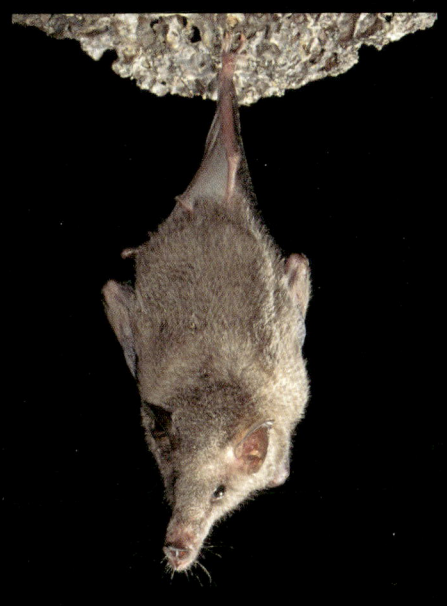

Family Mormoopidae

Ten species of New World bats make up the family Mormoopidae, which includes the mustached, naked-backed, and ghost-faced bats, which are variously distributed from the southern United States through Mexico, Central America, and South America to southern Brazil. Characterized by fleshy appendages on the snout and chin and a short tail protruding dorsally from the interfemoral membrane, these bats are abundant in the tropics as well as semiarid and subtropical environments. The only species of this family to occur in the United States is the ghost-faced bat, *Mormoops megalophylla*, which has been recorded in Texas and Arizona.

Mormoops megalophylla Peters, 1864
Ghost-Faced Bat

Etymology. The generic name *Mormoops* comes from Greek mythology, the first part of which, *mormo*, means "ghost," "bugbear," "hobgoblin," or "she-monster," a female vampirelike creature or spirit that bit children (Rezsutek and Cameron 1993). The second part, *ops*, translates as "face." The specific name also is Greek, from the words *megas* and *phyllon*, and means "big or great leaf" (Stangl et al. 1993).

Subspecies. Texas specimens are referable to the subspecies *M. m. megalophylla* Peters, 1864, as indicated by the most recent taxonomic revision of the species (Smith 1972).

Description. This is a relatively large (forearm = 51–59 mm) bat with reddish-brown to dark brown pelage. The fur becomes increasingly reddish with age. The ears are large and rounded and join across the forehead. *Mormoops* is readily recognized by its distinctive facial ornamentations, sharply upturned rostrum, and tail. Numerous tubercles and folds cover the rostrum and lips, and the ears are shaped and connected in such a way that the animal's mouth, when opened, functions as an orifice at the bottom of this funnel-shaped group of structures. The tail protrudes dorsally from the interfemoral membrane with approximately 1.3 cm not attached to this membrane and lying above it. The long forearm is strongly curved. The skull is markedly shortened, and the cranium is high and domed. Dental formula: I 2/2, C 1/1, Pm 2/3, M 3/3 × 2 = 34. Average external measurements are as follows: total length, 90 mm; tail, 26 mm; hind foot, 10 mm; ear, 13 mm; length of forearm, 54 mm. Weight: 13–19 g.

Distribution. In Texas this bat is known from the Chihuahuan Desert region (Trans-Pecos), southern edge of the Edwards Plateau, and Gulf Coastal Plains. It typically is found in lowland areas (below 3,000 m in elevation; Rezsutek and Cameron 1993), especially desert scrub and riverine habitats, as at Big Bend Ranch, although it also has been captured in the mountainous country of the Apache, Chisos, Chinati, Davis, and Elephant mountains and in the Sierra Vieja range. It is a common winter (November 1 to March 15) resident of caves along the extreme southern edge of the Edwards Plateau, where it apparently reaches its northern distributional limits in the United States. However, its occurrence at specific localities is highly variable and unpredictable.

Ghost-faced bat (*Mormoops megalophylla*)

Mormoops has been collected or observed at Frio Cave (Uvalde County), Webb Cave (Kinney County), Valdina Farms Sinkhole (Medina County), Sorcerer's Cave (Terrell County), and Haby, Ney, and Rattlesnake Caves (Bexar County) in January, February, March, May, September, November, and December, indicating that it uses the Edwards Plateau caves as a winter retreat (Eads et al. 1957a). Many apparently suitable caves near those just listed are not used, although they are occupied by bats often associated with *Mormoops* (Raun and Baker 1958). In contrast to the winter records from the Edwards Plateau, those from the Trans-Pecos are from the warmer months of the year (March 16 to October 31). This is suggestive of sea-sonal migration between the 2 regions, although such movements have yet to be substantiated.

Life History. Ghost-faced bats typically roost in caves, tunnels, and mine shafts, but they also have been found in old buildings. In addition to the caves listed earlier, these bats have been taken in a railroad tunnel near Comstock in Val Verde County. Specimens also have been captured in January and February in the junior high school at Edinburg in Hidalgo County. Students found them hanging from the rough plaster ceiling in one of the halls.

Although they may congregate in large numbers at a roosting site (as many as 500,000 individuals; Barbour and Davis 1969), these bats do not form the

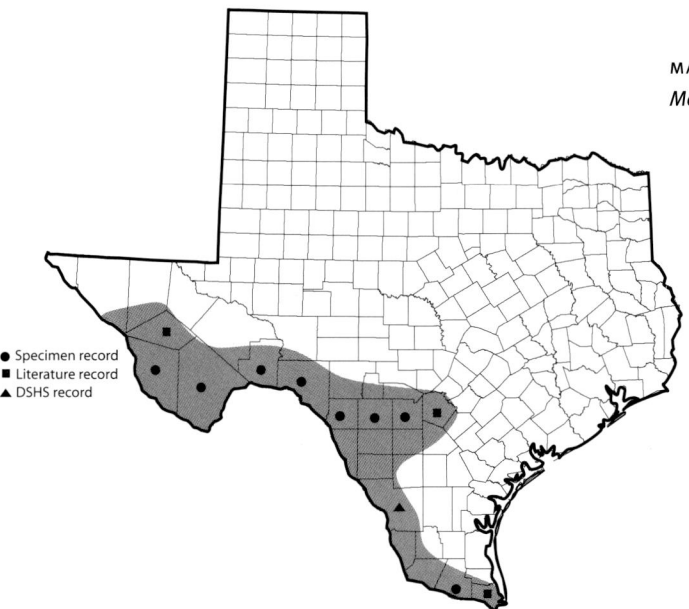

MAP 1. Distribution of the ghost-faced bat, *Mormoops megalophylla*.

- ● Specimen record
- ■ Literature record
- ▲ DSHS record

compact clusters typical of many other species; rather, they tend to space themselves approximately 15 cm apart across the roost ceiling (Raun and Baker 1958). Therefore, larger sites are required to house great numbers of these bats. One such chamber used by *Mormoops* at Frio Cave was described by Eads et al. (1956) as "10 to 20 feet (3.0–3.6 m) high, 30 to 40 feet (9.1–12.2 m) wide, and about the same length." Temperature and humidity in this chamber, which was called the "warm room," were higher than in the remainder of the cave. As many as 6,000 ghost-faced bats were observed at this site during winter, whereas the colony at Haby Cave was described as "small."

Lactating females roost in caves known as maternity roosts, separate from males and nonreproductive females. This is most likely the result of differences in the basal metabolic rate between males and females. Males have a higher metabolic rate and prefer relatively cooler caves (30°C, 86°F); repro-

ductively active females have a lower metabolic rate and prefer very warm caves (36°C, 97°F); and nonreproductive females are intermediate between the two (34°C, 93°F; Bonaccorso et al. 1992). However, ghost-faced bats have been found roosting in caves in Mexico at temperatures near 16°C (61°F; Avila-Flores and Medellín 2004). Other species that often cohabit roosting sites with *Mormoops* and generally greatly outnumber them include the cave myotis (*Myotis velifer*) and the Brazilian free-tailed bat (*Tadarida brasiliensis*).

Populations of this species generally have a highly skewed sex ratio in favor of one gender or the other depending on the time of year and the location of the roost within its geographic range. In Texas, the sex ratio is skewed in favor of females: In Brewster and Presidio counties, only 3–8% of captures of this species were male (Higginbotham and Ammerman 2002; Yancey 1997). However, the sex ratio is skewed in the opposite direction in South American populations: In Ecuador only 7% of captures were female (Boada et al. 2003).

Once out of the roost, individuals quickly fly to foraging sites along arroyos and canyons. These bats are relatively strong, fast flyers and travel at high altitudes to foraging areas (Bateman and Vaughan 1974). Foraging may occur over standing water, and these bats often are collected using mist nets positioned over ponds of water along arroyos and canyons. Clyde Jones and his graduate students at Texas Tech University captured many ghost-faced bats under such conditions in Big Bend State Ranch (Presidio County; Yancey 1997) and along Limpia Creek in the Davis Mountains (Jeff Davis County; DeBaca and Jones 2002).

Their food appears to consist entirely of insects that they capture in flight. Their echolocation calls are consistent with this foraging strategy (Smotherman and Guillén-Servent 2008). The stomachs of 2 individuals from Big Bend National Park were entirely filled with moths (Easterla 1973). At Big Bend Ranch State Park, the stomachs of 45 individuals included moths (100% occurrence), beetles (4.4%), true bugs (4.4%), and net-winged insects (2.2%; Yancey 1997). The bats begin returning to their caves and roosting sites, on average, about 7 hours after leaving them (Bateman and Vaughan 1974; Boada et al. 2003).

Few data on the breeding habits of *Mormoops* in Texas are available. In Big Bend National Park 2 pregnant females, each containing a single embryo, were captured in mid-June (Easterla 1973). At Big Bend Ranch State Park, 30 pregnant females were taken from April 29 to June 9, and all of them contained only a single embryo. In Coahuila and Nuevo Leon, 2 Mexican states bordering Texas, gravid females have been captured in March, April, and May. Each gravid female contained a single embryo (Villa-R. 1966). Farther south in Yucatan, Mexico, gravid females have been reported in February (Jones et al. 1973).

Lactating females have been captured at Big Bend Ranch State Park from mid-June to mid-September (Yancey 1997). Nursing appears to continue for at least 3 months, one of the longest durations known in bats (Hill and Smith 1984). Based on data collected in Central America and Mexico, it seems that mating begins in December. In Texas, Arizona, and northern Mexico, sexually mature females taken between February and June are likely to be gravid or lactating; no gravid females have been reported from late June through January in this portion of its range. Thus, it appears that the period of reproduction is confined to late winter and early spring in Texas and that each reproductively active female gives birth to only one offspring each year.

As is true of all species in the family Mormoopidae, males lack a baculum (os penis; Krutzsch 2000). In females, it appears that ovulation occurs only from the left ovary; thus, placental scars are asymmetrically found on the left side of the reproductive tract, although the number of females examined was small (6), so it is not possible to conclude anything about the functional capabilities of the right side of the tract (Rasweiler and Badwaik 2000).

Only 1 female was submitted to the DSHS for rabies testing from 1996 to 2000, and it was negative (Webb County, 2000VR7972). In addition, 42 free-flying individuals from Big Bend Ranch State Park (Presidio County) collected in 1994 were tested for rabies and found negative (Yancey et al. 1997). In Mexico, colonies of ghost-faced bats are known to suffer from epidemics of rabies that may be cyclic (Jiménez Guzmán 1982) and may cause mass mortality of this bat (Villa-R. 1955; Villa-R. and Jiménez Guzmán 1960).

Status. The IUCN 2011 status of the ghost-faced bat is "least concern." This bat is probably more common than previously thought (Scudday 1976a). In Presidio and Brewster counties it is often one of the most frequently captured species (Higginbotham and Ammerman 2002; Yancey 1997). The wintering cave populations that occur along the southern edge of the Edwards Plateau roost at just a few sites and could be susceptible to disturbance and disruption.

Specimens Examined. Go to www.batsoftexas.com for more detailed information about the total of 194 specimen records of *M. megalophylla* from Texas. Additional records: *Jeff Davis County,* (2) (DeBaca and Jones 2002), *Webb County,* (1) (DSHS #2000VR7972).

References. 23, 54, 60, 62, 75, 86, 92, 106, 118, 120, 145, 144, 155, 164, 178, 265, 269, 271, 323, 336, 338, 345, 368, 369, 381, 386, 486, 511, 564, 569, 575, 617, 619, 642, 682, 743, 767, 793, 830, 833, 834, 848, 869, 966, 971, 976, 979, 988, 1024, 1026, 1028, 1029, 1041, 1074, 1077, 1091, 1093, 1112, 1157, 1163, 1164, 1166, 1182, 1200, 1201, 1246, 1255

Family Phyllostomidae

The Phyllostomidae is a large family of New World bats primarily limited to tropical and subtropical areas, although a few species reach northward to subtemperate areas in the United States. The 160 species included in this family are characterized by a fleshy appendage, or nose leaf, that projects from the rostrum. Most of these bats feed on fruit or nectar, but this family also contains a few insectivores, carnivores, and the true vampire bats. Three species of phyllostomid bats have been recorded in Texas, including 2 nectarivores and 1 vampire, but none is widely distributed or very common.

Choeronycteris mexicana Tschudi, 1844
Mexican Long-Tongued Bat

Etymology. The generic name comes from 2 Greek words, *chorios* and *nykteris,* which mean "young pig" and "bat," respectively. The specific name denotes the location of the type locality and, fortuitously, the major area where this bat is distributed (Stangl et al. 1993).

Subspecies. *Choeronycteris mexicana* Tschudi, 1844, is a monotypic species with no subspecies recognized.

Description. This is a medium-sized bat (forearm = 43–45 mm) with a long, slender muzzle and prominent nose leaf approximately 5 mm tall. Pelage color ranges from sooty gray to brownish. A minute tail is present and extends less than halfway to the edge of the interfemoral membrane. The long tongue can

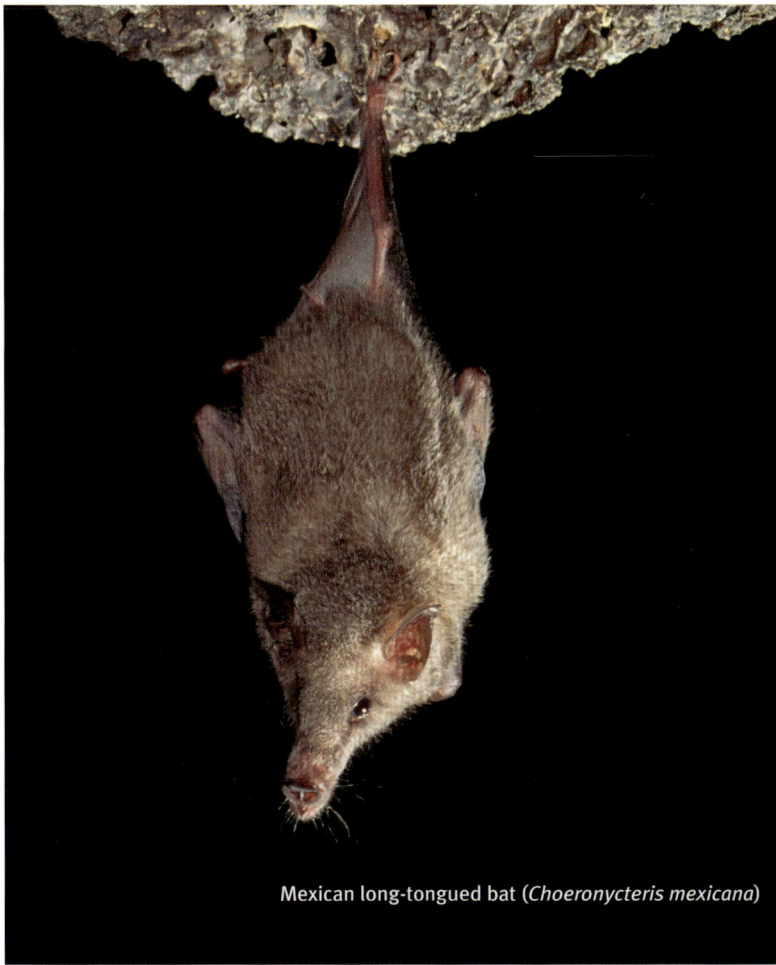
Mexican long-tongued bat (*Choeronycteris mexicana*)

extend to one-third its body length. Dental formula: I 2/0, C 1/1, Pm 2/3, M 3/3 × 2 = 30. Average external measurements are as follows: total length, 85 mm; tail, 10 mm; hind foot, 14 mm; ear, 16 mm; forearm, 44 mm. Weight: 10–20 g.

This bat is similar to the slightly larger Mexican long-nosed bat (*Leptonycteris nivalis*). The tail of *L. nivalis*, however, is not evident, and the interfemoral membrane is reduced to a mere fringe that lines the insides of the legs, in contrast to the better-developed interfemoral membrane of *C. mexicana*.

Distribution. This primarily Mexican species is rarely encountered in extreme southern Texas, New Mexico, Arizona, and California, although there are records of it as far north as Las Vegas, Nevada, and Grand Canyon National Park, Arizona (Constantine 1987; Cryan and Bogan 2003). The first documented record of the Mexican long-tongued bat in Texas was of a single specimen captured on December 11, 1970, at Santa Ana National Wildlife Refuge (Hidalgo County) in extreme South Texas. This bat was not preserved and is known

MAP 2. Distribution of the Mexican long-tongued bat, *Choeronycteris mexicana,* and Mexican long-nosed bat, *Leptonycteris nivalis.*

- Mexican long-tongued bat
- Mexican long-nosed bat

only from photographs deposited at the TCWC at Texas A&M University; however, the photographs clearly indicate the bat was *C. mexicana.* Subsequently, other scattered individuals have been obtained at Santa Ana Refuge and at Laguna Atascosa National Wildlife Refuge (Cameron County; Fernandez et al. 2000) and in El Paso County (Balin 2009). Additional specimen records from the DSHS document the occurrence of this species in Midland County in October 1995 and in Hays County in 2010. Thus, the distribution of this bat in Texas is enigmatic. It seems to be a southern species that is gradually making its way into Texas.

Life History. Mexican long-tongued bats inhabit deep canyons in the small, insular mountain ranges where they use caves and mine tunnels as day roosts. They occupy a wide range of habitats from arid desert to mixed oak-conifer forest; most records in Arizona and New Mexico are above 1,500 m (Fleming et al. 2003). Although they are widespread, they are relatively scarce throughout their range (Arita and Santos-del-Prado 1999). Their roosts are typically in relatively mesic areas with flowering agave in the vicinity (Cryan and Bogan 2003). In the northern part of their range they have been found roosting with Townsend's big-eared bats (*Corynorhinus townsendii*), cave myotis (*Myotis velifer*), big brown bats (*Eptesicus fuscus*), and Brazilian free-tailed bats (*Tadarida brasiliensis;* Arroyo-Cabrales et al. 1987). These bats also have been found in buildings.

Mexican long-tongued bats live in small colonies (< 50 individuals) in shallow, relatively well-lit and exposed roost sites. These bats do not cluster tightly together until temperatures drop below 21°C (70°F); instead, they roost spaced 2–5 cm apart. They hang by 1 foot so they can easily rotate 360° to watch intruders. They also are extremely wary and tend to take flight when disturbed (Cryan and Bogan 2003).

The main foods of these bats include pollen, nectar, fruit, and probably insects. They are obligate nectar and pollen consumers and important pollinators and seed dispersers for columnar cacti and about 60 species of *Agave* throughout their range (Valiente-Banuet et al. 1996). Because of their longer tongues, they may be able to recover nectar from a greater variety of night-blooming plants than the other nectar-feeding bat that occurs in Texas, the Mexican long-nosed bat (*Leptonycteris nivalis*). There are reports of this species feeding from hummingbird feeders (Cockrum and Petryszyn 1991), which may help sustain individuals when natural food sources are lacking. However, sugar water lacks the essential nutrients required for long-term survival.

Parturition occurs from June to early July in Arizona and New Mexico, and young have been reported as early as mid-April in Sonora, Mexico. Females give birth to 1 young per year. One of us (David J. Schmidly) collected a pregnant female (which gave birth shortly after capture) in May in the San Carlos Mountains of northern Tamaulipas, Mexico, which is no more than 240 km south of the Santa Ana Wildlife Refuge. Pregnant and lactating females have been recorded in March and June in Coahuila, Mexico, just south of the Texas border (Wilson 1979). Unlike many other bats, newborn *C. mexicana* are born in a remarkably advanced state

and are well furred (Barbour and Davis 1969). Females have been seen carrying their pup in flight, but probably carry them only from one shelter to another, not while foraging.

No evidence has been found that populations of this species in south-central Mexico migrate to follow the availability of food sources. However, it has been suggested that northern populations migrate to avoid the high metabolic costs associated with winter. This is further supported by the fact that columnar cacti that occur in the southwestern United States are not specialized for 1 type of pollinator, unlike the cacti that occur in the southern portion of this species' range (Valiente-Banuet et al. 1996). Three bats were submitted for rabies testing in the last 15 years, and all were negative.

Status. The IUCN 2011 status of the Mexican long-tongued bat is "near threatened." Recent surveys provide no evidence of significant population decline in New Mexico and Arizona (Cryan and Bogan 2003). For a long time this species was known in Texas on the basis of a single individual captured in Hidalgo County, which led to speculation that it was only of accidental occurrence in Texas. However, the recent documentation of specimens in Cameron, Hays, Midland, and El Paso counties in the western part of the state suggests there is a need to look for the species elsewhere in Texas. This species is listed as endangered by the Mexican government, and one could argue that it should be afforded the same status in Texas.

Mexican long-nosed bat (*Leptonycteris nivalis*)

Specimens Examined (o). Additional records: *El Paso County* (1) (Balin 2009); *Hays and Midland counties* (1) (DSHS).

References. 1, 62, 71, 92, 95, 104, 106, 261, 271, 280, 298, 419, 423, 432, 511, 575, 576, 730, 767, 833, 869, 932, 1026, 1028, 1029, 1091, 1148, 1164, 1232

Leptonycteris nivalis (Saussure, 1860) Mexican Long-Nosed Bat

Etymology. The generic name is derived from the Greek words *leptos*, meaning "small" or "slender," and *nykteris* for "bat." The specific name is from the Latin *nivis*, which means "snow." This probably refers to the location where the type specimen was collected, which was from the snowline of Mount Orizaba in Veracruz, Mexico (Hensley and Wilkins 1988).

Subspecies. *Leptonycteris nivalis* (Saussure, 1860) is a monotypic species, and no subspecies are recognized.

Description. This is a relatively large (forearm = 56–59 mm), sooty brown bat with a long muzzle and prominent nose leaf. The tips of the individual hairs are silvery, and the bases are white. The tail is so minute that it appears to be lacking, although it actually consists of 3 vertebrae, and the interfemoral membrane exists as a narrow fringe along the inside of each leg. The Mexican long-tongued bat (*Choeronycteris mexicana*) is similar in appearance but has a broader interfemoral membrane, a distinct tail, and a longer, narrower muzzle. Dental formula: I 2/2, C 1/1, Pm 2/3, M 2/2 × 2 = 30. Average external measurements are as follows: total length, 83 mm; hind foot, 17 mm; ear, 15 mm; forearm, 57 mm. Weight: 28 g.

Distribution. A highly colonial, cave-dwelling, migratory species, this bat has been recorded in the United States only in southwestern New Mexico (Hoyt et al. 1994) and Trans-Pecos Texas, where it has been captured in Big Bend National Park, Brewster County, and the Chinati Mountains of Presidio County (see distribution map in the account for *Choeronycteris mexicana*). Although occasionally found in the desert scrub habitats at lower elevations, these bats apparently prefer mountainous, pine-oak habitats at elevations of 1,000 to 2,300 m (Arita 1991). Mexican long-nosed bats occupy such areas in Texas and New Mexico from June to August, after which time they move out of the States to winter in Mexico. Few adult males have been recorded in Texas, which suggests that the sexes may segregate geographically, with males rarely appearing in the northernmost part of the species' range. However, work in southwestern New Mexico found a 50:50 ratio of males to females in populations there (Hoyt et al. 1994).

Although this bat occurs throughout much of Mexico, it is relatively rare over most of its range (Arita and Santos-del-Prado 1999). Several caves in central Mexico known to have housed considerable numbers of these bats in the past now contain only small colonies or lack bats altogether (Wilson et al. 1985).

However, there may be some indication that at least 1 colony is recovering (Medellín 2003). A colony of Mexican long-nosed bats found in the Chisos Mountains of Big Bend National Park (Emory Peak Cave) is the only known population of this species in Texas. Two caves housing both *L. nivalis* and the lesser long-nosed bat (*Leptonycteris yerbabuenae*) in Hidalgo County, New Mexico, were recently located in the Animas and Big Hatchet mountains (Paul Cryan, USGS, pers. comm.).

Life History. At Emory Peak Cave, *L. nivalis* form a large cluster with half-grown young and adult females intermingled. Adult males are rarely encountered. They share the cave with a colony of Townsend's big-eared bats (*Corynorhinus townsendii*) and fringed myotis (*Myotis thysanodes*); each colony roosts in a different part of the cave but not more than 6 m apart. Caves used as maternity roosts in Mexico are cooler than those used by many other species of cave-roosting bats (Moreno-Valdez et al. 2004). At Emory Peak Cave in July and early August, the average temperature in the main chamber, where the bats typically roost, was 16.4°C (61°F), and the average relative humidity was 87.7% (Ammerman et al. 2009). A distinct breeze blows through the cave at all times, and the air has a strong musky odor similar to that of the Brazilian free-tailed bat (*Tadarida brasiliensis*).

The colony size at Emory Peak Cave fluctuates annually; estimates range from zero to as many as 10,650 individuals (U.S. Fish and Wildlife Service 1994). The reasons for this instability are unknown, but Moreno-Valdez et al. (2004) has shown that the number of individuals at Infierno Cave in Nuevo León, Mexico, is highly correlated with the abundance of blooming *Agave*, its main food source. Thus, colony size of this nectar-feeding bat in Texas likely fluctuates with food availability in northern Mexico and Texas. A documented decrease in colony size at Emory Peak Cave in 2008 was probably due in large part to a decline in the number of flowering agave plants that year (Ammerman and Tabor 2008).

Population trends are difficult to assess, and recent observations in Emory Peak Cave cast doubt on the reliability of data in years when the population estimate was zero. Not only do the bats use the "main" room of Emory Peak Cave, but they also roost in deep passageways, where they are known to go undetected (Ammerman et al. 2009). In addition, the presence of guano is an unreliable indicator of cave use because water flows through portions of the main room and washes rocky surfaces clean after summer storms. Recently, heat-sensing cameras have been used successfully to more accurately count the number of bats emerging from Emory Peak Cave (Ammerman et al. 2009; Ammerman and Tabor 2008), but more work is necessary to understand how the cave is used throughout the summer months.

Nectar and pollen from 21 plant species across its range (Sánchez and Medellín 2007) constitute the diet of *L. nivalis*, but the flowers of century plants are their main food source in Texas. Century plants open their flowers at night

and attract bats with copious amounts of nectar. Mexican long-nosed bats are exceptionally strong, highly maneuverable fliers and are able to hover in flight while they feed, much as hummingbirds do. As the bats feed, their fur gets coated with pollen grains. When they fly to another plant in search of more food, they transfer the pollen to a new flower, assisting in cross-fertilization of the plants. Both the plant and the bat benefit from this mutual relationship. This association is the result of the coevolution of bats and plants; their interdependence is so strong that both the bats and the century plants need each other for survival. As the flower stalks of the agaves die in late summer, the bats disappear from the region because there is nothing left for them to eat.

The breeding season is restricted to April, May, and June (Wilson 1979). The only known mating roost of this species is in Morelos, Mexico, where mating occurs from September to February. It is generally thought that most males remain in south-central Mexico throughout the summer and that females migrate northward to establish maternity roosts (Téllez 2001). However, adult males are common occupants of roosts in southwestern New Mexico (Paul Cryan, USGS, pers. comm.).

Females typically give birth to a single pup prior to arriving in Texas. However, the capture of 2 pregnant females on April 25, 1996 (Higginbotham and Ammerman 2002), and others on April 23, 2006 (Brown 2008), at Emory Peak Cave suggests that some females give birth in the United States. The young are weaned in July or August at the peak of the rainy season and during the period of peak flower abundance. After weaning, up to 75% of the individuals captured at Emory Peak Cave are juveniles (Brown 2008).

Status. The IUCN 2011 status of Mexican long-nosed bats is "endangered." This species also is listed as endangered by the Texas Parks and Wildlife Department and the U.S. Fish and Wildlife Service. There is a high annual fluctuation in abundance of this bat at Emory Peak Cave in Texas, as well as throughout its range (Fleming et al. 2003; U.S. Fish and Wildlife Service 1994). Some studies suggest the population is increasing in parts of Mexico (Medellín 2003) but decreasing in Texas (Ammerman et al. 2009; Ammerman and Tabor 2008; Easterla 1972b). Others suggest it is decreasing everywhere (Arita and Humphrey 1988; U.S. Fish and Wildlife Service 1994; Wilson 1985) or that there is no evidence of a long-term decline in abundance (Cockrum and Petryszyn 1991). Such disparate claims point out the need for consistent monitoring of populations using unbiased methods, as one of us (Loren K. Ammerman) began doing in 2005.

Because this bat is highly specialized to consume nectar and pollen, destruction of its food sources could be detrimental to the population. This is particularly true for species of *Agave*, which are lost through agriculture, ranching, and human development. Reciprocally, the health of agave populations and other important plant species in desert ecosystems depends on these pollinators. Disturbance of the 1 known roosting site in Texas could also have a serious impact on the population.

Remarks. Another species of *Leptonycteris*, *L. yerbabuenae*, has been recorded from southwestern New Mexico, southern Arizona, and northern Mexico. Based on its current distribution, this species could occur in Texas, although it has not been recorded in our state to date. The 2 species can easily be distinguished by several characters. *Leptonycteris nivalis* is slightly larger (forearm, 56–59 mm vs. 52–56 mm), has a longer third finger (106–15 mm vs. 92–102 mm), has a conspicuous fringe of hair on the uropatagium, and has long, fluffy hair in contrast to the short, dense hair of *L. yerbabuenae*. In Mexico, this bat is known as the "tequila bat" because it is the major pollinator of the plant from which this beverage is made.

Specimens Examined. Go to www.batsoftexas.com for more detailed information about the total of 95 specimen records of *L. nivalis* from Texas.

References. 11, 54, 58, 61, 62, 69, 70, 71, 92, 95, 96, 106, 168, 192, 261, 271, 336, 337, 380, 381, 432, 497, 511, 564, 573, 591, 794, 830, 833, 834, 843, 932, 933, 976, 1011, 1024, 1026, 1028, 1029, 1064, 1089, 1112, 1115, 1139, 1164, 1200, 1232, 1233

Diphylla ecaudata Spix, 1823
Hairy-Legged Vampire

Etymology. The generic name *Diphylla* is Greek and means "two-leaved"; the species name *ecaudata* has a Latin root that means "without a tail" (Stangl et al. 1993).

Hairy-legged vampire (*Diphylla ecaudata*)

Subspecies. *Diphylla ecaudata* Spix, 1823, is monotypic, and subspecies are not recognized.

Description. This is a relatively large (forearm = 50–56 mm) dark brown bat with short, rounded ears and a short, "pug-nosed" snout. As with other phyllostomid bats, the hairy-legged vampire also has a nose leaf, although this feature is much reduced. The tail is absent, and a narrow, hairy interfemoral membrane extends from the body down the leg (thus, the common name). The dentition is highly modified with the bladelike middle upper incisors larger than the canines; the outer incisors are set close to the canines and are easily overlooked; premolars and molars are much reduced. This easily distinguishes this species from all others that are known to occur in Texas. Dental formula: I 2/2, C 1/1, Pm 1/2, M 2/2 × 2 = 26. Average external measurements are as follows: total length, 83 mm; hind foot, 18 mm; ear, 16 mm; forearm, 53 mm. Weight: 31 g.

Distribution. Primarily tropical in distribution, this bat is known from Texas on the basis of a single female collected on May 24, 1967, in an abandoned railroad tunnel 19 km west of Comstock in Val Verde County. This record extended the range of the hairy-legged vampire approximately 725 km to the northwest of Tamaulipas, Mexico, where it is more frequently encountered. The species is native to Central and South American tropical forests.

Life History. Only 3 species of true vampires exist in the world, and all are from the New World tropics and subtropics. These include the common (*Desmodus rotundus*) and white-winged (*Diaemus youngi*) vampires in addition to *Diphylla*. Most research on these unusual bats has been conducted with the common vampire, and very little information is available for the other 2 species.

Hairy-legged vampires primarily feed on bird blood, unlike the common vampire, which feeds upon the blood of other mammals—including humans on occasion. Vampires depend less upon their ability to fly and more upon stealth to approach their prey. All 3 species have special adaptations that allow them to "walk" up to prey animals. In the wild, *Diphylla* stalks birds in trees as they roost at night. The calcar is used as an opposable digit (like a thumb) for grasping branches as it climbs trees in pursuit of prey (Schutt and Altenbach 1997; Schutt et al. 1998). This is a unique use of the calcar among bats and is similar in its evolutionary development to the panda's "thumb."

Diphylla has been observed feeding from the legs and cloacal region of domestic chickens (Nowak 1999). Their enlarged and razor-sharp, bladelike upper incisors and canines are designed for inflicting a small, V-shaped wound on the prey animal. They do not have a lower lip cleft or grooves along the sides of the tongue, which are adaptations for blood feeding, as exhibited by the other 2 species of vampire bats. Instead, *Diphylla* possess a groove along the roof of the mouth, broad flexible lips, and greatly shortened rostrum, which led to speculation that this bat might form a seal over the wound with its lips and truly "suck" the blood instead of lapping up the fluid meal (Greenhall 1988). However, captive *Diphylla* have been observed feeding and clearly lapping blood in a manner similar to the other vampires (da Silva et al. 1998, Gerald Carter, pers. comm.). An anticoagulant secreted in the saliva keeps the wound open and the blood flowing.

As an adaptation to their peculiar diet, the digestive tract of all vampires has been highly modified. The stomach walls are thin and elastic, allowing a large volume of blood to be ingested at 1 meal (approximately 30 ml), and a network of capillaries around the stomach allows for rapid absorption and excretion of water contained in the blood meal. Although they are exceptionally agile when on the ground, vampires have difficulty flying after feeding and begin to urinate soon after commencing their meal in an effort to reduce their weight. *Diphylla* will share food with conspecifics that fail to find prey on a given night, much like the common vampire bat (Elizalde-Arellano et al. 2007).

Hairy-legged vampire bats seem to be reproductively active throughout the year. Pregnant females have been captured in March, July, August, October, and November in Mexico and Central America. Gestation lasts about 5.5 months, and the young are weaned after 223 days. Females will nurse and regurgitate blood to offspring other than their own (Delpietro and Russo 2002). Normally, *Diphylla* has 1 young per year, which weighs about 4.5 g at birth, but one was captured in Chiapas, Mexico, with 2 nearly full-term fetuses (crown-rump length = 34 mm). There is no indication that *Diphylla* breeds in Texas.

In Mexico, *Diphylla* are found in caves and abandoned mines with temperatures of 23–24°C (75°F; Ceballos and Oliva 2005). They also are rarely found roosting in hollow trees. In contrast to the common vampire, *Diphylla* is thought to be more solitary, and typically 1–3 individuals roost in the same cave (Dalquest 1955). However, as many as 500 individuals were found roosting in Las Vegas Cave in the state of Puebla, Mexico. Although they shared this cave with at least 12 other species, they did not associate with them and roosted individually on the walls (Medellín and López-Forment 1986).

Diphylla commonly roost in association with other bats, many of which are economically important insectivores and plant pollinators. Due to their unusual feeding habits, vampire bats (especially *Desmodus rotundus*) are regarded as a serious problem in much of Latin America because of the damage done to livestock herds through the occasional transmission of rabies. However, the indiscriminant poisoning of bats at roosts in an attempt to control vampires can be detrimental to the other, beneficial species of bats with which they roost.

A better understanding of the natural history of vampire bats has helped develop new methods to control their numbers and minimize the negative impact such control methods have on other species of bats. Upon returning to the roost, vampire bats clean themselves and their roostmates. Therefore, anything applied to the fur of a vampire, including poisons or hormones, will be carried back to the roost and ingested. Ranchers can capture common vampire bats at corrals and coat them with a petroleum product that contains anticoagulants. Upon returning to the roost, the vampires ingest the poison during the grooming process and die from internal hemorrhaging (Baer 1991). The use of a hormonal product to reduce fertility also has been proposed (Serrano et al. 2007). Prevention may be easier for protecting domestic fowl by simply keeping them in roofed wire pens to help prevent attacks by white-winged and hairy-legged vampires.

Status. IUCN 2011 status is "least concern." Although 1 specimen of the hairy-legged vampire bat is known from Texas, it is possible that a thorough search of the caves in the Hill Country and along the Rio Grande will reveal additional records of this species or the common vampire (*Desmodus rotundus*), which has been taken in northern Mexico less than 200 kilometers from the Texas border. However, *Diphylla* is probably of accidental occurrence in Texas.

Remarks. Although the common vampire (*Desmodus rotundus*) has not been recorded from Texas during Recent times, it is known from sub-Recent deposits in the Trans-Pecos, and specimens have been collected no more than 193 km south of the Rio Grande in northern Tamaulipas, Mexico. Under current climate conditions, vampires are unlikely to spread northward into cooler climates because they are constrained by meal size; they cannot meet the increased energy demands that are necessary to maintain their body temperatures in cool environments. However, it is distinctly possible that an occasional common vampire may periodically wander into the southern part of our state.

These 2 species of vampires can be distinguished by characters of the skull, pelage, and thumbs. In *Diphylla*, the first lower incisors are in contact, and the interfemoral membrane is lined with a distinct fringe of moderately long hairs. In *Desmodus*, the first lower incisors are not in contact, and the interfemoral membrane lacks a fringe of hair. *Desmodus* has distinct pads under the thumbs that are lacking in *Diphylla*. An additional feature found only in *Diphylla* is that the second lower incisor is comb-like, containing 7 lobes, and is more than twice as wide as the first lower incisor (Gardner 2007; Greenhall et al. 1983).

Specimens Examined. Go to www.batsoftexas.com for more detailed information about the single specimen record of *D. ecaudata* from Texas.

References. 9, 10, 54, 78, 95, 222, 307, 311, 317, 336, 347, 393, 457, 499, 511, 590, 767, 795, 833, 977, 1013, 1024, 1036, 1037, 1047, 1164

Family Vespertilionidae

Chiefly insectivorous bats, the Vespertilionidae constitute the largest family of bats (407 species) and are distributed worldwide with the exception of arctic regions. Consequently, these bats are found in nearly every conceivable habitat from tropical forests to desert and temperate regions, and most of the species in Texas belong to this family. Many are highly migratory and traverse great distances between their summer and winter ranges. Others, however, do not migrate and instead hibernate in their summer ranges.

Vespertilionids lack the facial adornments found in other families of bats and are commonly referred to as evening bats or vesper bats (vesper means "evening" in Latin). Several species have extremely large and complex ears, but most have small, simple ones. These bats typically have small eyes and a long tail completely enclosed by a well-developed interfemoral membrane.

Thirty-three species of vespertilionid bats range across the United States; of these, 25 are known from Texas.

Myotis austroriparius (Rhoads, 1897) Southeastern Myotis

Etymology. The generic name, *Myotis*, comes from the Greek words *mys*, meaning "mouse," and *ous* or *otis*, meaning "ear." The specific name comes from the Latin word for "southern," *auster*, and *riparius*, which refers to riparian habitats found along edges of streams (Stangl et al. 1993).

Subspecies. This bat is monotypic, and subspecies are not recognized, as indicated by the most recent taxonomic revision of the species (LaVal 1970).

Description. This is a small- to medium-sized bat (forearm = 33–40 mm) with thick, woolly fur and no keel on the calcar. They have relatively large feet, and the long toe hairs extend beyond the tips of the claws. Two color phases are found in Texas. The most common bats are a dull gray to gray-brown color, but

Southeastern myotis (*Myotis austroriparius*)

occasionally bright, orange-brown individuals are encountered. Individual hairs on the venter are bicolored with a black base and white tips, giving the belly a general white appearance that contrasts sharply with the dorsum. The facial skin is pinkish brown. The cranium is globose and normally has a low sagittal crest. Dental formula: I 2/3, C 1/1, Pm 3/3, M 3/3 × 2 = 38. Average external measurements are as follows: total length, 87 mm; tail, 34 mm; hind foot, 10 mm; ear, 13 mm; forearm, 37. Weight: 6 g.

The southeastern myotis is easily confused with 2 other species of *Myotis* that might occasionally occur in Texas— *M. septentrionalis*, which has relatively longer ears and tail, and *M. lucifugus*, which may be distinguished by its glossy, dark brown fur (Table 8).

Distribution. This bat occurs throughout the southeastern United States and is known from the South Central Plains (Pineywoods), East Central Texas Plains, and Western Gulf Coastal Plains in Texas. It reaches the westernmost part of its North American range in these areas of Texas and is relatively uncommon in the state. One specimen was collected in Comanche County in the Central Oklahoma/Texas Plains region, 240 km west of the previously known range (Higginbotham and Jones 2001). Collecting records are intermittent throughout the year, suggesting that this bat is a year-round resident of the state.

Life History. *Myotis austroriparius* predominantly roosts in caves in that part of its range where suitable caves occur. In East Texas, however, such caves are uncommon. Therefore, in Texas, these bats roost primarily in live, hollow, bottomland hardwood trees in proximity to slow-moving rivers and in structures

TABLE 8. *Trenchant morphological characters useful in distinguishing Texas species of Myotis (measurements in millimeters).*

Species	Ear Size (usual length)	Hind foot (usual length)	Overall (forearm length)	Pelage coloration	Special hair on body	Keel on calcar	Elevation from rostrum to braincase	Sagittal crest
M. austroriparius	small (11–16)	large (9–12)	medium (33–40)	dull gray or orange-brown	hairs on toes extend beyond tips of claws	none	abrupt	slight
M. californicus	small (9–13)	small (6–8)	small (29–34)	light brown	—	well developed	abrupt	none
M. ciliolabrum	small (12–14)	small (6–8)	small (30–34)	buff brown	—	well developed	gradual	none
M. occultus	small (11–15)	large (8–10)	medium (34–41)	shiny, dark brown	—	none	gradual	distinct
M. septentrionalis	large (14–19)	large (8–10)	medium (32–39)	dull, gray-brown	few, isolated hairs on edge of tail membrane	slight/none	gradual	none
M. thysanodes	large (16–20)	large (9–12)	large (40–45)	buff brown	thick fringe of hair on edge of tail membrane	none	abrupt	slight
M. velifer	intermediate (12–17)	large (8–11)	large (37–47)	dark, dull	bare patch on back between scapulae	none	abrupt	well developed
M. volans	small (12–14)	large (8–10)	medium (35–41)	dark brown	underside of wing furred to elbow	well developed	abrupt	well developed
M. yumanensis	small (12–14)	large (9–11)	small (32–36)	light, buff brown	—	none	abrupt	none

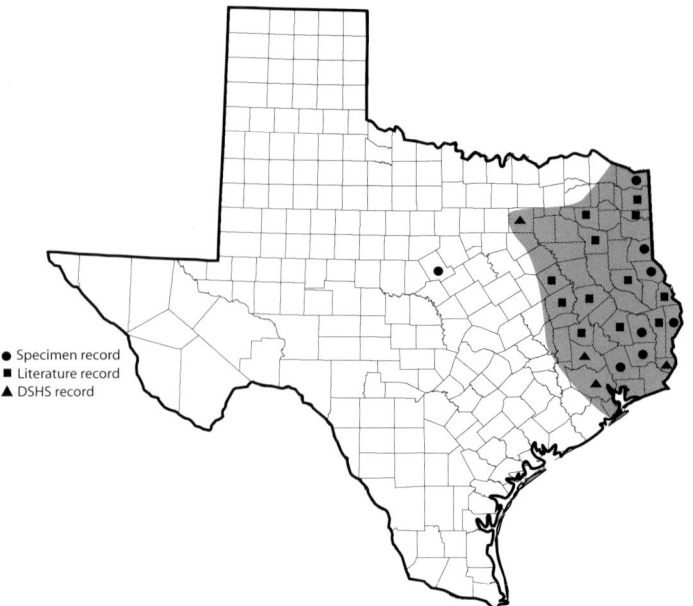

MAP 3. Distribution of the southeastern myotis, *Myotis austroriparius*.

● Specimen record
■ Literature record
▲ DSHS record

such as abandoned houses and culverts. Important species of trees used as roosts include black gum (*Nyssa sylvatica*), water tupelo (*N. aquatica*), and sweet-gum (*Liquidambar styraciflua;* Carver and Ashley 2008; Mirowsky et al. 2004).

Starting in March, large, hollow trees are used as maternity roosts and are occupied by several hundred females throughout the summer months. All maternity roosts and 61% of all roosts of this species in Texas have been found in trees (Horner and Maxey 1998). Occasionally Rafinesque's big-eared bats (*Corynorhinus rafinesquii*) are found sharing their maternity roosts. In winter, *M. austroriparius* roosts in small scattered groups, in contrast to larger maternity colonies found during the summer. These small groups are usually found either in smaller trees or cement culverts. They roost in association with American perimyotis (*Perimyotis sub-flavus*) and big brown bats (*Eptesicus fuscus*) in cement culverts in the winter.

They do not enter torpor during the winter in East Texas but instead are alert and fly when approached (Walker et al. 1996).

When they leave their diurnal roosts late in the evening (usually about dark), they fly to nearby ponds and slow-moving streams, over which they forage and from which they drink. They fly low over the water, usually within 60 cm of the surface, capturing insects. In Texas, this bat apparently is active throughout the year. Although the specific prey items are not well known in Texas, in the southeastern United States midges, mosquitoes, small moths, small beetles, and crane flies make up its diet (Whitaker and Hamilton 1998; Zinn and Humphrey 1981).

The southeastern myotis is unique among species of myotis in that it generally gives birth to twins; other species of myotis usually have only 1 young per year. Twinning in *M. austroriparius* is thought to be an adaptive response to the high mortality rate of young (about 12% during the first 4 weeks of life). However, one cost of twinning is that the young are more altricial (less developed) than those of other cave bats (Sherman 1930). Southeastern myotises prefer relatively high humidity in their maternity roosts; thus, they often roost in caves with water. Due to their relative weakness at birth, many young bats may fall and drown. Even at roosting sites with no water below, a fall for a young bat usually results in death (Foster et al. 1978; Hermanson and Wilkins 1986; Sherman 1930).

In Texas, pregnant females have been captured as early as April 2 and as late as May 11; the first pups are born in early May (Mirowsky et al. 2004). The breeding period is unknown, although in northern Florida, *M. austroriparius* is believed to mate in both the late fall and the early spring, with parturition occurring from late April to late May (Barbour and Davis 1969). Maternity colonies in caves can become quite large at this time, and the bats form dense clusters of as many as 90,000 individuals. Reportedly, these roosts are commonly found over pools of water (in caves or in trees in floodplains), and adult males constitute about 20% of such aggregations (Reynolds and Mitchell 1998). After the young mature, male may join these colonies in increasing numbers, but they leave again by October, when the populous nursery colonies begin to disperse.

Young bats weigh slightly over 1 gram at birth, but they grow quickly and are able to fly at 5–6 weeks. Both sexes reach sexual maturity before 1 year of age. Because they are active for most of the year, it has been suggested that the higher fecundity of these bats may be a response to their higher adult mortality rate compared to other northern species that hibernate and are inactive a large part of the year (Foster et al. 1978; Rice 1957). Both rat and corn snakes are known to frequent caves used by this species and may prey heavily upon these bats. Their habit of roosting over water, either in trees or in caves, is thought to lower the risk of predation (Foster et al. 1978). Opossums and owls also have been reported to prey upon them, and cockroaches are known to feed upon young bats that have fallen onto cave floors.

Four individuals submitted for rabies testing from 1984 to 1987 and 1 individual submitted between 1996 and 2000 were all negative.

Status. The IUCN 2011 status of the southeastern myotis is "least concern." It is considered a "species of concern" by the U.S. Fish and Wildlife Service. Although these bats may be more abundant in Texas and Arkansas than was once believed, major declines in populations (up to 50%) have been documented in other states over the past several decades, particularly in Florida, where most of the large maternity roosts are located (Cochran 1999; Gore and Hovis 1992; Humphrey and Gore 1992). This species is considered endangered in Illinois, Indiana, and Kentucky, threatened in South Carolina, protected in Alabama, and a species of special concern in North Carolina (Hofmann et al. 1999; Reynolds and Mitchell 1998). Clearing and draining of hardwood bottomland forests could reduce the amount of habitat available to this species in Texas (Clark 2003).

Specimens Examined. Go to www.batsof texas.com for more detailed information about the total of 25 specimen records of *M. austroriparius* from Texas. Additional records: *Cass, Houston, Jasper, Marion, Nacogdoches, Polk, Sabine, Smith, Wood counties* (Mirowsky et al. 2004) and *Freestone, Leon, Walker counties* (Walker et al. 1996).

References. 15, 102, 106, 219, 246, 251, 336, 421, 437, 495, 511, 524, 542, 568, 576, 586, 597, 632, 726, 760, 767, 813, 826, 833, 895, 981, 983, 1025, 1026, 1028, 1030, 1044, 1051, 1091, 1175, 1204, 1261

Myotis californicus (Audubon and Bachman, 1842)
California Myotis

Etymology. The generic name, *Myotis*, comes from the Greek words *mys*, meaning "mouse," and *ous* or *otis*, meaning "ear." The specific name refers to the original collection locality in California (Stangl et al. 1993).

Subspecies. Texas specimens are referable to the subspecies *M. c. californicus* (Audubon and Bachman, 1842), as indicated by the most recent taxonomic revision of the species (Bogan 1975).

Description. This is one of the smallest (forearm = 29–36 mm) species of *Myotis* in Texas. Pelage coloration is variable, but most are a light reddish brown. The bases of the hairs are black. The ears are black, and a dark facial mask extends across the eyes and rostrum from ear to

ear. The calcar is keeled, and, compared with other myotises, the feet are tiny and the tail and ears relatively long (Table 8). Dental formula: I 2/3, C 1/1, Pm 3/3, M 3/3 × 2 = 38. Females are slightly larger than males. Average external measurements are as follows: total length, 82 mm; tail, 36 mm; hind foot, 6 mm; ear, 13 mm; forearm, 32 mm. Weight: 3–5 g.

This bat is easily confused with the western small-footed myotis (*Myotis ciliolabrum*), from which it differs in having a sharply rising braincase, in contrast to the flattened skull of *M. ciliolabrum*. This cranial feature gives the California myotis a more prominent forehead than is evident in the western small-footed myotis (Bogan 1974; van Zyll de Jong 1984). The California myotis also has a shorter thumb (6–7.5 mm versus 8–8.5 mm; measurement includes wrist) and slightly shorter ears and a longer tragus than does *M. ciliolabrum* (Gannon et al. 2001).

The American parastrelle (*Parastrellus hesperus*) also resembles this species, although it may be easily distinguished by the lack of a keel on the calcar and the short, blunt tragus as opposed to the sharp-pointed tragus of *M. californicus*.

Distribution. The California myotis, which commonly occurs throughout the western United States and Mexico, is known in Texas predominately from the Chihuahuan Desert and the Arizona/ New Mexico Mountains (Trans-Pecos), where it has been found in desert, grassland, and wooded habitats. Apparently this is one of the few species that winters in the Trans-Pecos. Torpid individuals have been discovered in irrigation

California myotis (*Myotis californicus*)

tunnels in Presidio County in December (Young and Scudday 1975) and in the Franklin Mountains of El Paso County in January (Dooley 1974). Two disjunct records, one each from the Panhandle (High Plains region) and South Texas (Western Gulf Coastal Plains region), also exist. Based on the distribution of this species in Mexico, its occurrence in the southern portion of the state is not unexpected.

Life History. The California myotis typically roosts in a variety of crevicelike places such as rock fissures or behind signs and loose tree bark, and it is frequently found in buildings. This bat uses manufactured structures as roosts more than any other species of *Myotis* (Schmidly 1977). In winter, caves,

mine tunnels, and buildings are used as hibernacula. Colonies are usually small, consisting of up to 25 individuals. These bats apparently do not form the compact clusters typical of many other species. They are fairly active in winter, and records from the southwestern United States are relatively abundant during this season (Geluso 2007). In summer, they are quite transient and will use any suitable and immediately available site for shelter. They exhibit low roost fidelity in British Columbia (Barclay and Brigham 2001; Brigham et al. 1997). In Arizona and Nevada, they primarily use desert shrubs and trees as night roosts (Hirshfeld et al. 1977), but in California they preferentially use buildings (Krutzsch 1954).

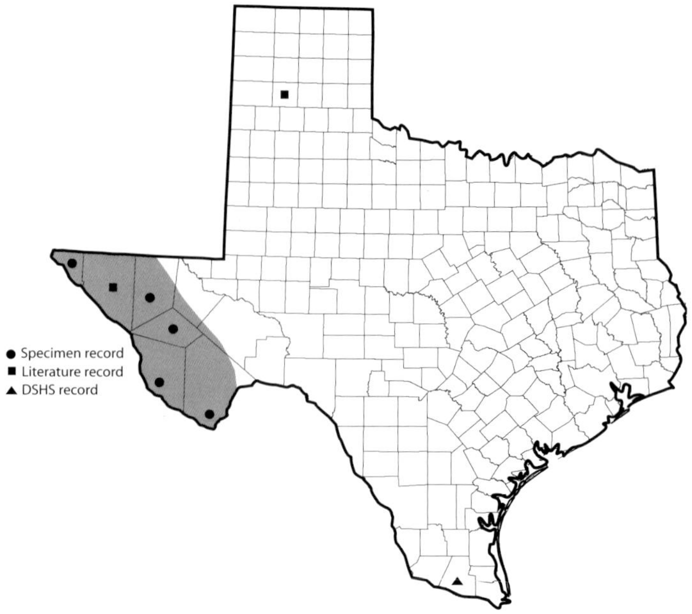

MAP 4. Distribution of the California myotis, *Myotis californicus*.

● Specimen record
■ Literature record
▲ DSHS record

California myotises appear on the wing much later in the evening than most species of *Myotis*. Their flight is relatively slow, fluttery, and highly erratic. Foraging occurs near vegetation and relatively close to the ground. They feed rapidly and fill their stomachs in a short time, after which they retire to a night roost near their foraging site. Feeding bouts continue throughout the night until almost dawn (Krutzsch 1954).

In Texas, specific food items are unknown. In other parts of its range, *M. californicus* is known to feed primarily on small moths, flies, beetles, and true bugs that occur between, within, or below the vegetative canopy (Black 1974). At more northerly latitudes this species partitions food resources spatially in areas where it is sympatric with the western small-footed myotis (*Myotis ciliolabrum*; Gannon et al. 2001; Woodsworth 1981). However, in Texas, the 2 species overlap substantially and, in Big Bend National Park, one of us

(Loren K. Ammerman) has caught both species at the same location on the same night on multiple occasions.

Breeding occurs in the fall and is followed by delayed fertilization. Males with descended testes were captured on November 22 in Big Bend National Park. Pregnant and lactating females have been captured in Big Bend National Park on May 20 (Ammerman, unpublished data). The date of birth can vary considerably, any time from late spring to early summer, at which time a single young is born (Krutzsch 1954). Maternity colonies are typically small, and, as with many other species, growth is quite rapid. By mid-July, the young are of adult size and volant.

The California myotis is well adapted for life in arid environments; its kidney has a greater ability to concentrate urine than that of the western small-footed myotis (*Myotis ciliolabrum*), thus allowing it to conserve water (Geluso 1978). They also are relatively long lived and have a potential reproductive lifespan of 15 years (Duke et al. 1979). Only 1 specimen was submitted to the DSHS from 1996 to 2000 for testing and was not rabid.

Status. The IUCN 2011 status of the California myotis is "least concern." This is a common bat in the Trans-Pecos and does not appear to need special conservation status at this time.

Remarks. The genetic divergence between *M. californicus* and *M. ciliolabrum* does not mirror morphological differences between the 2 species. In fact, they are more distinct morphologically than genetically, which suggests that they diverged quite recently (Rodriguez and Ammerman 2004).

Specimens Examined. Go to www.batsof texas.com for more detailed information about the total of 176 specimen records of *M. californicus* from Texas. Additional records: *Hudspeth County* (2) (Higgin-botham et al. 2002), *Randall County* (1) (Choate and Killebrew 1991).

References. 26, 54, 60, 62, 86, 92, 102, 106, 111, 140, 147, 157, 158, 168, 189, 230, 323, 336, 339, 354, 355, 362, 381, 405, 420, 421, 423, 443, 455, 466, 470, 475, 511, 564, 566, 570, 574, 575, 625, 626, 679, 727, 767, 830, 833, 994, 1001, 1024, 1026, 1028, 1029, 1091, 1112, 1152, 1153, 1164, 1243, 1256

Myotis ciliolabrum (Merriam, 1886) Western Small-Footed Myotis

Etymology. The generic name, *Myotis*, comes from the Greek words *mys*, meaning "mouse," and *ous* or *otis*, meaning "ear." The specific name comes from the Latin words *cilium*, which means "eye-lid" or "eyelash," and *labrum*, for "lip" (Stangl et al. 1993).

Subspecies. Texas specimens are refer-able to the subspecies *M. c. melanorhi-nus* (Merriam, 1890), which is slightly darker in color than *M. c. ciliolabrum* (Hall 1981; Holloway and Barclay 2001; van Zyll de Jong 1984). Recently it has been recommended that *M. c. melano-rhinus* be elevated to full species status (Simmons 2005). However, in our opinion, additional research is needed to support this recommendation. Until such support is forthcoming, we will continue to recognize the subspecies *M. c. melanorhinus* in Texas.

Description. This small *Myotis* (forearm = 30–34 mm) has buff brown pelage with dark roots and flaxen tips. The ears, face, and wing membranes are dark and contrast with the lighter pelage. A dark facial mask is present across the rostrum and eyes from ear to ear. Underparts are whitish to pale buff in color. The foot is less than half the length of the tibia, and the calcar is keeled. This bat is easily confused with the California myotis (*Myotis californicus*) and the American parastrelle (*Parastrellus hesperus*), from which it differs as described in the species account of *M. californicus*. Dental formula: I 2/3, C 1/1, Pm 3/3, M 3/3 × 2 = 38. Females are generally slightly larger than males. Average external measure-ments are as follows: total length, 80 mm; tail, 38 mm; hind foot, 7 mm; ear, 13 mm; forearm, 33 mm. Weight: 4–5 g.

Distribution. The western small-footed myotis is known in Texas primarily from the mountainous regions of the Chi-huahuan Desert and the Arizona/New

Western small-footed myotis (*Myotis ciliolabrum*)

Mexico Mountains ecoregion (Trans-Pecos). It occurs principally in wooded areas, although a few individuals have been taken in grassland and desert scrub habitats. Single specimens from Palo Duro Canyon, Armstrong County (Hollander and Jones 1987), and Canyon, Randall County (DSHS record) document its presence in the High Plains and Southwestern Tablelands of the Texas Panhandle. A small, resident population may occur in this area in the vicinity of Palo Duro Canyon.

Apparently this bat does not winter in Texas as specimens have been taken only from March through July. Also, the sparse collecting records suggest it may be fairly rare in our state. Outside

● Specimen record
■ Literature record
▲ DSHS record

Texas, the species is found throughout western North America from Canada into Mexico.

Life History. In the western United States, these bats are inhabitants of deserts, semideserts, and desert mountains. Over much of its range in North America, the western small-footed myotis is believed to hibernate within its summer range, but this has yet to be substantiated in Texas. Bats observed in winter are often found wedged deeply into narrow cracks and crevices in the rock ceilings of old mines. They are usually found hibernating alone (Kuenzi et al. 1999; Perkins et al. 1990; Szewczak et al. 1998). This is reportedly one of the last bats to enter torpor in winter. When probed from these crevices, they are able to fly, which indicates they do not go into a deep winter sleep. In summer, their daytime roosts may be in crevices and cracks in canyon walls, caves, and mine tunnels; behind loose tree bark; or in abandoned houses. Maternity colonies are small

and often comprise a lone female and her offspring, although a colony of up to 35 individuals, including juveniles, has been reported. Maternity roosts have been found in abandoned houses, barns, and cracks and crevices in rocks and vertical banks (Koford and Koford 1948; Tuttle and Heaney 1974).

In Arizona, a dietary analysis found that *M. ciliolabrum* ingests moths (100% frequency occurrence), beetles (88%), flies (63%), true bugs (38%), lacewings (25%), and ants (25%; Warner 1985). In regions where it occurs with the California myotis (*Myotis californicus*), it tends to forage in different areas, often over rocky arroyos, and on different species of prey. There is some suggestion that 1 species might be a "beetle strategist" and the other a "moth strategist," but no comparative dietary analysis has been conducted where they occur sympatrically (Black 1974). They forage at heights of 1–3 m along cliffs and rocky slopes with slow, maneuverable flight.

They may feed over water and close to the ground over desert chaparral. This bat is strong enough to take off from the surface of water if it is knocked into a pond or tank (Davis 1974).

This bat follows a typical reproductive pattern for hibernating bats: They mate in the fall, store sperm through the winter, and ovulate in the spring, at which time the egg is fertilized. Records indicate that the single young born annually appears in late May to early July and begins to fly about 1 month later (Kuenzi et al. 1999). Maximum lifespan has been recorded as 12 years (Wilson and Ruff 1999). Two specimens reported to the DSHS were not infected with rabies.

Status. The IUCN 2011 status of the western small-footed myotis is "least concern." This is a relatively rare bat in the western part of Texas, but there is not enough information available to accurately determine its status. For that reason, it is a species that should be monitored in the future.

Remarks. Scientists have had a difficult time determining the correct name of this species, as explained in the following account taken from *The Smithsonian Book of North American Mammals* (Wilson and Ruff 1999, 87): "It was known for years as *Myotis keenii* and was believed to occur throughout much of the United States and Canada. Later, it was called *M. subulatus* and was divided into three races: *leibii, subulatus,* and *melanorhinus.* Subsequently, van Zyll de Jong showed that there were 2 species of small-footed bats: *M. leibii* in eastern North America and *M. ciliolabrum* in the west. *M. ciliolabrum* is divided into 2 subspecies: *ciliolabrum* and *melanorhinus.* Because of this confusion, one must be careful to determine whether a given author is discussing eastern or western small-footed myotis."

Myotis ciliolabrum differs from the eastern small-footed myotis (*Myotis leibii*) in divergent cranial characters and in having a paler pelage, which varies from flaxen to yellowish brown, as compared with the pronounced sheen characteristic of *M. leibii*. Early editions of *The Mammals of Texas* referred to these bats as either *M. subulatus* or *M. leibii*.

Specimens Examined. Go to www.batsof texas.com for more detailed information about the total of 42 specimen records of *M. ciliolabrum* from Texas.

References. 13, 62, 102, 106, 140, 157, 166, 306, 334, 336, 339, 355, 381, 405, 420, 421, 423, 475, 484, 511, 540, 575, 578, 635, 640, 667, 686, 727, 756, 767, 830, 833, 914, 1024, 1026, 1028, 1031, 1066, 1091, 1109, 1112, 1134, 1152, 1153, 1179, 1235

Myotis occultus Hollister, 1909
Southwestern Little Brown Myotis

Etymology. The generic name, *Myotis*, comes from the Greek words *mys*, meaning "mouse," and *ous* or *otis*, meaning "ear." The specific name, *occultus*, is Latin for "mysterious" or "hidden," which seems appropriate for this confusing species.

Subspecies. Once considered a subspecies of *Myotis lucifugus* (Le Conte, 1831), authorities now regard *occultus* as specifically distinct from *lucifugus* (Piaggio et al. 2002). No subspecies are recognized.

Southwestern little brown myotis (*Myotis occultus*)

Description. This is a medium-sized (forearm = 34–41) myotis with long, full pelage that varies from brown to reddish brown dorsally and has a conspicuous sheen. The underparts are somewhat paler, almost tannish, in color. The interfemoral membrane is sparsely haired above, about to a line joining the knees. The tail is relatively short (tail length:head and body length < 0.8). The foot is relatively large and a little more than half the length of the tibia. The calcar is not keeled. Dental formula: I 2/3, C 1/1, Pm 2/3, M3/3 × 2 = 36. Most individuals of *M. occultus* have only 2 upper premolars, whereas most little brown myotis (*Myotis lucifugus*) have 3 upper premolars (Findley and Jones 1967). *Myotis occultus* also has a prominent sagittal crest and a relatively broad rostrum compared to other species of *Myotis*. Average external measurements are as follows: total length, 84 mm; tail, 34 mm; hind foot, 11 mm; ear, 14 mm; forearm, 39.5 mm. Weight: 8 g.

This bat may be confused with several other species of *Myotis*. For instance, *M. yumanensis* is smaller (forearm = 32–36) and has pale brown to yellowish pelage and smaller, lighter-colored ears. Also, *M. velifer* is larger (forearm = 40–44) and usually grayer than *M. occultus* and has a sparsely furred region between the scapulae on the back. Both *M. californicus* and *M. ciliolabrum* are smaller than *M. occultus* and have a keeled calcar. Also similar is *M. volans*, but it is larger and has a keeled calcar, and the underside of the wing is furred to the elbow and knee (Table 8).

Distribution. *Myotis occultus* occurs from southern California eastward into Arizona, New Mexico, and Colorado, as well as southward through the Mexican Plateau to Mexico City. This species is relatively uncommon throughout its range. Only a single specimen of *M. occultus* has been collected in Texas. This specimen, consisting of a skin and skull (NMNH 21083/36121), was collected near Fort Hancock, Hudspeth County, in the Trans-Pecos in June of 1893. It was probably a migrant individual, and it is doubtful that a resident population of this bat occurs in Texas.

Life History. Virtually nothing is known about the natural history of the southwestern little brown myotis in Texas; more information is available from the nearby states of Arizona, Colorado, and New Mexico. In Arizona, it is usually found in ponderosa pine or pine-oak woodland, although nursery colonies are known from much lower elevations (e.g., along the Colorado River in California and the Verde River in Arizona; Hoffmeister 1986). However, in New Mexico, vegetation zone seems unimportant in determining its distribution (Findley et al. 1975). Fort Hancock, Hudspeth County, where the bat has been taken in Texas, is in a region of desert scrub vegetation.

It is known to use attics, crevices in canyons, spaces beneath exfoliating bark on ponderosa pine snags, and caves as day roosting sites. In a survey of bridges in the Rio Grande Valley in southern New Mexico, more than 4,000 southwestern little brown myotises were documented using the crevices in bridges constructed from timbers as roosting sites. The bridges (9 in total) were occupied from April to September, and 2 bridges were used as maternity roosts (Geluso and Mink 2009).

The winter habitats of *M. occultus* are unknown although it is possible it hibernates in hollow tree cavities located within its summer range. One foraging bat was captured on March 30 in southwestern New Mexico, but netting during other winter months (November—February) did not result in any captures of this species. This individual likely represented an early arousal from hibernation (Geluso 2007).

Slightly before dark, *M. occultus* emerges to forage primarily around and over water. Its flight is erratic and relatively slow. There is limited information on the diet of *M. occultus*. In Arizona, it has been known to consume moths, beetles, flies, lacewings, wasps, ants, true bugs, and mayflies (Warner 1985). The diet of individuals from New Mexico and southern Colorado is similar and includes beetles, moths, parasitic wasps, and flies (Valdez 2006). However, hard-bodied prey items (e.g., beetles) constituted a larger component of the diet in southern latitudes, whereas soft-bodied prey items (e.g., flies) were more common in the diet of bats from northern latitudes; details of these food habits are being prepared for publication by Ernest Valdez (USGS).

In Arizona, the young are born in late June (Hoffmeister 1986). On June 29, 1960, a maternity colony of these bats was discovered in an abandoned house near the Verde River in Arizona. The river is located in a half-mile-wide valley and is lined with cottonwoods (*Populus* spp.), sycamores (*Platanus* spp.), and willows (*Salix* spp.). Sixty-seven adult females were captured and banded, but no adult males were found. Most of the females had already given birth. This site was revisited on June 4, 1961, and although 41 adult females were seen, there were no young present (Hayward 1963).

Although mating occurs in the fall, the gestation period is only 50–60 days after delayed fertilization. The weight of the young ranges from 1.8 g (an almost hairless individual) to 6.6 g (an individual not yet able to fly). On average, juvenile males weighed almost 0.5 g more than juvenile females of the same age (Hayward 1963). Other species of bats present in the maternity colony included the cave myotis (*Myotis velifer*) and the Yuma myotis (*Myotis yumanensis*).

Mites are the most common ectoparasites of this species (Valdez et al. 2009b). A new zoonotic coronavirus (the cause of severe acute respiratory syndrome, or SARS) has been isolated from *M. occultus* captured in Colorado (Dominguez et al. 2007), which is the first such virus isolated from bats in North America.

Status. The IUCN 2011 status of the southwestern little brown myotis is "least concern." This is one of the rarest species of bats in Texas. Populations have declined sharply in much of its historical range in California as a result of pesticides, control measures in nursery colonies, and disturbance at hibernation sites. It needs to be looked for elsewhere in the Trans-Pecos of Texas before its status can be determined.

Remarks. Taxonomists have long debated whether *M. occultus* is a distinct species or a subspecies of the wider-ranging *M. lucifugus*. Indeed, in the previous edition of *The Bats of Texas* it was treated as the latter—a subspecies of *lucifugus*. Historically, they were considered distinct species. Then Findley and Jones (1967) and Barbour and Davis (1970) showed that considerable intergradation in cranial and dental characters exists in bats from the southwestern United States. Subsequently, Hoffmeister (1986) reexamined the specimens used to unite the 2 species but was unable to conclude that the 2 are conspecific. He tentatively gave specific status to *M. occultus*.

Recent molecular genetic studies suggest that *occultus* is separated from *lucifugus* by sufficient genetic distance to be considered a separate species (Piaggio et al. 2002). Even so, morphological characteristics in *M. occultus* are highly variable geographically. Morphometric analysis of cranial, jaw, and dental features of *M. occultus* from Colorado and New Mexico demonstrates elevational and latitudinal variation in size and shape among populations. It has been suggested that these differences could be due to changes in diet in different habitats or that temperature might contribute to the observed morphological variation (Valdez 2006).

Specimens Examined. Go to www.bats oftexas.com for more detailed information about the single specimen record of *M. occultus* from Texas.

References. 5, 62, 93, 102, 106, 107, 140, 159, 221, 254, 260, 306, 322, 336, 353, 405, 420, 421, 423, 426, 443, 466, 468, 511, 512, 521, 524, 529, 571, 575, 635, 767, 823, 830, 832, 833, 834, 851, 878, 936, 957, 1001, 1024, 1026, 1028, 1030, 1040, 1044, 1112, 1142, 1143, 1146, 1153, 1179, 1236

Northern long-eared myotis (*Myotis septentrionalis*)

Myotis septentrionalis (Trouessart, 1897)
Northern Long-Eared Myotis

Etymology. The generic name, *Myotis*, comes from the Greek words *mys*, meaning "mouse," and *ous* or *otis*, meaning "ear." The specific name comes from the Latin word *septentrional*, which, literally translated, means "seven stars of the Great Bear" and also means "northern" in English (Stangl et al. 1993).

Subspecies. *Myotis septentrionalis* (Trouessart, 1897) is monotypic, and subspecies are not recognized.

Description. This is a small- to medium-sized (forearm = 32–39 mm) bat with long, dull, gray-brown pelage that is dark at the base. Membranes are similarly colored. Facial skin around the eyes and at the base of the ears is pinkish. Compared with other species of *Myotis*

in Texas, its ears are relatively long and extend past the end of the nose when pushed forward (Table 8). As with all *Myotis,* the tragus is long, narrow, and sharply pointed. The calcar is unkeeled (or weakly keeled), and the tail is relatively long (tail length:head and body length > 0.8). Dental formula: I 2/3, C 1/1, Pm 3/3, M 3/3 × 2 = 38. Average external measurements are as follows: total length, 78 mm; tail, 37 mm; hind foot, 9 mm; ear, 16 mm; forearm, 36 mm. Weight: 7 g.

Distribution. Although *M. septentrionalis* is widely distributed over eastern and northern North America, it is known in Texas on the basis of a single specimen collected at Winter Haven, Dimmit County, in South Texas. This specimen (comprising a skin and skull) was sent by S. E. Jones to the Division of Insects,

U.S. Bureau of Entomology, United States National Museum, on August 19, 1942. The nearest known locality of this species outside of Texas is the Kisatchie National Forest, Winn Parish, Louisiana (Crnkovic 2003), which is more than 800 km northeast of Dimmit County. Two males and 1 lactating female were collected at this location, suggesting a resident population there.

Life History. The northern long-eared myotis commonly hibernates in caves and mine tunnels of eastern Canada and in the United States from Vermont to Nebraska and southward to Louisiana and Mississippi. This bat is known to share hibernacula with the little brown myotis (*Myotis lucifugus*), big brown bat (*Eptesicus fuscus*), and American perimyotis (*Perimyotis subflavus*), and it generally chooses cooler sites with high humidity for hibernation (Barbour and Davis 1969). They are more solitary than other species of *Myotis* and prefer small out-of-the-way places in which to roost, such as tight crevices and cracks and hollow stalactites, which may result in underestimating their numbers in winter roosts (Caire et al. 1979; Griffin 1940).

The greatest number of northern long-eared bats found roosting in 1 cave (in Quebec, Canada) was estimated to be 300, but typically far fewer are observed (Thomas 1993). A large number of hibernating northern bats was found in a hydroelectric dam in Michigan. The number of bats (mostly little brown bats [*Myotis lucifugus*]) roosting in the dam was estimated to be 15,000. No estimate of the number of northern bats was given (Kurta and Teramino 1994). In Indiana, a few individuals have been found flying throughout the winter, and activity increases dramatically in mid-March, when feeding first begins after hibernation (Whitaker and Rissler 1992).

In the summer, *M. septentrionalis* roosts in the hollows of living and dead trees, behind tree bark, and occasionally in buildings or under bridges. Solitary bats, usually males, roost most frequently in the hollows of living hardwood trees, whereas maternity colonies are found predominately in the hollows or under the exfoliating bark of snags. These colonies range in size from 2 (a mother and her offspring) to 88. Individuals, including lactating females with nonvolant young, switch roost trees every 2–5 days. Trees are located from 6 m to 2 km (but generally less than 500 m) apart (Carter and Feldhamer 2005; Ford et al. 2006b; Foster and Kurta 1999; Johnson et al. 2009; Lacki and Schwierjohann 2001; Menzel et al. 2002b; Owen et al. 2002; Sasse and Pekins 1996). The average diameter at breast height was 30 cm, and the roost height ranged 3.7–10.8 m above the ground (Lacki et al. 2009a). In Arkansas, pine snags were preferred to hardwood snags (Perry and Thill 2007b).

These bats are highly maneuverable and prefer roosts in highly cluttered forested habitats with a dense, relatively closed canopy. The roost entrance is typically located below or within the canopy (Carter and Feldhamer 2005; Foster and Kurta 1999; Lacki and Schwierjohann 2001; Sasse and Pekins 1996). The average size of the home range in forested areas was 65 hectares in West Virginia (Owen et al. 2003) and 60 hectares in Kentucky (Lacki et al. 2009b). These bats typically forage within forests, along forest edges, over forest clearings, and occasionally over ponds.

Myotis septentrionalis has been known to consume moths, beetles, flies, true bugs, wasps, ants, caddis flies, crickets, net-winged insects, and spiders—in essence, anything that is small enough to capture and consume. Of these, the first 3 groups compose the bulk of its diet (Brack and Whitaker 2001; Carter et al. 2003a; Griffin and Gates 1985; Lacki et al. 2009b; Whitaker 2004). As a gleaning insectivore, *M. septentrionalis* is specialized to pluck insects from substrate surfaces such as tree bark, branches, foliage, and the forest floor, but it is capable of capturing airborne prey as well (Ratcliffe and Dawson 2003). The echolocation calls used by *M. septentrionalis* while gleaning are difficult for prey items, particularly moths, to detect (Faure et al. 1993).

The reproductive habits of this species are poorly known. Mating has been observed from the end of July to the end of October in Indiana (Whitaker and Rissler 1992), suggesting that females store sperm over winter and that fertilization occurs in the spring. A small amount of mating may occur in the spring, as a mating pair was seen in a cave in Missouri in April (LaVal and LaVal 1980). Reproductively active males have been taken from late July through September in Missouri (Caire et al. 1979). A single young is typical (Cope and Humphrey 1972). Parturition occurs from mid-May to mid-June in the southern portion of their range (Caceres and Barclay 2000), and the young are nursed for roughly 1 month (Kunz 1971). By late August, young of the year are difficult to distinguish from adults (Caire et al. 1979).

Status. The IUCN 2011 status of the northern long-eared bat is "least concern." Although no individuals have been taken in East Texas, a resident population of this species is possible there, as a population exists just across the border (approximately 120 km) to the east in Louisiana (Winn Parish). More fieldwork needs to be done to determine its status in northeastern Texas, though it is common in other parts of its range in eastern North America.

Remarks. Although previously considered a subspecies of *Myotis keenii* (*M. k. septentrionalis*), *M. septentrionalis* was elevated to full species status based upon cranial, dental, and external characters (van Zyll de Jong 1979). This taxonomic rearrangement created 2 monotypic species, *M. keenii* of the northwestern United States and Canada, and the paler *M. septentrionalis* of eastern North America.

Specimens Examined. Go to www.batsof texas.com for more detailed information about the single specimen record of *M. septentrionalis* from Texas.

References. 3, 54, 102, 106, 115, 177, 191, 204, 211, 216, 254, 285, 295, 306, 405, 407, 409, 6, 436, 438, 443, 500, 502, 511, 571, 621, 635, 665, 687, 709, 716, 717, 719, 729, 767, 807, 851, 893, 894, 917, 967, 1015, 1026, 1028, 1044, 1091, 1118, 1151, 1153, 1164, 1197, 1209, 1216

Fringed myotis (*Myotis thysanodes*)

Myotis thysanodes G. S. Miller, 1897
Fringed Myotis

Etymology. The generic name, *Myotis*, comes from the Greek words *mys*, meaning "mouse," and *ous* or *otis*, meaning "ear." The specific name comes from the Greek words *thysanos* and *odes*, which mean "tassel" or "fringe" and "resemblance," respectively (Stangl et al. 1993).

Subspecies. Texas specimens are referable to the subspecies *M. t. thysanodes* G. S. Miller, 1897 (O'Farrell and Studier 1980).

Description. This is the most easily recognized species of *Myotis* that occurs in Texas. It is a large myotis (forearm = 40–45) with a conspicuous fringe of short hairs lining the free edge of the interfemoral membrane, from which the species gets its common name. Pelage coloration is buff brown above and dull white below; the tips of individual hairs on the dorsal side are shiny, and the bases are blackish. The calcar is not keeled. Compared with other species of *Myotis*, the ears are long (projecting beyond the rostrum when laid forward) and dark in color; the feet are relatively large as well (Table 8). Dental formula: I 2/3, C 1/1, Pm 3/3, M 3/3 × 2 = 38. Females are slightly larger than males. Average external measurements are as follows: total length, 85 mm; tail, 37 mm; hind foot, 10 mm; ear, 17 mm; forearm, 43 mm. Weight: 9 g.

Distribution. This species is distributed throughout the western United States and is known in Texas from the Chihuahuan Desert and the Arizona/New Mexico Mountains (Trans-Pecos) during the summer months. Two specimens were captured in the High Plains

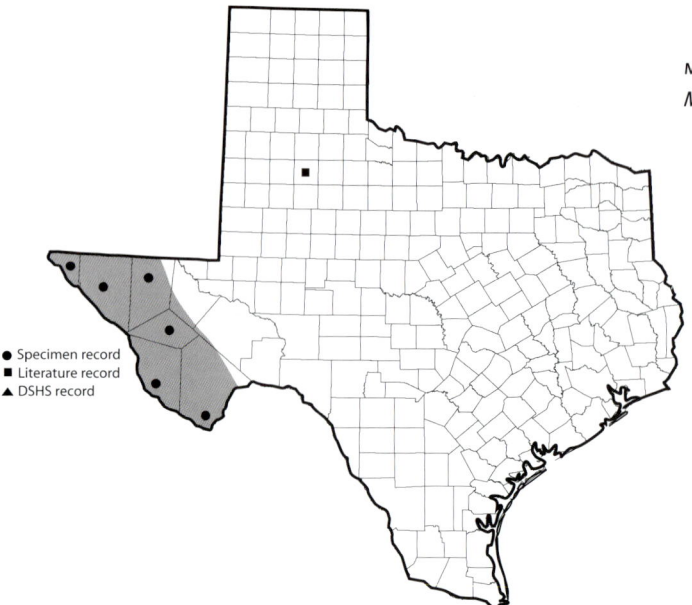

MAP 6. Distribution of the fringed myotis, *Myotis thysanodes*.

● Specimen record
■ Literature record
▲ DSHS record

(Crosby County; Jones et al. 1987b), but these may have been seasonal migrants. The fringed myotis has been collected in a wide range of habitats from mountainous pine, oak, and pinyon-juniper to riparian lowlands and desert scrub, but it seems to prefer grasslands at intermediate elevations. A specimen has been reported from spruce-fir forest at 2,850 m in New Mexico, but the fringed myotis is typically encountered at lower elevations of 1,200–2,150 m (Barbour and Davis 1969). No winter records are available for this bat in Texas, and its winter habits in the state remain unknown.

Life History. In the Trans-Pecos, *M. thysanodes* is a migratory species that arrives by May, at which time it begins forming nursery colonies. They have been collected in Texas as early as March 31. It is a colonial bat, and maternity roosts may contain several hundred individuals. It has been found inhabiting caves, mine tunnels, rock crevices, pine snags, and old buildings (Lacki and Baker 2007). It tends to roost in more open areas of caves and mines and not in hidden crevices. It is commonly found roosting in association with the cave myotis (*Myotis velifer*) and Townsend's big-eared bat (*Corynorhinus townsendii*) in the Trans-Pecos.

This species appears late in the evening to forage. They fly slowly and are highly maneuverable, allowing them to forage close to the vegetative canopy or about the face of small cliffs. No data are available on the specific food habits of this species in Texas. In New Mexico, they consume mostly small beetles (73% of their diet) and moths (Black 1974). In Arizona, beetles are the most common prey, followed by moths, true flies, and lacewings (Warner 1985). In Oregon, moths and spiders constitute the majority of their diet (Whitaker et al. 1977). Banding studies indicate a life span of at least 11 years (O'Farrell 1999).

Its reproductive biology in Texas is

poorly known. Unlike other species of *Myotis*, it appears that copulation, fertilization, and ovulation occur in the spring (O'Farrell 1999). Parturition times in New Mexico were found to occur during a 2-week period—from late June to early July—after a gestation period of 50 to 60 days (O'Farrell and Studier 1973, 1975). In Arizona, pregnant individuals were captured from mid-June until mid-July, and lactating females from late June until late July (Morrell et al. 1999). However, in the Trans-Pecos, the date of parturition appears to vary greatly even within a colony. Juveniles were observed at various stages of development (newborn pups to volant juveniles) on July 10 within a single maternity colony in Big Bend National Park (Easterla 1973). Congruous with this observation, lactating females were captured by one of us (Loren K. Ammerman) in Big Bend National Park as early as May 12. Apparently the young are born in Texas over a period of several weeks, from mid-May to early July.

The single pup is large and precocial and at birth often weighs more than 25% of the weight of its mother. Young bats are able to fly at 16–17 days and, until that time, are guarded at night roosts by "nurse bats"—adult females that remain with the nursery colony to protect and watch over the baby bats while the rest of the adults are out foraging. Maternity colonies retain group integrity through roost changes, and nonvolant young are carried to new roost sites by their mothers (Cryan et al. 2001). Mortality rate is low among neonates, less than 1%, partially because the adults retrieve the fallen young (O'Farrell and Studier 1973). Males are usually absent from maternity roosts, and these colonies begin to disperse by October. In the fall, the latest date on which this species has been collected in Texas is October 19. Hibernacula are unknown in Texas.

Status. The IUCN 2011 status of the fringed myotis is "least concern." There are no known threats to this species, as it is one of the commonest bats in the Trans-Pecos region during the summer. This species is easily disturbed by human presence and seems particularly sensitive to disturbance just prior to parturition (O'Farrell and Studier 1980).

Specimens Examined. Go to www.batsof texas.com for more detailed information about the total of 122 specimen records of *M. thysanodes* from Texas. Additional records: *Crosby County* (2) (Jones et al. 1987b).

References. 22, 60, 62, 102, 106, 140, 147, 168, 234, 300, 306, 336, 355, 381, 405, 420, 421, 423, 443, 475, 511, 564, 574, 575, 625, 626, 633, 637, 640, 647, 715, 727, 767, 830, 833, 846, 848, 871, 878, 879, 976, 1024, 1026, 1028, 1091, 1112, 1153, 1164, 1179, 1186, 1200, 1207

Myotis velifer (J. A. Allen, 1890)
Cave Myotis

Etymology. The generic name, *Myotis*, comes from the Greek words *mys*, meaning "mouse," and *ous* or *otis*, meaning "ear." The specific name comes from the Latin words *velum* and *ferre*, which can be translated as "bearing a sail" (Stangl et al. 1993).

Subspecies. Texas specimens are referable to 2 subspecies according to the most

Cave myotis (*Myotis velifer*)

recent taxonomic revision of the species in Texas (Dalquest and Stangl 1984a; Parlos 2008). The first of these, *M. v. incautus* (J. A. Allen, 1896), is found in the Chihuahuan Desert, Edwards Plateau, Texas Blackland Prairies, East Central Texas Plains, Western Gulf Coastal Plains, and Southern Texas Plains regions, while *M. v. magnamolaris* Choate and Hall, 1967, occupies the High Plains, Southwestern Tablelands, and Central Great Plains of the Texas Panhandle.

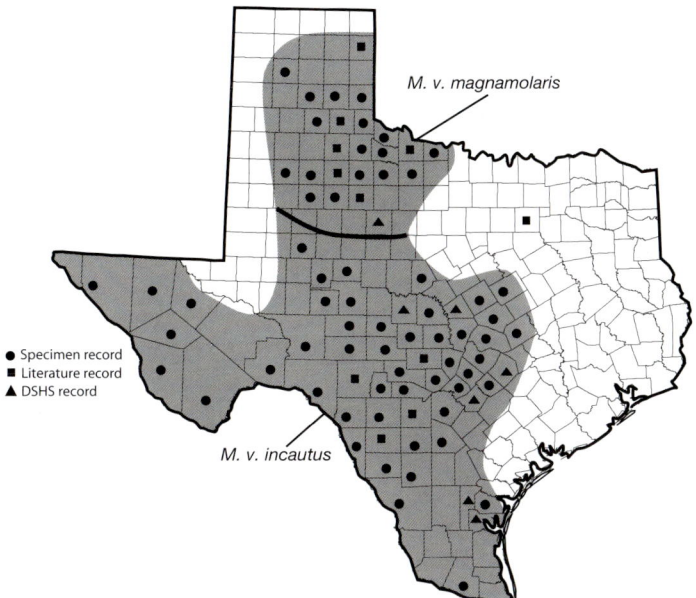

MAP 7. Distribution of the two subspecies of the cave myotis, *Myotis velifer*.

M. v. magnamolaris

M. v. incautus

● Specimen record
■ Literature record
▲ DSHS record

The subspecies *M. v. magnamolaris* is paler in coloration, is on average 5 mm longer in total length, and has slightly greater skull measurements than *M. v. incautus* (Choate and Hall 1967). Interestingly, genetic data do not support the recognition of geographically distinct subspecies. Nonetheless, because morphological and behavioral differences between the 2 subspecies have been well documented, we continue to recognize 2 subspecies in Texas (Parlos 2008).

Description. This is the largest (forearm = 37–47 mm) species of the genus *Myotis* in Texas. Males are usually slightly smaller than females. This bat ranges in color from gray-brown to dark, dull brown dorsally with paler underparts. The dorsal fur is monochromatic and does not have contrasting tips. There is typically a sparsely furred region between the scapulae. The facial skin is pinkish, and when the ears are laid forward, they reach to or slightly beyond the nostril. In the west, the cave myotis is sympatric with several other species of *Myotis*, but its larger size, dark, dull brown coloration, and lack of a keeled calcar serve to distinguish it from the others (Table 8). Dental formula: I 2/3, C 1/1, Pm 3/3, M 3/3 × 2 = 38. Average external measurements are as follows: total length, 94 mm; tail, 42 mm; hind foot, 10 mm; ear, 15 mm; forearm, 42 mm. Weight: 12 g.

Distribution. The cave myotis is a year-round resident of Texas, although it exhibits a varied seasonal distribution in the western two-thirds of the state. During summer months (March 16 to October 31) it occupies every ecological region in Texas except for the South Central Plains (Pineywoods), whereas in winter (November 1 to March 15) the species is apparently restricted to the central and north-central parts of the state. Only 1 winter record exists from the Trans-Pecos (Presidio County in February; Yancey and Jones 1996a), and none exist from the Rio Grande Valley.

Seasonal variation in the distribution of the sexes is not evident on the High Plains or Rolling Plains, but significantly more males than females are known from the Edwards Plateau during the winter. In summer, males and females appear equally distributed across the Trans-Pecos, Edwards Plateau, and South Texas Plains. However, most high-elevation records (Chisos and Davis mountains) are males.

Life History. The cave myotis is the most abundant myotis of the Edwards Plateau (Reddell 1994), where it is second in abundance only to the Brazilian free-tailed bat (*Tadarida brasiliensis*). As its name suggests, it usually roosts in caves and tunnels, typically in clusters numbering into the thousands. Seasonal uses of the caves vary: some sites are used as maternity roosts, and others as important hibernacula (Elliot 1993; Pekins 2006). In winter they are known to hibernate in Central Texas caves, where colony size can range from 1,000 to 40,000 individuals. The cave myotis also hibernates in the gypsum caves found throughout the Panhandle and rarely make major movements during winter, although short flights between hibernacula occasionally occur. In Oklahoma, sex ratios of hibernating populations are often highly skewed in favor of females (up to 18.3:1, F:M), but in Kansas, the sex ratio slightly favors males (0.6:1, F:M), suggesting that more males hibernate in the northern part of their range (Loukes and Caire 2007).

Cave myotises return to the same caves to hibernate and raise their young each year (Hayward 1970; Tinkle and Patterson 1965; Twente 1955a, 1955b). Caves that have high humidity and/or standing water are often used as winter roosts by *M. velifer* in Texas (Tinkle and Patterson 1965), and they tend to hibernate in relatively warm, humid areas within a cave (Jagnow 1998). Although these bats prefer very warm and humid caves as maternity roosts, they will abandon caves in which the walls become saturated due to heavy rains and return after such episodes (Pekins 2007). During migration they have been found in rock fissures, carports, occasionally attics and old buildings, and even abandoned Cliff or Barn Swallow nests (Buchanan 1958; Hayward 1970; Howell et al. 2009; Manning et al. 1987b; Ritzi et al. 1998).

Several other bats, including Townsend's big-eared bats (*Corynorhinus townsendii*), Brazilian free-tailed bats (*Tadarida brasiliensis*), big brown bats (*Eptesicus fuscus*), Yuma myotises (*Myotis yumanensis*), and ghost-faced bats (*Mormoops megalophylla*) often hibernate or roost at the same sites as *Myotis velifer*, although the bats usually segregate by species: Different species inhabit separate areas or rooms of the roosting site. For example, at the Eckert James River Bat Cave Preserve in Mason County, cave myotises (about 20,000 bats) are located closest to the entrance, whereas Brazilian free-tailed bats (*Tadarida brasiliensis*) roost deeper in the cave. Although cave myotises roost closer to the entrance, they emerge later in the evening than do the Brazilian free-tailed bats (*Tadarida brasiliensis*).

In summer, these bats typically leave

the roost 20–25 minutes after sunset to forage (Pekins 2006), and they leave to forage again before dawn (Kunz 1974a). They have been observed coming to pools of water and open tanks early in the evening to drink. Their larger size and stronger, more direct flight relative to other species of *Myotis* help to distinguish them from others feeding in the same area. Also, cave myotises generally fly approximately 4–12 m above ground while foraging and tend to fly close to vegetation (Fitch et al. 1981). Their minimum home range size has been estimated to be 932 to 1,619 km^2 in Arizona (Hayward 1970).

As with most other insectivorous bats, the cave myotis is an opportunistic feeder and takes a wide variety of insects depending upon what is available on a particular evening. A study of the species in Arizona revealed that small moths make up the largest part of its diet, although weevils, ant lions, and several species of small beetles also are taken (Hayward 1970). They also have been observed feeding on large concentrations of flying ants (Vaughan 1980). Banding studies have shown that these bats do not always return to the same roosting sites on succeeding nights; in combination with their relatively powerful flying ability, this may explain why they are able to forage farther afield than other species of *Myotis* (Tinkle and Patterson 1965; Twente 1955b).

Data on their reproductive habits are sparse. As with many other vespertilionids, *M. velifer* typically mates in the fall; ovulation and fertilization are delayed until spring. However, cave myotis

have occasionally been observed copulating during the winter and in the spring (Kunz 1973b). Equal numbers of males and females are present at maternity roosts until parturition, after which time the number of males decreases, but they do not disappear entirely (Dunnigan and Fitch 1967). Adult females disperse from maternity colonies shortly after weaning, perhaps to reduce competition with juveniles for food (Kunz 1973b).

In Arizona, embryo implantation is believed to occur by the first of May, with parturition occurring in the last week of June, resulting in a gestation period of 45–55 days (Hayward 1970). In Texas, gravid females have been observed as early as April 12, which suggests that individuals in our state may give birth somewhat earlier than in other areas. On the Edwards Plateau, lactating females have frequently been captured in May, which supports a parturition date as early as the first 2 weeks of May in Texas (Land 2001; Pekins 2009b).

One young per year is born to the female. Newborn bat pups are lightly haired and pink skinned and have dark membranes and ears. Baby bats are "hung" in a nursery colony with other newborns, where they are nursed and protected by the adult females. Such nursery colonies can be quite large—up to 40,000 individuals—but generally contain between 5,000 and 10,000 bats (Angelo 2009). The young are capable of flight at about 3 weeks of age and, until that time, may be moved to a different location by the mothers if the nursery colony is disturbed.

Known predators include snakes, hawks, owls, raccoons, ring-tailed cats, skunks, and gray foxes (Angelo 2009). Of 254 cave myotises reported to the DSHS (82 from 1984 to 1987 and 172 from 1996 to 2000), only 6 tested positive for rabies (2.4%).

Status. The IUCN 2011 status of the cave myotis is "least concern." Capture records suggest that the species is abundant throughout its known distribution in Texas, especially in areas where caves are common. However, due to the abandonment of historic roosts for unknown reasons, some scientists have suggested that there could be cause for concern about the conservation of the cave myotis (O'Shea and Vaughan 1999; Pekins 2009b). This species is apparently quite sensitive to anthropogenic roost disturbances (Angelo 2009). In Arizona, it was found that the breeding behavior of a cave population was negatively impacted by disturbance caused by regular cave tours, and the negative impacts increased as the maternity season progressed (Mann et al. 2002). Pesticides also have been implicated as a threat to this species, but recent studies conducted in Coryell and Mason counties has found that organochlorine residue levels are well below that which would cause symptoms of pesticide poisoning (Land 2001; Thies and Thies 1997). So although roost disturbances could negatively impact the population, we do not appear to have any reason to be overly concerned about the long-term status of this bat in Texas.

Specimens Examined. Go to www.batsof texas.com for more detailed information about the total of 1,371 specimen records of *M. velifer* from Texas. Additional records: *Briscoe County* (5), *Dickens County* (7), *Hall County* (3), *Motley County* (15) (Howell et al. 2009); *Kimble County* (104) (Goetze et al. 1996); *Gillespie County* (2) (Johnson et al. 2005); *Cottle County* (5) (Milstead and Tinkle 1959).

References. 7, 37, 54, 59, 62, 63, 77, 86, 92, 102, 106, 144, 145, 146, 159, 194, 205, 229, 240, 254, 262, 264, 266, 313, 319, 321, 322, 323, 336, 356, 365, 369, 370, 381, 394, 405, 420, 421, 423, 443, 475, 478, 487, 511, 527, 530, 541, 549, 564, 574, 575, 588, 611, 614, 615, 617, 623, 625, 633, 635, 637, 668, 684, 689, 690, 722, 727, 758, 764, 767, 768, 774, 790, 820, 823, 830, 832, 833, 884, 903, 909, 910, 911, 912, 913, 944, 968, 971, 976, 978, 988, 989, 999, 1001, 1024, 1025, 1026, 1028, 1029, 1045, 1071, 1091, 1100, 1105, 1107, 1112, 1117, 1122, 1123, 1136, 1137, 1138, 1154, 1156, 1164, 1200, 1246, 1249

Myotis volans (H. Allen, 1866)
Long-Legged Myotis

Etymology. The generic name, *Myotis*, comes from the Greek words *mys*, meaning "mouse," and *ous* or *otis*, meaning "ear." The specific name *volans* comes from the Latin word *volare*, which means "to fly" (Stangl et al. 1993).

Subspecies. Texas specimens are referable to the subspecies *M. v. interior* G. S. Miller, 1914.

Description. The long-legged myotis is a medium-sized myotis (forearm = 35–41mm), which may be distinguished by a combination of features, including a keeled calcar, a relatively long tail, and

Long-legged myotis (*Myotis volans*)

short, rounded ears (Table 8). As its common name suggests, this species has long legs (tibia) relative to foot length. The hind foot is typically about 41% of the length of the tibia, whereas it is generally more than half the length of the tibia in other large species of myotis (such as *M. thysanodes* and *M. velifer*). Most important, the underside of the wing membrane is lightly furred to an imaginary line that connects the elbow with the knee. Fur on the back is russet to dark brown, and the ears are almost black. Dental formula: I 2/3, C 1/1, Pm 3/3, M 3/3 × 2 = 38. Females are larger than males. Average external measurements are as follows: total length, 95 mm; tail, 43 mm; hind foot, 8 mm; ear, 11 mm; forearm, 37 mm. Weight: 7 g.

Distribution. Although distributed throughout western North America, the long-legged myotis apparently is relatively rare in Texas. It is known primarily from the Chihuahuan Desert and the Arizona/New Mexico Mountains ecoregions, where it seems to prefer high (above 2,000 m), open woods and mountainous terrain in the Guadalupe, Davis, Chinati, Chisos, and Sierra Vieja mountain ranges. A single, enigmatic specimen has been taken from Knox County in the Southwestern Tablelands. Originally reported as *M. lucifugus*, this bat was found in a region of much local relief, in keeping with this species' habitat preferences (Baker 1964; Mollhagen and Baker 1972). This was probably a wandering individual, and resident populations are not believed to inhabit this region of Texas.

No winter records are available for this species in Texas, and it is most likely

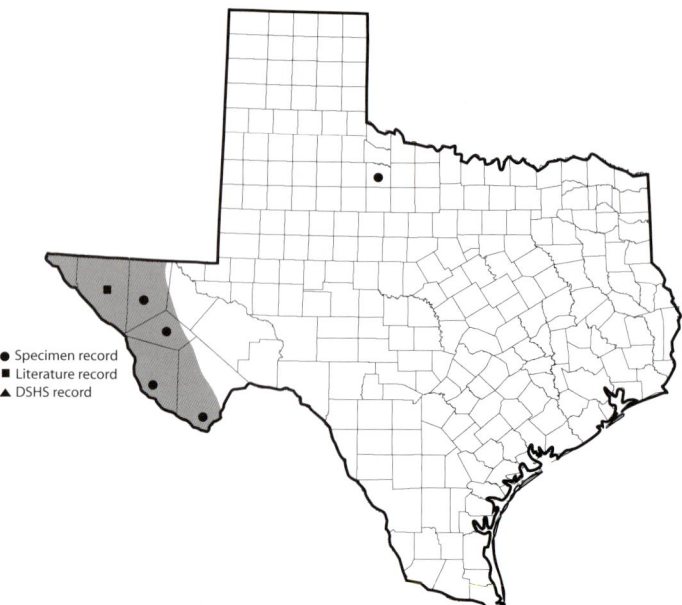

MAP 8. Distribution of the long-legged myotis, *Myotis volans*.

● Specimen record
■ Literature record
▲ DSHS record

only a summer resident of the state, having been captured from March 31 to October 13.

Life History. Over much of their range, long-legged myotis are forest inhabitants that prefer high, open woods and mountainous terrain. Roost sites for this species include dead hollow trees, rock or cliff crevices, and buildings. In Arizona and New Mexico, they prefer ponderosa pine snags with long vertical cracks or loose bark (Chung-MacCoubrey 1996; Rabe et al. 1998a). Snags that contained the largest colonies of bats are larger and taller and retain more bark than random snags (Baker and Lacki 2006; Ormsbee and McComb 1998). Bats frequently change roosts sites; in fact, they use the same roost for only 2.16 consecutive days on average (Ormsbee 1996). These bats apparently do not use caves as day roosts, although they may use such sites at night (Barbour and Davis 1969). The winter range and habits of the long-legged myotis are poorly understood, but

they are known to use caves and mines as hibernacula.

These bats emerge early in the evening to forage around cliffs, over water, and in forest clearings. Evidence from New Mexico and Arizona suggests they forage mainly in the open and prey primarily on small moths and other soft-bodied insects (Black 1974; Warner 1985). In Idaho, their diet comprises moths (49.2%), beetles (31.2%), flies (2.8%), and leafhoppers (0.1%, Johnson et al. 2007; Lacki et al. 2007). This differs from their diet in Oregon, where about 78% of their diet is moths (Whitaker et al. 1977, 1981b).

Only 1 young is born annually to the females. Copulation occurs in late summer, and females store the sperm until ovulation occurs in the spring. In Texas parturition probably takes place in June or early July. Females have been captured carrying newborn young (Hoffmeister 1986). Their home range size in coniferous forests in Idaho was 304 hectares for lactating females, 448 hectares for pregnant females, and 647 hectares for males, although the differences in average home range size between genders and reproductive conditions were not statistically significantly different (Johnson et al. 2007).

The maximum life span for *Myotis volans* is 21 years based on recapture of banded individuals (Warner and Czaplewski 1984). Consistent with their occurrence in cool climates at high elevations, this species is able to fly at low body temperatures—as low as 25°C (77°F; O'Farrell and Bradley 1977).

Status. The IUCN 2011 status of the long-legged myotis is "least concern." Although it has been considered rare in the state, more recently it has been found to be fairly common in appropriate habitats (Bradley et al. 1999; Higginbotham et al. 2002).

Specimens Examined. Go to www.batsoftexas.com for more detailed information about the total of 40 specimen records of *M. volans* from Texas. Additional records: *Hudspeth County* (1) (Higginbotham et al. 2002), *Jeff Davis County* (62) (Bradley et al. 1999).

References. 29, 62, 89, 93, 102, 106, 115, 140, 142, 147, 178, 234, 306, 318, 336, 381, 405, 420, 421, 423, 443, 475, 511, 566, 574, 575, 622, 625, 626, 714, 718, 727, 767, 830, 833, 835, 873, 887, 888, 957, 976, 988, 1024, 1026, 1028, 1091, 1112, 1153, 1164, 1179, 1200, 1206, 1207

Myotis yumanensis (H. Allen, 1864)
Yuma Myotis

Etymology. The generic name, *Myotis*, comes from the Greek words *mys*, meaning "mouse," and *ous* or *otis*, meaning "ear." The specific name, *yumanensis*, refers to Fort Yuma, California, where the type specimen was collected (Stangl et al. 1993).

Subspecies. Texas specimens are referable to the subspecies *M. y. yumanensis* (H. Allen, 1864), based on the most recent taxonomic revision of the species (Harris 1974).

Description. This is a small (forearm = 32–36 mm), buff brown bat with an unkeeled calcar, relatively large feet, and short ears. The underparts range from white to pale buff brown. The membranes, ears, and facial skin are usually pale brown and similar in color to the fur. Color is somewhat variable, and some individuals have darker fur and ears. Dental formula: I 2/3, C 1/1, Pm 3/3, M 3/3 × 2 = 38. Average external measurements are as follows: total length, 80 mm; tail, 34 mm; hind foot, 9 mm; ear, 13 mm; forearm, 33 mm. Weight: 4 g.

The combination of large feet and small forearm should distinguish it from similar species. The feet are more than half as long as the tibia in *M. yumanensis*. Other *Myotis* species sympatric with this one include *M. velifer*, which is larger and darker in coloration; *M. volans*, *M. californicus*, and *M. ciliolabrum*, all of which have keeled calcars; *M. thysanodes*, which is larger and has a fringe of hairs on the edge of the uropatagium; and *M. occultus*, which has longer hairs tipped with dark brown, which imparts a glossy appearance (Table 8).

Distribution. The Yuma myotis is a summer resident of the southern Chihua-

Yuma myotis (*Myotis yumanensis*)

huan Desert ecoregion and the area just east of the Pecos River in Val Verde County. There is one disjunct record from Starr County in the Southern Texas Plains region. This bat is most commonly encountered in lowland habitats near open water, where it prefers to forage. Most specimens collected in Texas are from lowland habitats near the Rio Grande, although several specimens also have been obtained from the Chisos Mountains.

Recently, 3 adult females were captured in southwestern Cimarron County, Oklahoma, only 8 km from the Texas border. Of the 3 females, 2 were pregnant (Roehrs et al. 2008). This strongly suggests that a population of this species could occur in the northwestern Texas Panhandle. This species likely remains in the Trans-Pecos during winter months

as indicated by capture records from January through November. However, their winter habits are largely unknown.

Life History. Like many other species of *Myotis*, these bats roost in a variety of places, including caves, cliff crevices, bridges, tunnels, and buildings. They also use abandoned Cliff Swallow nests as day roosts (Vaughan 1980). In a suburban area in coastal California, they prefer to roost in large live trees that are close to water (Evelyn et al. 2003). Night roosts of several hundred bats have caused damage to historic structures along the Rio Grande in Big Bend National Park, leading to the use of exclusionary screens.

The Yuma myotis is closely associated with open water, especially along the Rio Grande in the Big Bend region of Texas, where it comes to drink just after

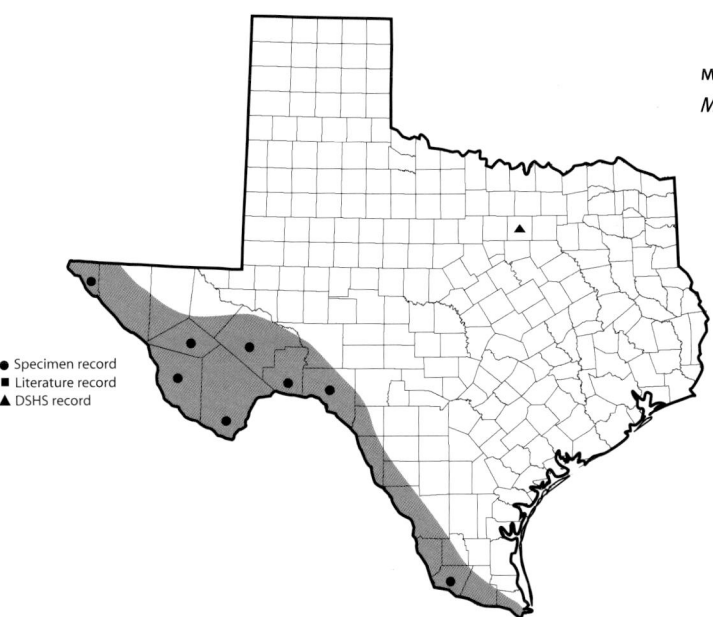

MAP 9. Distribution of the Yuma myotis, *Myotis yumanensis*.

● Specimen record
■ Literature record
▲ DSHS record

sundown. Most commonly found in desert areas, this bat typically forages just above the surface of streams and ponds or other open, uncluttered habitats over land, typically for whatever insects are available (Brigham et al. 1992). Their diet seems to be more general than that of other species of *Myotis,* and they do not specialize in moths or beetles but instead consume a larger variety of insects (Black 1974; Whitaker et al. 1977, 1981b). In Big Bend National Park, the stomach contents of 14 *M. yumanensis* were found to comprise the following: moths (39.5%), midges (12.5%), June beetles (4.6%), ground beetles (2.5 %), muscid flies (2.5 %), caddis flies (2.1%), froghoppers and leafhoppers (1.4 %), crane flies (1.4%), and unidentified insects (33.0%; Easterla 1973).

The reproductive biology of *M. yumanensis* in Texas is poorly known. As with many of the other myotises, it is thought that this species breeds in the fall and that the sperm is retained until spring.

However, males with descended testes have been collected in March in Brewster County, which suggests that some breeding occurs in the spring. Parturition is believed to occur from late May to early June, with the female giving birth to 1 young per year. One female specimen from Brewster County carried an 18 mm embryo on May 23. Lactating females have been documented on May 30 in Val Verde County (Manning et al. 1987b).

Large nursery colonies may form in buildings, caves, and mine tunnels and under bridges. In Nevada, a large nursery colony of about 5,000 bats inhabited a church belfry and attic (Dalquest 1947b). Maternity roost sites in mines in Hell's Canyon, Idaho, were selected on the basis of higher humidity and lower levels of anthropogenic disturbance when compared to mines without maternity colonies; ambient temperature differed little among the mines (Betts 1997). As with many other bats, the males take no part in the care of the young and are not usually found near nursery colonies; instead, they typically scatter and lead somewhat solitary lifestyles (Dalquest 1947b). Nursery colonies are very sensitive and quickly abandoned if disturbed. The nursery colonies are eventually vacated in the fall, although the destination of the bats is unknown. A few individuals may overwinter in Texas.

Status. The IUCN 2011 status of the Yuma myotis is "least concern." This species is relatively common in the desert lowland habitats near the Rio Grande. Not enough is known about its abundance and natural history to identify possible threats to the species, although pesticide use and changes in vegetation caused by invasive species of plants (i.e., salt cedar, giant cane) along the Rio Grande could potentially impact it in Texas.

Remarks. *Myotis yumanensis* and *M. occultus* are particularly difficult to distinguish, especially where the dark form of *M. yumanensis* occurs. When clean skulls are available, the abrupt rise of the forehead and smaller mastoid breadth (< 7.5 mm) can be used to readily identify *M. yumanensis*. The apparent intergradation of characters has led some authors to propose hybridization between the 2 species (Harris 1974), although electrophoretic and molecular data indicate distinctness consistent with species designation (Herd 1987). A combination of echolocation call frequencies and forearm length has been used to aid in field identification of these species in the Pacific Northwest, although broad overlap in these characters between *M. yumanensis* and *M. lucifugus* has been documented (Weller et al. 2007).

Specimens Examined. Go to www.batsof texas.com for more detailed information about the total of 149 specimen records of *M. yumanensis* from Texas.

References. 60, 62, 86, 92, 102, 106, 133, 140, 147, 159, 168, 187, 309, 336, 355, 381, 386, 403, 405, 420, 423, 427, 443, 449, 505, 511, 521, 540, 564, 574, 575, 625, 626, 767, 768, 830, 833, 842, 884, 976, 988, 995, 1024, 1026, 1028, 1091, 1112, 1153, 1156, 1164, 1185, 1200, 1206, 1207, 1254

Lasiurus blossevillii (Lesson, 1826) Western Red Bat

Etymology. The generic name, *Lasiurus*, comes from the Greek words *lasios*, meaning "shaggy," and *oura*, meaning "tail," which refers to the well-furred uropatagium. The species name honors the French explorer J. P. de Blosseville (Stangl et al. 1993).

Subspecies. Two subspecies of *L. blossevillii* are recognized. The first, *L. b. blossevillii* (Lesson, 1826), is found only in South America. The second, *L. b. frantzii* (Peters, 1870), occurs in Central and North America and has been documented in Texas (Morales and Bickham 1995).

Description. This medium-sized (forearm = 38–42 mm) bat is similar in overall appearance to the eastern red bat (*Lasiurus borealis*). It has short, rounded ears and a relatively long tail. The pelage is rusty red to brownish and lacks the white-tipped hairs that give the frosted appearance so characteristic of *L. borealis*. The posterior third of the interfemoral membrane is bare or only sparsely haired. In addition, *L. blossevillii* is slightly smaller than *L. borealis*, and most cranial mea-

Western red bat (*Lasiurus blossevillii*)

surements (i.e., greatest length of skull, zygomatic breadth, mastoid breadth, and length of maxillary toothrow) are significantly smaller (Schmidly and Hendricks 1984). Dental formula: I 1/3, C 1/1, Pm 2/2, M 3/3 × 2 = 32. Average external measurements are as follows: total length, 103 mm; tail, 49 mm; hind foot, 10 mm; ear, 13 mm; forearm, 40 mm. Weight: 10–15 g.

Distribution. *Lasiurus blossevillii* ranges across the southwestern and far western areas of the United States southward into Mexico and Central America. Only 1 specimen has been recorded from Texas (Genoways and Baker 1988). It was captured on July 15 in the Sierra Vieja Mountains of Presidio County, in the Chihuahuan Desert ecoregion. Specimens of the eastern red bat (*Lasiurus borealis*), also have been reported from Presidio County, suggesting that the geographic ranges of the 2 species overlap somewhat in northern Chihuahua and western Texas (Higginbotham et al. 2002; Jones and Bradley 1999; Jones and Lockwood 2008).

In the southwest, the western red bat has been documented from scattered localities in New Mexico (Findley et al. 1975) and Arizona (Hoffmeister 1986). These writers indicate that *L. blossevillii* is occasionally captured in riparian habitats dominated by cottonwoods, oaks, sycamores, and walnuts and is rarely found in desert habitats. This bat has been captured in riparian, xeric thorn scrub, and pine-oak forest habitats of the San Carlos Mountains of Tamaulipas, Mexico, which is approximately 160 km south of the Texas border. Evidence suggests that the 2 species of red bats might interbreed in northeastern Mexico (Schmidly and Hendricks 1984). However, no data suggest that such intergradation occurs in Texas (Genoways and Baker 1988).

Life History. Western red bats seem to prefer riparian areas, where they usually roost solitarily in the foliage of deciduous trees. However, up to 15 bats have been observed roosting in an apricot tree in an orchard in the North Bay area, which is a subregion of the San Francisco Bay area of California. They roost on the underside of overhanging leaves, where they are sheltered, and typically there is an open area just below the roost site to facilitate takeoff. In the spring, they prefer roosts with a southern exposure, whereas no preference in exposure is exhibited during the summer (Constantine 1959). In Arizona, 1 specimen was found hanging from a fig leaf about

2.1 m above the ground near a pond and a grove of cottonwood trees (Hargrave 1944). Roost heights up to 10–12 m have been documented (Carter and Menzel 2007; Constantine 1959; Pierson et al. 2006).

Lasiurus blossevillii is associated with mature stands of cottonwood and sycamore in Arizona and New Mexico, where they also have been found roosting in black walnut trees (*Juglans nigra*) and leafy shrubs (Findley et al. 1975; Hoffmeister 1986). In California, these bats have been observed roosting in fig, apricot, orange, almond, and walnut orchards, as well as in salt cedar (*Tamarix* sp.; Pierson et al. 2006). The specimen from Texas was an adult female captured over permanent water in desert scrub habitat in ZH Canyon of the Sierra Vieja. Riparian areas within xeric landscapes are important for both roosting and foraging and likely are the limiting factor in the distribution of this species in desert environments (Pierson et al. 2006; Snow 1996).

In the southwestern United States *L. blossevillii* is probably migratory. In Arizona, specimens have been captured only in July and August (Hoffmeister 1986), and in New Mexico, they have been taken from May 15 to August 4 (Findley et al. 1975). These authors suggest that a winter withdrawal of red bats from this region into Mexico seems likely. Winter records of this species from central Mexico support this suggestion, although migratory routes could not be determined because of a lack of museum specimen records (Cryan

2003). Western red bats occur year-round in California. During the spring and summer months females primarily occupy lower-elevation riparian habitats, while males are more often found at higher elevations. This difference is not apparent during the winter (November–February; Cryan 2003; Pierson et al. 2006).

The food habits of *L. blossevillii* in Texas are essentially unknown. In California, they usually forage over water and along river edges in riparian corridors, but they also have been observed foraging in clearings at least 500 m from a riparian area (Pierson et al. 2006). The diet of the western red bat probably is similar to that of the eastern red bat, which includes mainly small moths, crickets, and grasshoppers (Ross 1961). The echolocation call of western red bats is distinctive: The end of the call abruptly increases in frequency, which is not characteristic of the calls of the eastern red bat (Pierson et al. 2006).

As with many other vespertilionids, mating occurs in the late summer and fall, and sperm is stored until spring, when fertilization occurs. In Tamaulipas, Mexico, a pregnant female carrying 3 fetuses was captured on May 23 (Schmidly and Hendricks 1984). Pregnant individuals have been captured in New Mexico from May 15 to June 29, and lactating females have been collected from June 13 to July 12 (Findley et al. 1975). A female captured on June 7 gave birth to 2 young that were 43 mm long with forearms 11.5 mm in length (Mumford et al. 1964). These observations suggest that *L. blossevillii* may raise as many as 3 young annually and that parturition occurs from mid-May through late June (Snow 1996).

Status. The IUCN 2011 status of the western red bat is "least concern." It is uncommon and patchily distributed in the western part of its range. More work is needed to determine whether a resident population of this species occurs in the Trans-Pecos and the extent to which it is sympatric with the eastern red bat (*Lasiurus borealis*).

Remarks. Based on genetic studies, Baker et al. (1988) combined and elevated the subspecies *Lasiurus borealis teliotis* and *L. b. frantzii* to specific status under the scientific name *Lasiurus blossevillii*, which is distinct from the eastern red bat, *Lasiurus borealis*. Phylogenetic analysis indicates that *L. borealis* is more closely related to the Seminole bat (*Lasiurus seminolus*) than to *L. blossevillii*. Fossil evidence from Arizona suggests that *L. blossevillii* and *L. borealis* have been distinct species for at least 3.5 million years (Czaplewski 1993a). The authorship of the species name has been corrected—Lesson, 1826, should be sole author of the species name and not Lesson and Garnot (Gardner and Handley 2008).

Specimens Examined. Go to www.batsoftexas.com for more detailed information about the single specimen record of *L. blossevillii* from Texas.

References. 62, 101, 106, 159, 212, 270, 297, 303, 423, 458, 474, 511, 519, 520, 566, 575, 627, 629, 638, 767, 834, 841, 850, 940, 998, 1024, 1026, 1028, 1029, 1073, 1078, 1091, 1164

Eastern red bat (*Lasiurus borealis*)

Lasiurus borealis (Muller, 1776)
Eastern Red Bat

Etymology. The generic name, *Lasiurus*, comes from the Greek words *lasios*, meaning "shaggy," and *oura*, meaning "tail," which refers to the well-furred uropatagium. The species name comes from the Greek word *boreas*, which means "northern" (Stangl et al. 1993).

Subspecies. *Lasiurus borealis* is a monotypic species with no subspecies rec-

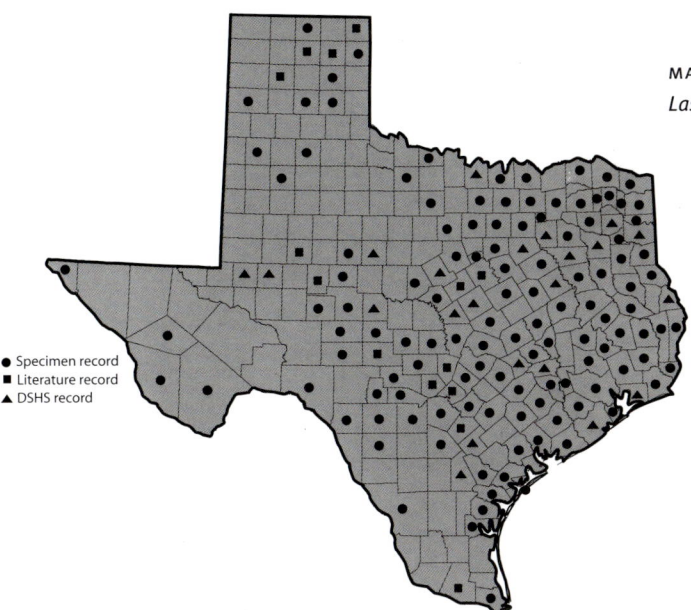

MAP 10. Distribution of the eastern red bat, *Lasiurus borealis*.

● Specimen record
■ Literature record
▲ DSHS record

ognized (Morales and Bickham 1995; Simmons 2005).

Description. The eastern red bat is a medium-sized (forearm = 35–45 mm), distinctly reddish or orange-red bat with short, rounded ears and a relatively long tail. The interfemoral membrane is densely and completely furred dorsally. The tragus is triangular in shape, and the tip has a slight forward bend. Pelage coloration is sexually dimorphic: Females have reddish fur with white hair tips that produce a frosted appearance; males lack the white hairs and are more uniformly red. Also, females are slightly larger than males. Dental formula: I 1/3, C 1/1, Pm 2/2, M 3/3 × 2 = 32. Average external measurements are as follows: total length, 106 mm; tail, 45 mm; hind foot, 8 mm; ear, 13 mm; forearm, 40 mm. Weight: 10–15 g.

The eastern red bat is similar in appearance to the Seminole bat (*Lasiurus seminolus*), but coloration will distinguish between the 2 species inasmuch as

L. seminolus has a rich, mahogany-brown pelage in contrast to the brick-red fur of *L. borealis*. The eastern and western red bat (*Lasiurus blossevillii*) also are very similar and may be confused in the Trans-Pecos, where they could occur together. In coloration, *L. blossevillii* is rusty red to brownish and lacks the frosted appearance of *L. borealis*. Also, *L. blossevillii* is significantly smaller in most cranial measurements (Schmidly and Hendricks 1984), and the posterior third of the interfemoral membrane is only sparsely haired.

Distribution. The eastern red bat is a tree-roosting, forest-dwelling species. It is widely distributed throughout Texas and is one of the commonest bats of the eastern part of the state (Schmidly et al. 1977). There are fewer records from the Chihuahuan Desert region, where the species seems to be limited primarily to mountainous areas. Locations in this region include El Paso County (Jones and Lee 1962); the Chisos and the

Rosillos mountains, Brewster County (Anderson 1972; Smith 1975); the Davis Mountains, Jeff Davis County (Jones and Bradley 1999; Mollhagen 1973); Presidio County (Higginbotham et al. 2002; Jones and Bradley 1999; Jones and Lockwood 2008); and northwestern Chihuahua, Mexico (Anderson 1972; Bogan and Williams 1970). The Presidio County specimens were captured at a riparian site dominated by cottonwoods. The western red bat (*Lasiurus blossevillii*) also has been documented in Presidio County, which suggests the geographic ranges of the 2 species overlap somewhat in northern Chihuahua and western Texas (Jones and Bradley 1999).

Lasiurus borealis is highly migratory and has been found 105 km out to sea during fall migration (Carter 1950). Although it is considered a year-round resident of eastern Texas, collecting records decline sharply in the winter months. During the winter it occurs throughout the southeastern United States and northeastern Mexico, but concentrations are highest in coastal Atlantic and Gulf of Mexico regions. Females are more common in the southern part of its range, and males more common in the north during the winter (Davis and Lidicker 1956). In the summer, its range expands northward, with no clear differences in the distribution of males and females (Cryan 2003). In northern Texas, the population comprises solely reproductive females that arrive from parts unknown in April and depart with their young in August (Spradling et al. 2003).

Eastern red bats are thought to be only a summer migrant in the Trans-Pecos region. The status of the species in Big Bend National Park is an enigma. Two specimens have been reported from that area, but both records are suspect (Easterla 1975). However, 1 specimen has been taken just north of the park boundary (Smith 1975), and several pregnant females have been captured nearby at Big Bend Ranch State Park (Higginbotham et al. 2002).

Life History. Eastern red bats are solitary in their habits except for small family groups comprising an adult female and her offspring. This tree-roosting species is almost never found using caves, mines, buildings, or similar sites frequented by other bats (but see Saugey et al. 1998). Roosting sites are commonly in tree foliage, where the bats are concealed because they closely resemble dead leaves. They have been observed roosting in a large variety of tree and shrub species, although they seem to prefer hardwoods over conifers (Carter and Menzel 2007). In only 2 studies, conducted in Mississippi (Elmore et al. 2004) and Missouri (Mormann and Robbins 2007), were a substantial number of roosts located in conifers, including 30% in loblolly pine (*Pinus taeda*) and 44% in eastern red cedar (*Juniperus virginiana*), respectively. Eastern red bats demonstrate some selectivity in their choice of roosting sites, as some species of trees are used more or less often than would be expected based on their availability. This preference changes geographically: Most commonly used are sweetgum and black oak (*Quercus nigra*) in Georgia (Menzel et al. 1998); tulip poplar

(*Liriodendron tulipifera*), sweetgum, and black gum in Maryland (Limpert et al. 2007); tulip poplar, white oak (*Quercus alba*), and pignut hickory (*Carya glabra*) in South Carolina (Leput 2004); white oak and hickories in Arkansas (Perry et al. 2007a); and oaks and sweetgum in Illinois (Mager and Nelson 2001).

Trees that are selected as roosts are usually taller and larger in diameter (> 30 cm diameter at breast height) than are nearby trees (Hutchinson and Lacki 2000b; Leput 2004; Mager and Nelson 2001; Menzel et al. 1998, 2000; Perry et al. 2007a), although they have been found roosting in smaller trees in Mississippi (Elmore et al. 2004) and North Carolina (O'Keefe et al. 2009). Individuals have been found roosting in trees from 1.5 to 19 m above the ground and tend to cling to leaf petioles or the tips of small branches (Hutchinson and Lacki 2000b; Mager and Nelson 2001; Menzel et al. 1998, 2000; O'Keefe et al. 2009). Juvenile bats and females with young tend to inhabit roosts that are higher than those occupied by solitary adults (Constantine 1966a). Red bats select roost sites that provide shade, protection from wind, and cover from above and both sides. The area directly below the roost is usually unobstructed to allow the bats an easy drop for takeoff (Constantine 1966a; Hutchinson and Lacki 2001; Saugey et al. 1998).

Roost trees are frequently close to linear "edge" habitat, such as streams, roads, and trails (Constantine 1966a; Hart et al. 1993; Leput 2004; Limpert et al. 2007; Mumford 1973; O'Keefe et al. 2009). Such corridors could pro-vide either convenient foraging sites or unobstructed flyways to preferred forag-ing sites. Eastern red bats preferentially roost in mature forest, most likely due to the availability of preferred roost trees there, but they also have been found roosting in intensively managed for-ests in Mississippi and North Carolina (Hutchinson and Lacki 2000b; Leput 2004; Menzel et al. 2000; O'Keefe et al. 2009; Perry et al. 2007a). It is interest-ing to note that all of the bats found roosting in younger forest were males or juveniles (no pregnant or lactating females). Roost sites in mature forest generally had higher canopy density and lower ground cover and tree density, although not all of these variables were met at all sites (Hutchinson and Lacki 2000b; Leput 2004; Limpert et al. 2007; Menzel et al. 2000; Perry et al. 2007a).

As is typical for a foliage-roosting species, roost fidelity is low, with summer roosts normally used for 1–2 consecutive days, although 1 pregnant female occupied the same roost for 14 consecutive days in North Caro-lina, and a lactating female used the same roost on 11 out of 12 consecutive days in Kentucky (Elmore et al. 2004; Hutchinson and Lacki 2000a, 2000b; Leput 2004; Menzel et al. 1998, 2000; O'Keefe et al. 2009). Individuals have been observed using the same roost on nonconsecutive days (Mager and Nelson 2001). Alternate roost sites are usually within 100 m of each other, and the total area encompassed by the roost sites ranged from 0.01 to 14.2 hectares in forest habitat (Elmore et al. 2004, 2005; Leput 2004; Menzel et al. 1998). In an urban landscape in Illinois, the mean roost area was much larger, 90 hectares, which was attributed to a lower density of suitable trees in the area (Mager and Nelson 2001). It has been suggested that frequent roost switching is an adaptation to variation in microclimate at the roost site, changes in resource availability, or predator and ectoparasite avoidance (Lewis 1995).

Although eastern red bats are migratory, they have been found overwintering in several states. Winter roosting sites frequently are located in leaf litter (Boyles et al. 2003; Moorman et al. 1999; Mormann and Robbins 2007). The coloration of their fur provides remarkable camouflage among fallen leaves. They also have been found in a woodpecker hole and other tree roosts in the winter (Fassler 1974; Hutchinson and Meisenburg 2004; Koontz and Davis 1991). In Missouri, these bats switch from tree roosts to leaf litter when the ambient temperature approaches or falls below freezing. They have been found an average of 62 mm below the surface of the litter, which was dominated by oak leaves. Such roosts are occupied for 12–18 days, longer than is typical for summer roosts but not unexpected since the bats are hibernating during this time (Mormann and Robbins 2007; Saugey et al. 1998). Their dense fur and ability to lower their bodily functions dramatically during cold weather allow them to survive rigorous winter weather in relatively open roosting sites. They are often found with their well-furred uropatagium pulled up over their face for additional insulation. In fact, the red bat

is so adapted to withstanding inclement weather that it may not be able to arouse spontaneously from torpor in caves and thus may not survive hibernation in such sites (Barbour and Davis 1969; Myers 1960; Quay and Miller 1955; Saugey et al. 1998).

Eastern red bats typically follow a specific territory while feeding and generally forage from near the forest canopy to just above the ground (Barbour and Davis 1969; Menzel et al. 2005). These bats seem to prefer foraging in relatively open habitats, such as young, thinned stands of pine, clearings, and aquatic habitats (Carter et al. 2004; Elmore et al. 2005; Hutchinson and Lacki 1999; Loeb and Waldrop 2008). Sizes of foraging areas are highly variable, from 62 to 925 hectares (Elmore et al. 2005). The largest foraging areas are of males in hardwood forests in Kentucky. At the same location, the foraging areas of adult females increase in size from the time of pregnancy to weaning. Diurnal roosts are usually located within the foraging area (Hutchinson and Lacki 1999).

Eastern red bats maximize their foraging efforts in the first 2–3 hours after sunset and often hunt around street lamps in towns (Hickey et al. 1996; Hickey and Fenton 1990; Kunz 1973a; Stangl et al. 1996). They are known to "eavesdrop" on the calls of conspecifics, particularly feeding buzzes, to alert them to the presence of vulnerable prey (Balcombe and Fenton 1988). Eastern red bats are less specialized than many other bats in their prey preferences. The majority of prey items include moths and beetles; the relative abundance of each of these insects in the diet changes geographically and seasonally. In South Carolina, moths increased from 5% of the diet in early summer to 63% in late summer. Concurrently, beetles decreased from 75% to 21% of the diet (Carter et al. 2004). In Illinois, beetles constituted nearly 70% of the volume of food consumed (Feldhamer et al. 1995). However, in West Virginia, Virginia, and North Carolina, moths made up 50–65% of the diet (Carter et al. 2003a; Whitaker et al. 1997). Other insects that are sometimes consumed include flies, true bugs, ants, wasps, net-winged insects, crickets, grasshoppers, and caddis flies (Whitaker 2004).

Unlike most North American bats, eastern red bats are commonly active during winter months, when they have been observed feeding at temperatures as low as 1°C (34°F; Boyles et al. 2003, 2006; Dunbar and Tomasi 2006; Dunbar et al. 2007; Easterla 1967; Padgett and Rose 1991; Whitaker et al. 1997). Since they do not put on additional fat before hibernation, they presumably must feed to supplement the energy supplies they lose during the winter (Milam-Dunbar 2005). They are more likely to arouse on mild winter days, when insects are more likely to be available.

At northern latitudes mating takes place in August and September, sperm is stored in the uterus and oviducts through the winter, and fertilization and parturition occur in spring (Shump and Shump 1982b). Limited mating also takes place in the spring (Saugey et al. 1998). A similar reproductive chronol-ogy most likely happens in eastern Texas. Mating is initiated in flight (Saugey et al. 1989; Shump and Shump 1982b). In fact, there are several accounts of copulating red bats falling from the sky (Glass 1966; Love 2009; Saugey et al. 1998; Stuewer 1948). Pregnant and lactating females have been reported from Arkansas in May (Gardner and McDaniel 1978).

One to 5 young (average 2.3) are born in May or June after a gestation period of 80–90 days (Jackson 1961; LaVal and LaVal 1979; Shump and Shump 1982b). This is one of the few species of bats that have more than 2 teats (they have 4), which enables the females to success-fully raise 3 or 4 young. Genetic evi-dence suggests that litters may be sired by more than 1 male (Spradling et al. 2003). Multiple paternity could explain the large variation observed in sizes and stages of development of some siblings.

Each baby bat weighs in excess of a gram at birth. The young are able to fly at 3–4 weeks but continue to nurse until they achieve adult size. Fully weaned and adult-sized young have been found clinging to their grounded mother even though the young were independently volant, suggesting a longer maternal association than previously considered (Stangl et al. 1996).

Juvenile red bats have a high mortal-ity rate (as high as 30%); predation, flying accidents, and bad weather are the most common causes of death. In fact, violent weather may be the single most important mortality factor for red bats in North Texas (Stangl et al. 1996). Mothers are known to carry nonvolant young to new roosts, and it is possible that they sometimes fall from their mothers in transit, although pups have remarkably well-developed thumbs and hind feet that enable them to cling to their mother with great tenacity (Barbour and Davis 1969; Davis 1970; Stangl et al. 1996). Predators include opossums, Sharp-shinned Hawks, American Kestrels, Merlins, Loggerhead Shrikes, Great Horned Owls, Roadrunners, and Blue Jays. This last bird is probably the most important predator of red bats in eastern North America (Mumford 1973; Sarkozi and Brooks 2003; Shump and Shump 1982b).

The eastern red bat is the second-most frequent bat reported to the DSHS (second only to the Brazilian free-tailed bat, *Tadarida brasiliensis*). This is no doubt due to the species' wide range, abundance, and propensity to roost and forage around human habitations. From 1984 to 1987 and from 1996 to 2000, 1,340 *L. borealis* were reported to the DSHS, of which 94 tested positive for the rabies virus. This represents about a 7% infection rate.

Status. The IUCN 2011 status of the eastern red bat is "least concern." The conservation status of this bat is good, as it is fairly common in the eastern part of the state. However, there are serious concerns over the welfare of this species in areas where wind turbines have been erected. A summary of fatalities at wind-energy facilities throughout the United States in 2005 found that, out of 2,486 bats killed at such facilities, 580 (23%) were eastern red bats (Kunz et al. 2007). The majority of bats killed were male, and most were killed in late sum-mer and early fall, when this species is migrating (Arnett et al. 2008). Declines in the populations of red bats also have been documented in Arkansas (Carter et al. 2003b), Indiana (Whitaker et al. 2006), and Michigan (Winhold et al. 2008). Texas has the largest installed capacity of wind energy on the continent, but no data on wildlife fatalities at any of these facilities are available (Arnett et al. 2008). More research is necessary to determine whether the fatality rate of eastern red bats at wind facilities in Texas is similar to that of other regions.

Specimens Examined. Go to www.batsof texas.com for more detailed information about the total of 852 specimen records of *L. borealis* from Texas. Additional records: *Brown County* (1) (Yancey and Jones 1996b), *Comal County* (14) (Goetze 1998), *Kimble County* (29) (Goetze et al. 1996), *Lamar County* (2) (Edwards and Johnson 2007), *Presidio County* (4) (Higginbotham et al. 2002; Jones and Lockwood 2008), *Bosque County* (1) and *Hamilton County* (1) (Goetze et al. 2003), *Jeff Davis County* (2) and *Presidio County* (1) (Jones and Bradley 1999), *Tom Green County* (1) (Yancey and Jones 1996b).

References. 25, 43, 48, 54, 59, 62, 74, 77, 86, 92, 98, 101, 102, 103, 106, 136, 145, 144, 146, 159, 172, 175, 212, 214, 216, 217, 218, 225, 231, 268, 274, 283, 297, 306, 313, 315, 318, 327, 336, 343, 363, 364, 374, 382, 388, 389, 399, 400, 405, 406, 410, 434, 443, 459, 482, 486, 487, 489, 511, 519, 522, 524, 561, 562, 566, 574, 579, 604, 605, 606, 607, 608, 610, 613, 625, 626, 627, 629, 635, 638, 640, 641, 669, 688, 696, 702, 714, 728, 742, 746, 749,

Hoary bat (*Lasiurus cinereus*)

Lasiurus cinereus (Palisot de Beauvois, 1796)
Hoary Bat

Etymology. The generic name, *Lasiurus*, comes from the Greek words *lasios*, meaning "shaggy," and *oura*, meaning "tail," which refers to the well-furred uropatagium. The specific name comes from the Latin word *cinereus*, meaning

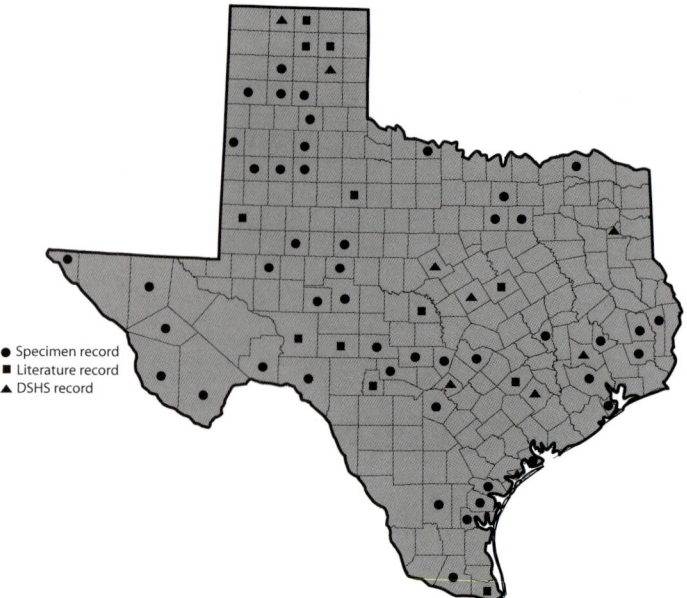

MAP 11. Distribution of the hoary bat, *Lasiurus cinereus*.

● Specimen record
■ Literature record
▲ DSHS record

"grayish" or "ash" colored, referring to the frosted tips of the pelage (Shump and Shump 1982a; Stangl et al. 1993).

Subspecies. Populations of this bat in the continental United States, Mexico, and Guatemala are referable to the subspecies *L. c. cinereus* (Palisot de Beauvois, 1796). Different subspecies occur in Hawaii (*L. c. semotus*) and South America (*L. c. villosissimus;* Hall 1981).

Description. This is a large (forearm = 46–58 mm) and heavily furred bat with short, rounded ears and striking pelage coloration. Overall, the fur is a dark mahogany brown, but the hair tips are white, giving the bat a distinctive frosted appearance. The ears are furred and rimmed with black. Fur on the throat and face is yellowish, and the dorsal surface of the interfemoral membrane is densely furred, as in other *Lasiurus* species. Yellowish fur extends onto the ventral wing membrane from body to wrist. The forearms appear pinkish.

Due to its large size and beautiful,

distinctive fur, the hoary bat is not easily confused with any other bat. In flight, this bat produces a chattering noise similar to that of the big brown bat (*Eptesicus fuscus*). Dental formula: I 1/3, C 1/1, Pm 2/2, M 3/3 × 2 = 32. Females are larger than males. Average external measurements are as follows: total length, 136 mm; tail, 57 mm; hind foot, 12 mm; ear, 18 mm; forearm 52 mm. Weight: 20–35 g.

Distribution. This is a forest-dwelling, transcontinental species that has been recorded from scattered localities throughout Texas. It has been reported from all ecological regions in the state. Males are more common in mountainous regions of the western United States, and females are more abundant in the eastern United States (Findley and Jones 1964).

Hoary bats are migratory, and the sexes appear to segregate geographically in summer. Females are known to migrate through Texas in spring and fall; males first appear in spring and remain throughout the summer in small numbers. Hoary bats are thought to prefer forested areas but also have been captured in lowland desert and along the Rio Grande in Texas. Although the overall pattern for this species in Texas is one of a spring-fall migrant, some hoary bats may also overwinter in the state. Museum specimen records suggest that some populations overwinter in coastal California or Mexico. Individuals have been observed in Arkansas, Indiana, and Pennsylvania in the winter (Doutt et al. 1966; Saugey et al. 1988; Whitaker 1967). Nonetheless, where the majority of individuals overwinter is largely unknown (Cryan 2003).

Life History. The hoary bat typically roosts singly in deciduous or coniferous tree foliage, 3–19 m above the ground, and often near the edge of a clearing (Constantine 1966a; Gruver 2003; Perry and Thill 2007a). They prefer trees that are taller and larger than average in diameter and tend to roost in stands dominated by more mature trees (Jung et al. 1999; Perkins and Cross 1988; Perry and Thill 2007a). In Arkansas, hoary bats seem to preferentially roost on the east side of the canopies of pine and oak trees (Perry and Thill 2007a), whereas in Saskatchewan, Canada, they were found to roost on the southeast-facing branches of spruce trees (Willis and Brigham 2005). Roost sites also tend to have a relatively open flyway to facilitate the departure and arrival of these fast-flying bats, which have relatively low maneuverability compared to other forest species (Perkins and Cross

1988; Perry and Thill 2007a). Although this species generally has low roost-site fidelity, females have been documented using the same roost for 23–26 consecutive days during the maternity season (Perry and Thill 2007a).

Hoary bats have infrequently been found roosting in sites other than tree foliage. These include a squirrel's nest, the side of a building, underneath driftwood, in a woodpecker hole, under a bridge, and in caves (Bowers et al. 1968; Carter and Menzel 2007; Hendricks et al. 2005; Neill 1952). Individuals in caves are rarely found alive; those that are alive are often extremely weak and emaciated and are unable to fly. Presumably they become disoriented and are unable to find an exit once they enter a cave (Myers 1960).

These bats are migratory and exhibit an interesting seasonal distribution as a result. In summer, females move to the northern, eastern, and central United States to give birth and raise their young. Males, however, remain in the western states, generally in montane areas (Cryan 2003). This pattern of migration and sexual segregation is illustrated in Texas by the distribution of this species in the Chisos Mountains of Big Bend National Park. Here, hoary bats are a rare summer resident, and only males occur in the mountains at this time. In the spring and fall, however, both males and females are found (Easterla 1973; Higginbotham and Ammerman 2002).

Interestingly, females are believed to precede males in migratory movements, which contrasts with the pattern found in migratory birds (Valdez and Cryan

2009). Hoary bats breed on the winter range prior to migration, whereas birds breed on the summer range (Findley and Jones 1964). Unlike migratory birds, male hoary bats have no need to set up territories or otherwise prepare for breeding prior to the female's arrival and simply do not need to make the long journey north. In northern New Mexico, they have been found migrating in dispersed groups that fly below the tree canopy along streams (Valdez and Cryan 2009). Hoary bats cover great distances during their seasonal migrations, and distances of 2,000 km and 1,800 km have been documented for a male and a female, respectively (Cryan et al. 2004).

During spring migration in New Mexico, male and female hoary bats appear to differentially use torpor to overcome thermoregulatory challenges. Male bats readily enter torpor at temperatures below 25°C (77°F), whereas females maintain normothermic body temperatures. The differences in thermoregulation between males and females can be attributed to different energetic demands associated with reproduction. Maintenance of a higher body temperature by females speeds embryonic development, whereas males have no such energetic constraints (Cryan and Wolf 2003). However, the use of torpor by pregnant hoary bats has been documented in Saskatchewan, Canada (Willis et al. 2006a).

Hoary bats are powerful fliers and emerge relatively late in the evening to forage (Caire et al. 1986; Shump and Shump 1982b). They forage on the wing, cover large areas, and also forage considerable distances from their diurnal roost sites (Black 1974). In South Carolina, they were more active above than below the forest canopy (Menzel et al. 2005). They have a strong preference for moths, which often constitute more than 90% of their diet (Black 1972, 1974; Carter et al. 2003a; Ross 1967; Valdez and Cryan 2009; Warner 1985). Indications are that the bat approaches a flying moth from the rear, engulfs the prey's abdomen-thorax in its mouth, and then bites down, allowing the sheared head and wings to drop to the ground. Hoary bats preferentially feed on larger moths (wingspan of 8–31 mm) when compared to eastern red bats (*Lasiurus borealis*; Hickey et al. 1996; Shump and Shump 1982a). Hoary bats also are known to eat beetles, flies, true bugs, grasshoppers, termites, dragonflies, and wasps (Ross 1967; Shump and Shump 1982a; Warner 1985; Zinn and Baker 1979). Although moths also dominate the diet of hoary bats in northern New Mexico during their spring migration, their eating habits shift to include more beetles later in the spring (Valdez and Cryan 2009).

Female hoary bats change their foraging behavior based on their stage of reproduction. Foraging time increases by 73% between early lactation and fledging and then declines as the young become independent. Presumably this is to meet increased energy demands while raising young. Females with 2 pups forage for longer periods than do females with a single pup. As young bats become volant, females forage in habitats that are different from those they used before their pups could fly, although the reasons for this difference are not known (Barclay 1989).

The diet of juvenile hoary bats (up to 2 weeks postfledging) in Manitoba, Canada, is significantly different from that of older juvenile and adult bats, with youngsters generally consuming smaller prey items. In particular, they eat substantially fewer dragonflies than do adults. During the juvenile hoary bats' first week of flight, small flies of the family Chironomidae are a major component of their diet. Differences in diet can be attributed to several potential factors, including higher echolocation call frequencies, lower wing loading (thus higher maneuverability), earlier time of foraging, or simply the inability of inexperienced young bats to handle large prey items such as dragonflies (Rolseth et al. 1994).

Echolocation calls of hoary bats have been examined throughout their range in the United States, including Hawaii, where it is the only native terrestrial mammal. The call structure of the mainland and island populations differs, but it is not known whether this is due to endogenous factors or to the fact that the bats forage in structurally dissimilar habitats (habitats are more cluttered in Hawaii than on the mainland; O'Farrell et al. 2000).

Mating begins in early September and potentially continues through the winter and into spring based on the males' ability to store viable sperm (Drueker 1972). Parturition seems to range from mid-May to early July; the usual number of young per litter is 2 and ranges from 1 to 4 (Caire et al. 1986; Shump and Shump 1982a; Willis and Brigham 2005). Little is known about the reproduction of hoary bats in Texas, but pregnant individuals have been captured in May and June in New Mexico and Arkansas (Bogan 1972; Cook 1986; Drueker 1972; Gardner and McDaniel 1978; Saugey et al. 1989), and females with young have been observed in roosts in mid-June in Arkansas (Perry and Thill 2007a). Lactating females have been captured from mid-June to early August in Oklahoma, Colorado, and Iowa (Caire et al. 1986; Everette et al. 2001; Kunz 1971). The young typically are left at the roost while the female forages (Davis 1970).

The development of newborn hoary bats has been studied in the wild in Manitoba, Canada (Koehler and Barclay 2000), and in captivity in New Mexico (Bogan 1972). Newborn young are covered with fine, silver-gray hair on the back of the head, shoulders, uropatagium, and feet; otherwise, they are naked (Bogan 1972). They have an average forearm length of 19 mm and weigh 4.7 g—more than an adult American parastrelle (*Parastrellus hesperus*)! Nursing continues until the young are 7 weeks old and have been flying for nearly 3 weeks. Growth is slower in cool, wet years. Lactating females go into torpor on nights when the temperature drops below 13°C (55°F; Hickey and Fenton 1996). Growth of the young may be faster in the southern portion of their range (Koehler and Barclay 2000). Fledging occurs 30–34 days after birth, and strong, controlled flight is attained by day 44.

No important predators are known for hoary bats, although American Kestrels and rat snakes have been reported to prey upon them. They also have been found accidentally impaled on barbed wire fences (Shump and Shump 1982b). Hoary bats have been observed attacking other smaller bats, although it was not clear whether the hoary bats were pursuing the other bats as prey or attacking them to defend a feeding territory (Bishop 1947; Orr 1950).

Ninety-seven bats were reported to the DSHS in the years 1984–1987 and 1996–2000, and 25 of these (25.7%) proved to be rabid. This is the highest incidence of rabies for any Texas bat, but the sample is too small to warrant any definitive conclusions regarding the species' susceptibility to the rabies virus. Nationwide, this species accounted for 2.3% of all rabies-positive bats in 2006 (Blanton et al. 2007).

Status. The IUCN 2011 status of the hoary bat is "least concern." This bat commonly occurs in parts of Texas, and there appear to be no threats to the species at this time. However, its abundance seems to be declining in Indiana, possibly due to the loss of woodlands to development (Whitaker et al. 2006). In addition, some evidence suggests that their population may be declining in Arizona (Morrell et al. 1999). Wooded habitats also have been impacted by development in Texas, so continued loss of this type of habitat could impact the availability of roosting sites for hoary bats.

Although there are no apparent threats to this species in Texas, recent studies on the impact of wind-energy facilities on volant wildlife raise concerns. In 2005, *L. cinereus* represented the highest proportion of bats killed at these facilities throughout the United States. An average of 41% (1,023/2,486 bats) of all bat fatalities were hoary bats, and a startling 89% of all bats killed at wind-energy facilities in the Rocky Mountain region were hoary bats (Kunz et al. 2007). Most were adult males killed in late summer and fall during their seasonal migration (Arnett et al. 2008). At one site in southeastern Wyoming, the number of hoary bats found dead under turbines was 4 times that of the number netted alive during the same time period (Gruver 2003). It has been suggested that hoary bats fly at lower altitudes during the spring migration and at higher ones during the fall migration (Valdez and Cryan 2009). Thus, the likelihood of being struck by wind turbine blades would be higher in the fall, which is consistent with the pattern of fatalities observed. The number of bats killed at these facilities increases exponentially with turbine tower height (Barclay et al. 2007).

It is unclear whether migration "routes" exist for bats as they do for many migratory birds. At 7 sites across southern Alberta, Canada, hoary and silver-haired bats (*Lasionycteris noctivagans*) concentrated their migratory activities at sites closer to the Rocky Mountains, suggesting the possibility of such a route. Researchers hypothesize that tree bats avoid the treeless prairies because of the paucity of roost sites and instead use the foothills of the Rockies, where there are more predictable roosting options (Baerwald and

Barclay 2009). Information such as this can potentially be used to minimize the negative impacts of wind-energy facilities on bats if sites for new facilities are chosen where there is less bat activity during migration.

Specimens Examined. Go to www.batsof texas.com for more detailed information about the total of 174 specimen records of *L. cinereus* from Texas. Additional records: *Fayette County* (1) (Johnson et al. 2005), *Kimble County* (6) (Goetze et al. 1996), *Real County* (1) (Goetze 1998), *San Saba County* (Davis 1966a), *Stonewall County* (1) (Ruhl and Stangl 1997).

References. 24, 46, 48, 54, 59, 62, 74, 83, 86, 92, 98, 101, 102, 106, 108, 110, 136, 138, 139, 140, 142, 145, 144, 147, 148, 156, 159, 168, 169, 172, 203, 212, 216, 225, 231, 274, 283, 297, 299, 301, 302, 306, 313, 318, 319, 323, 327, 335, 336, 355, 357, 360, 372, 381, 404, 405, 423, 424, 443, 459, 475, 486, 487, 506, 511, 524, 537, 561, 563, 564, 575, 610, 620, 623, 625, 626, 635, 640, 641, 650, 666, 687, 696, 727, 760, 767, 768, 774, 798, 823, 833, 846, 848, 851, 855, 859, 860, 874, 889, 915, 916, 970, 976, 988, 997, 999, 1003, 1017, 1019, 1024, 1025, 1026, 1028, 1029, 1030, 1040, 1044, 1071, 1091, 1102, 1112, 1144, 1153, 1164, 1179, 1193, 1198, 1200, 1218, 1228, 1229, 1239, 1250, 1260

Lasiurus ega (Gervais, 1856)
Southern Yellow Bat

Etymology. The generic name, *Lasiurus*, comes from the Greek words *lasios*, meaning "shaggy," and *oura*, meaning "tail," which refers to the well-furred uropatagium. The specific name refers to the location where the type specimen was collected in Ega, Brazil (Stangl et al. 1993).

Subspecies. Texas specimens are referable to the subspecies *L. e. panamensis* (Thomas, 1901) (Baker et al. 1988; Morales and Bickham 1995).

Description. This is a medium-sized bat (forearm = 42–48 mm) with relatively long wings and short, rounded ears. Pelage coloration is a dull sooty yellow washed with flecks of gray, brown, or black. The hairs are black at the base. Fur on the venter does not contrast with that on the dorsum. The dorsal surface of the interfemoral membrane is incompletely furred, and only the anterior half is covered by hair. Wing membranes are dark, and ears are pinkish brown. Dental formula: I 1/3, C 1/1, Pm 1/2, M 3/3 × 2 = 30. Females are larger than males. Average external measurements are as follows: total length, 118 mm; tail, 51 mm; hind foot, 9 mm; ear, 13 mm; forearm, 47 mm. Weight: 12 g.

The southern yellow bat may be confused with the northern yellow bat (*Lasiurus intermedius*). These 2 bats, which may occur together in extreme South Texas, can be separated on the basis of size: *L. ega* is much smaller (average total length 118 mm) than *L. intermedius* (average total length 133 mm). Their skulls differ in the length of the maxillary toothrow—in *L. ega* it is less than 6 mm, while in *L. intermedius* it is more than 6 mm. It is also similar to the western yellow bat (*Lasiurus xanthinus*), from which it can be distinguished by the lack of contrasting bright yellow fur on the interfemoral membrane. How-

Southern yellow bat (*Lasiurus ega*)

ever, *L. ega* and *L. xanthinus* have not yet been found in sympatry in Texas.

Distribution. *Lasiurus ega* is a neotropical bat that has been recorded in the United States only from the Southern Texas Plains and the Gulf Coastal Plains ecoregions of Texas. It also occurs along the Gulf Coast of Mexico. It has been suggested that this bat is expanding its range in the United States because of the increased usage of ornamental palm trees in landscaping. Palms are the

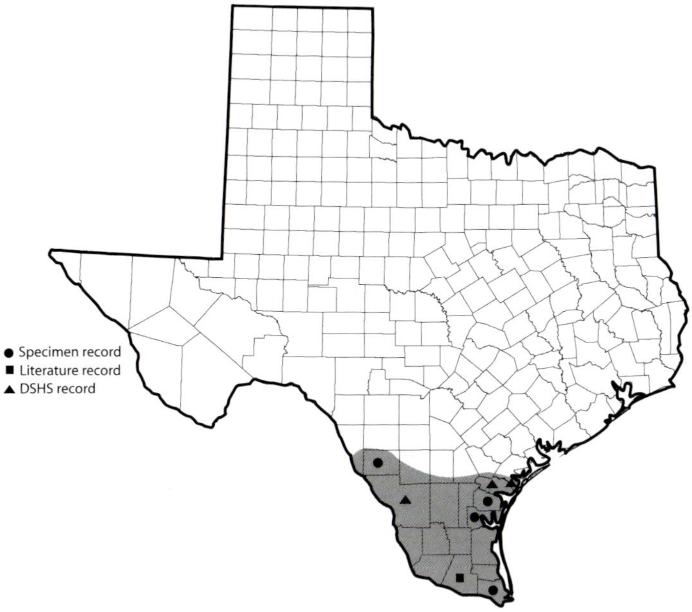

MAP 12. Distribution of the southern yellow bat, *Lasiurus ega*.

- ● Specimen record
- ■ Literature record
- ▲ DSHS record

preferred roosting site for this species in Texas (Carter and Menzel 2007; Spencer et al. 1988).

Most Texas specimens have been collected along the Rio Grande near Brownsville, where this bat is known to inhabit a natural grove of palm trees. The northernmost record in the state is from the Chaparral Wildlife Management Area in Dimmit County (Suchecki et al. 2003). This species appears to be a year-round resident of the Brownsville area, where it has been collected in 6 different months, including December (Baker et al. 1971).

Life History. This is one of the rarest and most restricted bats in Texas, and very little is known about its biology. It is a tree-roosting species that commonly roosts in the dead fronds of palm trees (*Sabal texana* and *Washingtonia robusta*). The bats' coloration provides a perfect camouflage against the yellow-brown color of these fronds (Mirowsky 1997; Spencer et al. 1988). Although lasiurine bats are generally considered to be solitary in their roosting habits, southern yellow bats have been found roosting with northern yellow bats (*Lasiurus intermedius*) in ornamental palms in Nueces County (Chapman and Chapman 1990; Spencer et al. 1988). The southern yellow bat may be migratory in parts of its range, but it appears to be a permanent resident of Texas (Baker et al. 1971; Van Deusen 1961).

Its food habits are not well documented, although small to medium-sized, night-flying insects are probably the main prey items. In Mexico it has been found to consume beetles more frequently than other potential prey items (Kurta and Lehr 1995). Reproductive data are equally scarce. In Texas, *L. ega* is believed to breed in winter; 1 male with enlarged testes was collected in September (Spencer et al. 1988). Pregnant females have been collected in April and June. These individuals carried 2, 3, or 4 embryos. Gestation lasts 3–3.5 months.

Lactating females have been reported from mid-June to mid-July (Baker et al. 1971; Kurta and Lehr 1995). As with other lasiurine bats, the females have 4 mammae, not 2 as is seen in most North American bats.

Eighty-one specimens of *L. ega* have been reported to the DSHS (in the years 1984–1987 and 1996–2000), and 2 tested positive for the rabies virus. Most of the bats (80) were submitted between 1996 and 2000. The drastic increase in human encounters with this species may be related to the rapid development and population growth in South Texas since 1990.

Status. The IUCN 2011 status of the southern yellow bat is "least concern." The southern yellow bat is listed as threatened by Texas Parks and Wildlife in the most recent Texas Wildlife Action Plan due to its limited distribution in Texas, although the range of the subspecies found in Texas encompasses Mexico, Central America, and northern South America. The primary threat to this species is the removal of potential roost sites (occasionally with unfledged pups) by trimming dead palm fronds in residential areas where this species is known to occur (Mirowsky 1997).

Remarks. Specimens from Texas were formerly assigned to the subspecies *L. e. xanthinus*. In genetic studies of *Lasiurus*, however, it has been shown that this subspecies should be elevated to specific status (*Lasiurus xanthinus*) and that *L. ega* from Texas belongs to the subspecies *L. e. panamensis* (Baker et al. 1988; Morales and Bickham 1995).

Specimens Examined. Go to www.batsoftexas.com for more detailed information about the total of 24 specimen records of *L. ega* from Texas.

References. 20, 54, 92, 98, 99, 101, 102, 106, 136, 212, 225, 226, 336, 423, 443, 511, 514, 519, 575, 767, 825, 833, 841, 999, 1026, 1028, 1029, 1082, 1091, 1105, 1112, 1149, 1164

Lasiurus intermedius H. Allen, 1862
Northern Yellow Bat

Etymology. The generic name, *Lasiurus*, comes from the Greek words *lasios*, meaning "shaggy," and *oura*, meaning "tail," which refers to the well-furred uropatagium. The specific name, *intermedius*, is Latin for "intermediate" (Stangl et al. 1993).

Subspecies. Texas specimens are referable to 2 subspecies, according to the latest evaluation of the species' taxonomy (Hall and Jones 1961). *Lasiurus i. floridanus* (G. S. Miller, 1902) is known from Bexar and Travis counties eastward, and *L. i. intermedius* H. Allen, 1862, from San Patricio County southward. *Lasiurus i. floridanus* is generally smaller and slightly less yellow in pelage coloration than *L. i. intermedius*. The average total length of *L. i. floridanus* is less than 135 mm, whereas it is more than 135 mm (up to 164 mm) in *L. i. intermedius* (Ceballos and Oliva 2005; Webster et al. 1980). However, there is substantial overlap in other measurements, and females are generally larger than males.

Description. This is a large bat (forearm = 45–56 mm) with short, pointed ears and long wings. Pelage coloration is yellowish

Northern yellow bat (*Lasiurus intermedius*)

brown to yellowish gray, and only the anterior half of the interfemoral membrane is furred. The calcar is slightly keeled. Dental formula: I 1/3, C 1/1, Pm 1/2, M 3/3 × 2 = 30. Average external measurements are as follows: total length, 133 mm; tail, 55 mm; hind foot, 10 mm; ear, 17 mm; forearm, 53 mm. Weight: 14–31 g.

This bat is distinguished from other members of the genus *Lasiurus* by its pointed, rather than rounded, ears and yellowish pelage. The northern yellow bat may be confused with its southern relative (*L. ega*), and one must be careful with identification in South Texas, where the 2 species are sympatric.

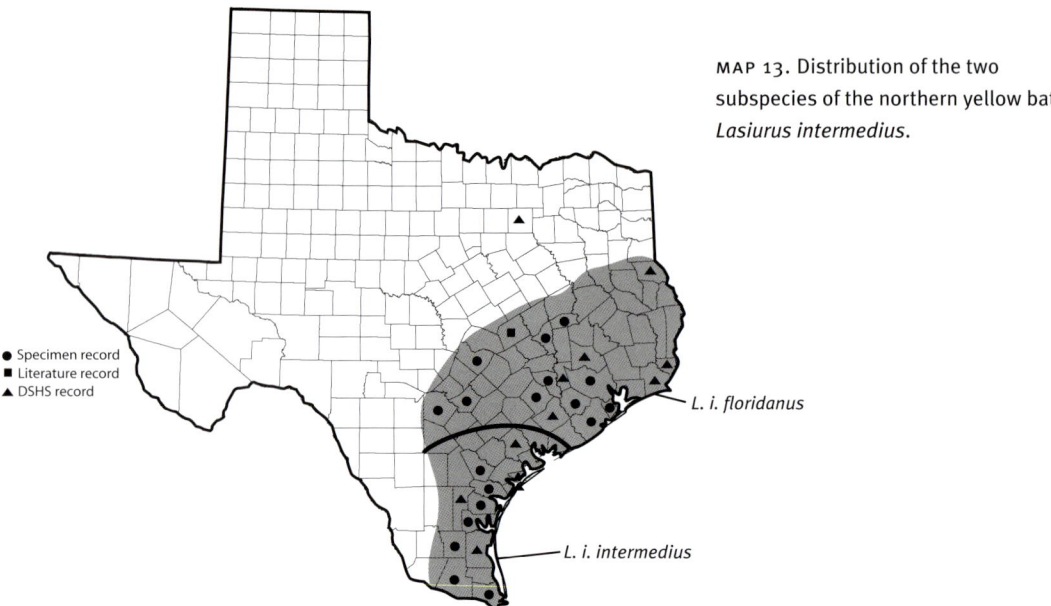

MAP 13. Distribution of the two subspecies of the northern yellow bat, *Lasiurus intermedius.*

● Specimen record
■ Literature record
▲ DSHS record

L. i. floridanus

L. i. intermedius

Distinguishing features are listed in the account of *L. ega.*

Distribution. The northern yellow bat occurs primarily along the Gulf Coast in the Gulf Coastal Plain ecoregion from Jefferson County south to Cameron County. Inland records are uncommon but include specimens from the South Central Plains, East Central Great Plains, and the Blackland Prairies ecoregions of Texas. A specimen from Dallas County submitted to the DSHS was probably an accidental occurrence.

Migration and winter habits are poorly known. In Texas *L. i. floridanus* has been reported in late winter, but *L. i. intermedius* has been recorded only from January to September in southern Texas (Chapman and Chapman 1990). The extent to which they use torpor at low ambient temperatures is not known. A northern yellow bat was discovered in a flower shop in Illinois in January, but it is believed to have arrived with a shipment of Spanish moss (*Tillandsia usneoides*) from Florida (Pinto 2006).

Life History. The distribution of this bat in the United States closely coincides with that of Spanish moss, which is its preferred roosting site throughout the southeastern United States (Carter and Menzel 2007; Kern 1992). In South Texas, however, *L. intermedius* roosts in the dead leaves of the petticoat of palm trees, where individual bats are well concealed behind the large fronds. In these sites, they have been found with southern yellow bats (*Lasiurus ega*; Chapman and Chapman 1990). A single roosting site may contain several bats, and such groups are often quite noisy, especially when young are present; their bickering quickly gives them away (Davis 1974). A relatively large group (45 females and juveniles) of northern yellow bats was found roosting among dried corn stalks leaning against a building in the state of Veracruz, Mexico (Baker and Dickerman 1956).

In Florida and Georgia, roosts typically are found near permanent bodies of freshwater in live oak (*Quercus virginiana* and *Q. geminata*) stands. Roosts are 2.23 m above the ground in Florida and more than 10 m high in Georgia (Hutchinson 2006; Menzel et al. 1999) and are shaded for much of the day. Their home range is relatively small (10.47 hectares) when compared to 3 other species of bats on Sapelo Island, Georgia, but this is based on 1 male bat. This bat was tracked in live oak stands (73%) and loblolly/slash pine (*Pinus taeda* and *P. elliottii*) stands (25%; Krishon et al. 1997).

Northern yellow bats prefer to forage over open, grassy areas such as pastures, lake edges, airports, and golf courses, as well as along forest edges (Kern 1992). They emerge about 15–20 minutes after sunset and fly a straight, steady course about 4.5–6 m above the ground (Marks and Marks 2006). In Florida, they often form "feeding aggregations" when foraging. Such groups are segregated by gender; males are rarely found in these groups, and they seem to be more solitary in their habits than females. Specific prey items include leafhoppers, dragonflies, true bugs, flies, diving beetles, beetles, ants, and mosquitoes (Sherman 1939).

Mating occurs in the fall, when males are actively producing sperm; copulating bats have been captured in late November in Florida (Kern 1992; Sherman 1944). Females carry 3 to 4 embryos in spring, and litter size is usually 2 or 3. In Florida, the mean number of embryos was 3.5, with a mean postpartum litter size of 2.7. Parturition occurs in late May or June in Florida, and the timing is probably similar in Texas (Kern 1992). Lactating females have been documented in June (Webster et al. 1980). One female with 3 pups was found in a fan palm that was cut down on Galveston Island on June 14, and several other bats (presumably northern yellow bats) were seen flying from the palm as it was felled (Nedbal et al. 1994).

Out of 279 northern yellow bats reported to the DSHS from 1984 to 1987 and 1996 to 2000, 25 (9%) tested positive for rabies.

Status. The IUCN 2011 status of the northern yellow bat is "least concern." Little is known of this species' population in Texas. Although records indicate it is a year-round resident, specimen records and reports suggest it is relatively rare in the state. It has not been found in large numbers anywhere in Texas.

Remarks. The closest relative of *L. intermedius* is *L. ega*. This was first predicted by examining karyotypic data (Baker et al. 1988; Baker and Patton 1967; Bickham 1987) and has been confirmed by genetic analysis (Morales and Bickham 1995).

Specimens Examined. Go to www.batsof texas.com for more detailed information about the total of 114 specimen records of *L. intermedius* from Texas.

References. 32, 54, 88, 94, 98, 101, 102, 106, 136, 137, 144, 212, 222, 225, 226, 268, 336, 369, 511, 514, 519, 603, 662, 676, 760, 767, 770, 781, 803, 833, 841, 859, 942, 1025, 1026, 1028, 1029, 1030, 1053, 1054, 1082, 1091, 1112, 1164, 1250

Seminole bat (*Lasiurus seminolus*)

Lasiurus seminolus (Rhoads, 1895)
Seminole Bat

Etymology. The generic name, *Lasiurus*, comes from the Greek words *lasios*, meaning "shaggy," and *oura*, meaning "tail," which refers to the well-furred uropatagium. The species name refers to the Native American Indian tribe, the Seminoles, that historically occupied the region of Florida where the type specimen was collected (Stangl et al. 1993).

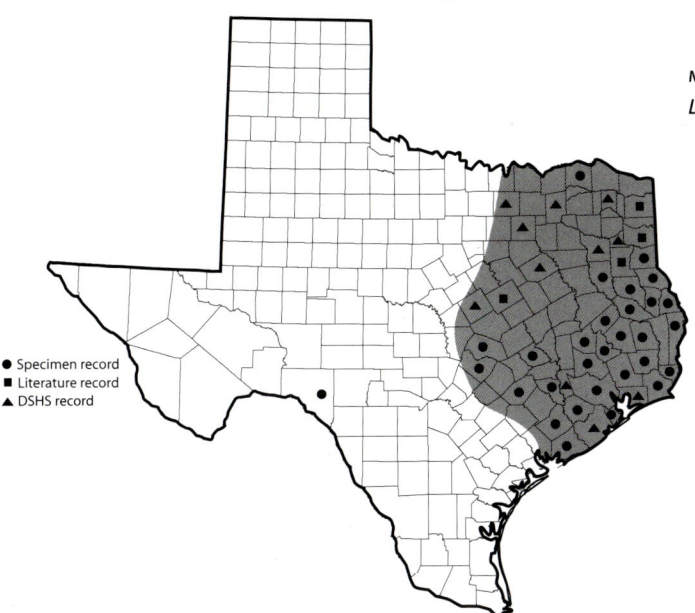

MAP 14. Distribution of the Seminole bat, *Lasiurus seminolus*.

● Specimen record
■ Literature record
▲ DSHS record

Subspecies. *Lasiurus seminolus* (Rhoads, 1895) is a monotypic species, and no subspecies are recognized.

Description. This is a medium-sized bat (forearm = 35–45 mm) with short, rounded ears. As with other bats of the genus *Lasiurus,* the interfemoral membrane is densely furred dorsally. The pelage is a dark reddish brown (or mahogany), and the hair tips are sometimes white, giving the fur a frosted appearance. *Lasiurus seminolus* may be confused with the eastern red bat (*Lasiurus borealis*), from which it is distinguished as described in the account of the latter. Dental formula: I 1/3, C 1/1, Pm 2/2, M 3/3 × 2 = 32. Average external measurements are as follows: total length, 103 mm; tail, 44 mm; hind foot, 8 mm; ear, 12 mm; forearm, 39 mm. Weight: 9–14 g.

Distribution. This bat is distributed across the southeastern United States, and in Texas it had been known from 5 physiographic regions (South Central Plains, Gulf Coastal Plains, East Central Great Plains, Blackland Prairies, Cross Timbers) until it was recently documented on the western edge of the Edwards Plateau as well (Brant and Dowler 2000). It is most commonly encountered in the pine-oak forests of eastern Texas. The species is a year-round resident of Texas, although winter records are rare, and none exist for the months of November and December.

Life History. For many years, it was thought that Seminole bats preferred roosting in clumps of Spanish moss. However, few bats actually were ever found in Spanish moss, and those were captured in the winter (Carter and Menzel 2007). Recently, several radio-telemetry studies have found that these bats most frequently roost in clumps of needles on small branches of pine trees, particularly slash and loblolly pine. Roosts in the foliage of pines constitute 85–95% of the roosting sites of this species in the summer in South Carolina,

Georgia, and Arkansas. Mature stands of pines are preferred roosting habitat, and trees used as roosts are taller and larger in diameter than the surrounding trees (Hein et al. 2008b; Menzel et al. 1998, 1999, 2000; Perry and Thill 2007d; Perry et al. 2007b). When roosting among pine needles, these bats hang motionless and appear to be a pine cone. During the winter, they use a wider variety of roosting sites, including Spanish moss, the canopy of overstory trees, hanging vines, shrubs, and leaf litter on the forest floor. Supposedly, these winter roosting substrates provide insulation against cold winter temperatures (Constantine 1958d; Hein et al. 2005, 2008a; Hutchinson and Meisenburg 2004).

Roosts typically are occupied by a single individual or a female with young. The height of roosts ranges from 1 to 19 m above the ground (Constantine 1958d; Menzel et al. 1999, 2000). These bats change roosts frequently and rarely use the same one for more than 1–2 days but tend to stay in the same general area (Hein et al. 2008b; Menzel et al. 1998). The mean size of the roosting home range is from 0.2 hectares in Georgia to 5.85 hectares in South Carolina, and reproductive females have the largest roosting home range size (Hein et al. 2008b; Krishon et al. 1997). Roosting home range sizes are substantially larger in winter than in summer, although roost switching is less frequent during colder weather, when bats may occupy the same roost for up to 12 consecutive days (Hein et al. 2005, 2008a). The Seminole bat is thought to remain active throughout the year, although it begins

to reduce its activity when ambient temperatures reach 21°C (70° F; Constantine 1958d), and generally they do not fly at temperatures below 18°C (64° F; Jennings 1958). In South Carolina, when temperatures fall below freezing, roosts are typically found beneath leaf litter on the forest floor. At temperatures between 0 and 4°C (32–39°F), they select roost sites in pine needle clusters < 5 m above the ground. Roosting at lower heights in cold weather likely provides protection from the wind (Hein et al. 2008a).

Although migratory movements have been documented for several tree bat species, the evidence of Seminole bat migration is largely circumstantial. Many of the extralimital records of Seminole bats are from the autumn, when young bats presumably are wandering (Wilkins 1987). For example, an adult female was captured outside a mineshaft in Arkansas in September (Heath et al. 1983); records from New York (Layne 1955) and Pennsylvania (Poole 1932, 1949) were in the autumn, and the single record of *L. seminolus* from West Texas was also in September (Brant and Dowler 2000). Specimens have been collected in East Texas from February through November (Schmidly et al. 1977).

Seasonal differences in the distribution of males and females have been documented. For example, in South Carolina, females outnumber males nearly 2:1 in the summer, but no females have been captured from January to March (Hein et al. 2008a, 2008b). In the summer, the ratio of females to males is even more highly skewed toward females in Mississippi, where they outnumber males 5:1 (Miller 2003).

Seminole bats feed early in the evening over watercourses and clearings, generally at treetop level. Their flight is swift and direct, and they may occasionally alight on vegetation to capture prey (Marks and Marks 2006). In North Carolina, they forage primarily in pine stands (55% of the time) but also use bottomland (35%) and upland hardwood habitats (11%; Carter et al. 2004). Foraging area was found to range from 79.2 hectares for a lactating female to 289.6 hectares for a juvenile female in Georgia. The foraging area was located in a pine stand that was spatially discrete from the roosting area (Krishon et al. 1997). In Georgia, 95% of the diet of Seminole bats comprises beetles, flying ants, and moths (Carter et al. 1998). Additional prey items include true bugs, flies, and even ground-dwelling crickets (Carter et al. 2004; Wilkins 1987). In urban areas, Seminole bats are known to frequent streetlights to feed on insects attracted to the light.

Seminole bats are known to mate in the fall; reproductively active males have been documented in February and April in Georgia, so mating likely occurs in the spring as well (Constantine 1958d). Females may carry from 1 to 4 embryos but usually give birth to 2 young (Davis 1974). Pregnant females have been taken in May and June (Wilkins 1987). In Texas, parturition occurs in late May or June. Lactating females have been documented in the Big Thicket of Texas in late June and in Mississippi in mid-July (Miller 2003). The young bats mature quickly and are able to fly at 3–4 weeks of age. Young bats tend to wander far from their natal roost after they have been weaned.

Out of a total of 200 Seminole bats submitted to the DSHS from 1984 to 1987 and 1996 to 2000, 19 (9.5%) tested positive for the rabies virus.

Status. The IUCN 2011 status of the Seminole bat is "least concern." This species is locally abundant throughout eastern Texas and may be expanding its range westward.

Specimens Examined. Go to www.batsof texas.com for more detailed information about the total of 129 specimen records of *L. seminolus* from Texas. Additional records: *Cass County* (1) (McAllister et al. 2004), *Harrison County* (1) (Yancey and Jones 1996b), *Travis County* (1-TTU76878).

References. 33, 59, 86, 98, 101, 106, 116, 136, 180, 212, 214, 215, 225, 268, 336, 511, 524, 532, 533, 534, 535, 608, 618, 661, 676, 702, 732, 734, 760, 767, 770, 780, 781, 799, 801, 803, 817, 833, 919, 923, 951, 952, 970, 1017, 1025, 1026, 1028, 1030, 1044, 1091, 1164, 1250

Lasiurus xanthinus Thomas, 1897
Western Yellow Bat

Etymology. The generic name, *Lasiurus*, comes from the Greek words *lasios*, meaning "shaggy," and *oura*, meaning "tail," which refers to the well-furred uropatagium. The specific name comes from the Greek word *xanthos*, for "pertaining to yellow" (Stangl et al. 1993).

Subspecies. *Lasiurus xanthinus* Thomas, 1897, is monotypic, and no subspecies are recognized.

Western yellow bat (*Lasiurus xanthinus*)

Description. This is a medium-sized bat (forearm = 43–47 mm) with pale yellow pelage. Hairs have a dark base. The interfemoral membrane is furred for about half its length, and the hairs here are brighter yellow and contrast with the rest of the pelage. The western yellow bat's wing membranes are dark, and it lacks a dark face mask. The ears are pinkish brown and have a rounded tip. Dental formula: I 1/3, C 1/1, Pm 1/2, M 3/3 × 2 = 30. Average external measurements are as follows: total length, 105 mm; hind foot, 10 mm; ear, 13.5 mm; forearm, 45 mm. Weight: 12–20 g.

The western yellow bat may be confused with its southern relative, the southern yellow bat (*L. ega*), and identification is very difficult. In *L. ega*, hairs of the interfemoral membrane do not contrast with the color of hairs on the dorsum, and the bats have a dark face mask. The northern yellow bat (*L. intermedius*) and *L. xanthinus* are similar in color, but the former is larger (forearm usually greater than 47 mm) and does not occur in the same areas where one would expect to find *L. xanthinus*.

Distribution. *Lasiurus xanthinus* occurs in arid regions of the southwestern United States, along the Mexican Plateau west of the Sierra Madre, and Baja California. It was first documented in Brewster County, Texas, in October 1996 (Higginbotham et al. 1999). It has since been reported from other sites in the Chihuahuan Desert region as far east as Val Verde County (Bradley et al. 1999; Jones et al. 1999; Weyandt et al. 2001). Most individuals have been captured in autumn in Black Gap Wildlife Management Area and the adjacent Big Bend National Park. However, they have been captured in Texas in March, June, July,

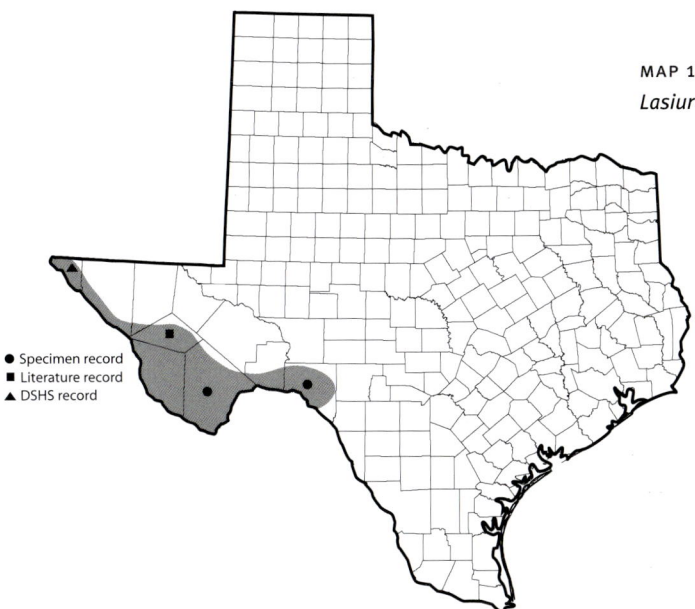

MAP 15. Distribution of the western yellow bat, *Lasiurus xanthinus*.

● Specimen record
■ Literature record
▲ DSHS record

September, October, and November. Outside of Texas (Nevada) they are somewhat active in the winter months (O'Farrell et al. 2004).

The presence of lactating females in Jeff Davis and Brewster counties indicates that at least some *L. xanthinus* raise their young in Texas. Recent captures by Raymond Matlack (pers. comm.) of West Texas A&M University confirm that the species still occurs in the state. Nine individuals (8 males, 1 female) were captured in September 2007 in Black Gap Wildlife Management Area, Brewster County. Furthermore, as recently as 2010 specimens have been documented from El Paso County (Tipps et al. in press).

Life History. This species represents a recent addition to the fauna of Texas. All evidence suggests that this is a result of an expansion of the species' range rather than the absence of previous collecting success. The reasons for such a range expansion are not well understood, but

the most probable explanation is the warmer temperatures recently recorded in West Texas. Vegetational changes in the region also may have facilitated the northward range expansion of the species. Cottonwood trees (*Populus* spp.) and other riparian vegetation in the Trans-Pecos region were almost exterminated early in the twentieth century, but the establishment of Big Bend National Park and other protected areas in the region have allowed the recovery of natural vegetation. This woody, riparian growth provides suitable roosting and foraging habitat for *L. xanthinus*.

Western yellow bats have been found using giant dagger yucca (*Yucca carnerosana*) as a roost site in Texas (Higginbotham et al. 2000), and Washington fan palms (*Washingtonia filifera*) are used in Arizona and Nevada (Cockrum 1961; O'Farrell et al. 2004). In both cases, the bats were found roosting among the dead leaves of the petticoat. In addition, *L. xanthinus* roosts in some deciduous tree species, including hackberry (*Celtis* spp.), sycamore (*Platanus* spp.), and cottonwood (Bond 1970; Brown 1996; Findley et al. 1975; Mumford and Zimmerman 1963).

Western yellow bats have been found in a wide variety of habitats in the United States at elevations of 550–1,750 m. In Nevada (Moapa Valley), these bats primarily use riparian woodlands dominated by *Washingtonia* fan palms (O'Farrell et al. 2004; Williams et al. 2006). In Big Bend National Park, Texas, they have been taken in lowland desert riparian zones (850 m) dominated by cottonwood, acacia (*Acacia* spp.), and willow (*Salix* spp.; Higginbotham et al. 1999). They have been encountered in the Davis Mountains over a creek at approximately 1,525 m among cottonwood and willow species (Jones et al. 1999). They also have been encountered in riparian areas dominated by sycamore or cottonwood and in woodland habitats with oak or pinyon-juniper vegetation in the Animas Mountains of New Mexico (Cook 1986). Farther south, on the Mexican Plateau, they have been captured in agricultural areas and mesquite/scrub habitat from 700 to 2,350 m (Baker 1956; Genoways and Jones 1968).

As one of us (Loren K. Ammerman) has documented, most captures of this species in mist nets in Texas tend to take place later in the evening—at least 1.5 hours after sunset, but more commonly 3–4 hours after sunset. In New Mexico this species was captured somewhat earlier in the evening, from 30 minutes to 2.25 hours after sunset (Mumford et al. 1964), but other researchers in New Mexico have taken them "relatively late at night" (Cook 1986). Discrepancies in time of capture could be influenced by several factors, including a late emergence time, foraging before drinking, or commuting long distances between roosting and foraging areas. Based on radio-tracking data, 1 individual in Big Bend National Park was captured more than 22 km away from its roost site, which lends some support to the third possible factor (Higginbotham et al. 2000).

Information about their diet is scarce. The diet of 1 male from Brewster County comprised true bugs, flies, wasps, moths, beetles, and crickets (Higginbotham et al. 1999). In addition to these food items, leafhoppers also have been documented in their diet in southern Nevada (O'Farrell et al. 2004).

Details about reproduction in this species are not well known. Its reproductive biology is probably similar to that of the southern yellow bat (*Lasiurus ega*), with one notable difference—litter size is typically 2, as opposed to 3–4 as seen in other lasiurines. In New Mexico, pregnant females, all carrying 2 embryos, were captured in May and June; lactating females have been taken in August (Cook 1986; Mumford and Zimmerman 1963). Pregnant females also have been reported from southern California in June (Bond 1970), and lactating females have been captured in Texas in late June (Jones et al. 1999). Juvenile western yellow bats have been captured in July and September in southern Nevada (O'Farrell et al. 2004).

Rabies was first reported in this species in 1979 in California, where 8 out of

23 (34.8%) bats submitted over a 9-year period tested positive for rabies (Constantine et al. 1979). In 2010, the first specimens were submitted to the DSHS for rabies testing, and 3 from El Paso County were found to be positive (Tipps et al. in press).

Status. The IUCN 2011 status of the western yellow bat is "least concern." In Texas and elsewhere the species has expanded its range northward (O'Farrell et al. 2004). There appears to be an established population in Brewster County, Texas, where it is encountered most often in the fall. Additional fieldwork is needed to ascertain the status of this species in other regions of Texas.

Remarks. Until 1988 *L. xanthinus* was recognized as a subspecies of *L. ega* (*L. e. xanthinus*), at which time it was elevated to specific status based on genetic evidence (Baker et al. 1988). Additionally, the species most morphologically similar to *L. xanthinus*, *L. ega,* is actually more closely related to *L. intermedius* than it is to *L. xanthinus* (Morales and Bickham 1995). Because of these taxonomic changes, older references that mention *L. ega* in the southwestern United States are usually referring to *L. xanthinus*. *Lasiurus ega* occurs in southern Texas, as described in its species account.

Specimens Examined. Go to www.batsof texas.com for more detailed information about the total of 8 specimen records of *L. xanthinus* from Texas. Additional records: *Brewster County* (1) (Bradley et al. 1999), *Jeff Davis County* (1) (Jones et al. 1999).

References. 92, 101, 165, 178, 193, 212, 256, 275, 282, 283, 423, 476, 565, 567, 575, 628, 767, 841, 850, 852, 880, 1026, 1028, 1091, 1124, 1190, 1224

Lasionycteris noctivagans
Le Conte, 1831
Silver-Haired Bat

Etymology. The generic name, *Lasionycteris,* comes from the Greek words *lasios,* meaning "shaggy" or "hairy," and *nykteris,* meaning "bat." The specific name comes from the Latin words for "night," *nox,* and "wanderer," *vagari* (Stangl et al. 1993).

Subspecies. *Lasionycteris noctivagans* Le Conte, 1831, is a monotypic species, and no subspecies are recognized.

Description. This medium-sized (forearm = 37–44 mm) bat is easily recognized by its distinctive pelage coloration. It is entirely black, but individual hairs of the dorsal surface are partially white near the ends, giving the pelage a frosted appearance. The ears are dark, short, and rounded with a paler patch at the base. The tragus is short and curved forward. The upper surface of the interfemoral membrane is furred, though not as heavily as in the tree bats of the genus *Lasiurus*. Other bats that have similar pelage include the hoary bat (*Lasiurus cinereus*), which is much larger and grayer, and the eastern (*Lasiurus borealis*) and western (*Lasiurus blossevillii*) red bats, which are reddish rather than black. Dental formula: I 2/3, C 1/1, Pm 2/3, M 3/3 × 2 = 36 (it is easy to overlook the upper incisors and first lower premolars because they are very small). Females are slightly larger than males. Average external measurements are as follows: total length, 96 mm; tail, 38 mm; hind

Silver-haired bat (*Lasionycteris noctivagans*)

foot, 9 mm; ear, 14 mm; forearm, 41 mm. Weight: 8 g.

Distribution. The silver-haired bat is broadly but erratically distributed across North America. It has been recorded from localities scattered throughout Texas and apparently is a fall-spring migrant in our state. Eight physiographic regions feature the silver-haired bat—they are absent only from the East Central Texas Plains, Blackland Prairies, Cross Timbers, and Southern Texas Plains regions of Texas, but this bat could be found all over the state. Although

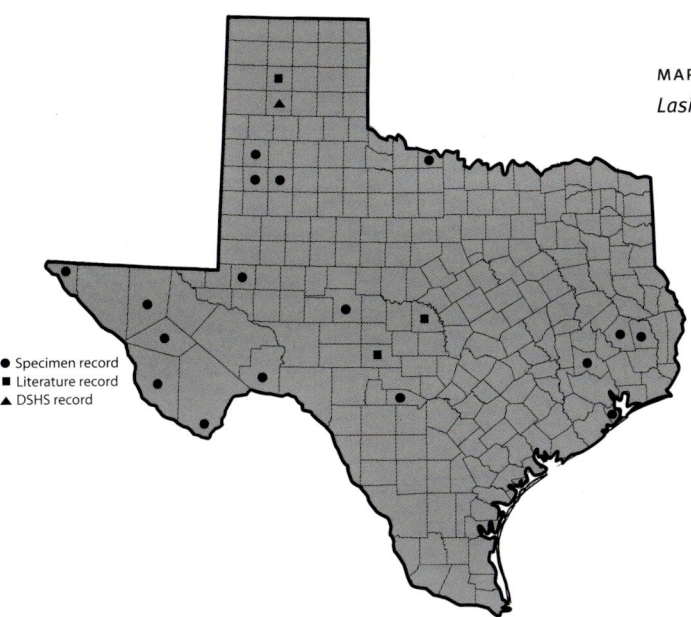

MAP 16. Distribution of the silver-haired bat, *Lasionycteris noctivagans.*

● Specimen record
■ Literature record
▲ DSHS record

primarily a species that inhabits forested areas, this bat occasionally may be found in xeric habitats during migration.

There are few midsummer records of silver-haired bats in Texas—these include 2 males taken in the Davis Mountains of the Trans-Pecos region in June and 2 males from the Chisos Mountains in late May (Ammerman 2005). Other mountain ranges, such as the Guadalupe Mountains, also may support a summer population. Interestingly, most summer records of this bat across the southwest are of males, which suggests that geographical segregation of the sexes may occur during the warmer months (Cryan 2003). Females appear to move north in spring and summer to bear young, whereas the males remain behind at more southerly locales.

Life History. Although *L. noctivagans* is generally considered a "tree bat," which roosts in hollow trees, woodpecker holes, and behind loose bark, it also has been found in buildings. Unlike tree bats

of the genus *Lasiurus*, silver-haired bats do not roost out in the open on branches or in foliage. In the Pacific Northwest, they prefer dead or dying trees that are larger in diameter and taller than other trees (Betts 1996; Campbell et al. 1996; Crampton and Barclay 1998; Jung et al. 1999). They have been found roosting in several species of coniferous and deciduous trees, with ponderosa pine (*Pinus ponderosa*), trembling aspen (*Populus tremuloides*), and peach-leaved willow (*Salix amygdaloides*) the most frequently used species (Barclay et al. 1988; Betts 1998a; Mattson et al. 1996; Vonhof 1996). Roosting colonies are small and transient, usually with fewer than 20 individuals, although a colony of 55 bats was found in a tree in South Dakota (Betts 1996; Crampton and Barclay 1998; Mattson et al. 1996; Vonhof and Barclay 1996).

This species is migratory, at least in part. There are oceanic records of migrating silver-haired bats nearly

161 km off the coast of Long Island, New York (Mackiewicz and Backus 1956). Although they may occasionally hibernate as far north as New York and British Columbia (Kunz 1982), most silver-haired bats are believed to overwinter in the southern part of the species' range. Surprisingly few winter records are available; thus, the mystery of just where these bats spend the winter is still not completely solved. Individuals have been reported from southeastern Arizona from November through March (Hoffmeister 1986) and from Missouri in winter (Dunbar et al. 2007).

Silver-haired bats typically forage in or near coniferous and/or mixed deciduous forests and adjacent to ponds or other bodies of water. Although its diet has not been studied specifically in Texas, it is considered to be a moth strategist based on data from New Mexico (Black 1974) and Arizona (Warner 1985). However, it consumes a large variety of small to medium-sized insects, including moths, true bugs, beetles, flies, and caddis flies (Kunz 1982; Lacki et al. 2007; Whitaker et al. 1981a). Captive specimens have eaten banana, bits of raw meat, and insects (Barbour and Davis 1969).

Reports on the temporal activity patterns of this species conflict. Kunz (1982) reports it is a relatively late flyer that often appears after other species have begun feeding. However, other authors have commented on their early activity (Adams 2003; Ammerman 2005; Hoffmeister 1986; Whitaker et al. 1977). In Iowa, they exhibit 2 peaks of foraging activity, one 2–4 hours after sunset and another 6–8 hours after sunset (Kunz 1973a).

Mating is thought to occur in the fall (possibly during migration), and fertilization is delayed until spring. Sperm production peaks in late August in New Mexico, copulation starts in late September, and ovulation peaks in late April and early May (Drueker 1972). Gestation is 50–60 days. Adult females typically raise 2 (or occasionally 1) young annually (Kunz 1982). Parturition typically occurs in June or July, although a pregnant individual was captured in Saskatchewan on August 7 (Parsons et al. 1986).

Females roost head up and use the tail membrane to safely catch the newborn during parturition (Kunz 1971). At birth, pups are hairless and pink with dark membranes. The young are able to fly at about 3 weeks of age. *Lasionycteris noctivagans* seems to have a shorter lifespan than that of other vespertilionid bats; the oldest individual known was estimated to be 12 years of age (Schowalter et al. 1978). Ten specimens of *L. noctivagans* reported to the DSHS were not infected with rabies (5 from 1984 to 1987 and 5 from 1996 to 2000).

Status. The IUCN 2011 status of the silver-haired bat is "least concern." This species is widely distributed in Texas, but it is not common. A new feature on the landscape in Texas, energy-producing wind turbine facilities, could pose a risk to this, as well as other, species. Although no data are available from Texas, *L. noctivagans* is 1 of the 3 bats with the highest fatality rates at such facilities in the United States. At 19 facilities located throughout the United States, silver-haired bats accounted for 1–56% of the bats killed. The highest fatality rates for this species were at facilities in the Pacific Northwest (Arnett et al. 2008). A study in southern Alberta, Canada, found that *L. noctivagans* migrated at a lower altitude than did hoary bats (*Lasiurus cinereus*) and that total bat fatalities increased with turbine tower height (Baerwald and Barclay 2009). As the number of wind-energy facilities in Texas continues to grow, it will be important to monitor their impact on migratory bat species such as *L. noctivagans.*

Specimens Examined. Go to www.batsof texas.com for more detailed information about the total of 53 specimen records of *L. noctivagans* from Texas. Additional records: *Kimble County* (4) (Goetze et al. 1996), *Presidio County* (2) (Brant et al. 2002).

References. 17, 37, 56, 60, 74, 83, 102, 106, 110, 112, 132, 135, 134, 136, 140, 145, 181, 208, 209, 223, 290, 297, 306, 315, 318, 336, 355, 359, 360, 364, 405, 423, 443, 475, 487, 511, 524, 574, 575, 594, 612, 625, 626, 633, 635, 648, 650, 687, 688, 718, 727, 760, 762, 767, 768, 772, 778, 823, 833, 851, 904, 1000, 1024, 1025, 1026, 1028, 1029, 1030, 1034, 1040, 1091, 1153, 1169, 1171, 1179, 1205, 1207, 1218, 1236, 1250

Parastrellus hesperus (H. Allen, 1864) American Parastrelle

Etymology. The generic name, *Parastrellus,* comes from the Greek words *para,* meaning "beside," and *strellus,* referring to *Pipistrellus,* derived from the Italian word for "bat," *pipistrello* (Hoofer et al. 2006). The specific name comes from the Greek word *hesperos,* which means "of evening" (Stangl et al. 1993).

Subspecies. Texas specimens are referable to the subspecies *P. h. maximus* Hatfield, 1936, as indicated by the latest taxonomic revision of the species (Findley and Traut 1970). It is the larger of the 2 subspecies in North America.

Description. This is the smallest North American bat (forearm = 29–33 mm). The pelage varies from light gray to yellowish, and the hairs are black at the bases. A dark and leathery facial mask is present across the rostrum and eyes from ear to ear. The wing and tail membranes also are very dark. The calcar of *P. hesperus* is subtly keeled; its ears are short and rounded, and the tragus is slightly curved and blunt at the tip. The eyes are fairly large in this species.

Parastrellus hesperus may be confused with the California myotis (*Myotis californicus*) and the western small-footed myotis (*Myotis ciliolabrum*), both of which have the narrow, straight, and sharply pointed tragus characteristic of the genus *Myotis.* The American perimyotis (*Perimyotis subflavus*) is also similar in appearance to *P. hesperus,* and the ranges of these 2 bats overlap in Texas. Characters that distinguish the 2 species are given in the account for *P. subflavus.* Dental formula: I 2/3, C 1/1, Pm 2/2, M 3/3 × 2 = 34. As with many vespertilionids, males are typically smaller than females of the same age. Average external measurements are as follows: total length, 78 mm; tail, 32 mm; hind foot, 6 mm; ear, 11 mm; forearm, 31 mm. Weight: 4 g.

American parastrelle (*Parastrellus hesperus*)

Distribution. The American parastrelle is particularly abundant in the mountain ranges and rocky canyon country of the Chihuahuan Desert region, but it also has been documented from scattered localities in the Southern Texas Plains, Central Great Plains, Southwestern Tablelands, High Plains, the northern and western edges of the Edwards Plateau, and the Arizona/New Mexico Mountains region. A nonreproductive female was collected from the wall of a building on the campus of Laredo Community College in September 2001 (Goetze et al. 2003). This is the southernmost record of this species in Texas.

American parastrelles have been captured in Texas at elevations ranging from near sea level on the lower Rio Grande to 2,100 m in the Chisos Mountains of Big Bend National Park, but they are most often encountered in the rocky canyon drainages in desert scrub habitat. In winter, females are thought to hibernate at higher elevations than males, possibly because their slightly larger size makes them less vulnerable to climatic extremes (Hayward and Cross 1979). However, recent work in New Mexico suggests that females might not move to higher elevations when overwintering there (Geluso 2007).

The American parastrelle is sporadically active during the winter in Texas. A few winter records (December, January, and February) have been documented,

● Specimen record
■ Literature record
▲ DSHS record

and there is no evidence that the species migrates. Apparently these bats do not enter a deep torpor in Texas and are capable of arousing and becoming active during warm spells in winter (Genoways et al. 1979). They have been captured in all months in southern Nevada and throughout the winter in southern New Mexico and southern Utah, although their activity peaks during the warmer months (Geluso 2007; O'Farrell and Bradley 1970; Poche 1981; Ruffner et al. 1979).

Life History. During the day, the American parastrelle roosts in cracks and crevices of canyon walls or cliffs, under loose rocks, and in buildings, desert shrubs, and caves (Blair and Miller 1949; Cross 1965; Hayward and Cross 1979; Milstead and Tinkle 1959; Moor et al. 1965). Similar sites also are used as night roosts (Hirshfeld et al. 1977). In winter, in the part of their range where they hibernate, these bats do so in mine tunnels and caves (Kuenzi et al. 1999).

American parastrelles tend to roost singly or in very small groups. A maternity colony of 12 individuals (adults and juveniles) is the largest group of these bats that has been found roosting together (Koford and Koford 1948).

These bats do not exhibit high fidelity to a particular crevice; instead, they tend to roost repeatedly in the same general area. In Arizona, occupied crevices were usually located on rocky outcrops that had a south to southeastern orientation. The crevices were vertical and about 2.5 cm wide and 13 cm to 1 m deep. Bats typically were found roosting within about 15 cm of the opening, regardless of the depth of the crevice (Cross 1965).

These are among the most diurnal of bats. They begin their foraging flights very early in the evening (often before sunset) and often remain active later into the morning hours than other bats. The peak of foraging activity is 30–60 minutes after sunset (Bradley and O'Farrell 1969; Cockrum and Cross

1964; O'Farrell and Bradley 1970; Szew-czak et al. 1998). In Arizona, lactating females were found to visit waterholes throughout the night during the month of July, but males did not (Cox 1965).

American parastrelles are slow bats and may be distinguished on the wing by their erratic, fluttery flight, which is restricted to small foraging circuits (Hayward and Davis 1964). These small bats are less active when wind speeds exceed 16 km per hour (O'Farrell and Bradley 1970). They forage from 2 to 15 m above ground, where they capture small, swarming insects, and consume about 20 percent of their body weight in insects per feeding (Hayward and Cross 1979; Mumford et al. 1964). Specific prey items include caddis flies, stone-flies, moths, small beetles, leaf and stilt bugs, leafhoppers, flies, mosquitoes, ants, and wasps. In Arizona, flying ants are their principal food during the sum-mer rainy months. Stomachs of indi-vidual bats were often found to contain only a single species of insect, or if more than 1 species was present, the remains were clumped together within the stom-ach, which suggests that these bats take advantage of swarming insects and feed intensely within such swarms (Hayward and Cross 1979). Spiders are occasion-ally eaten in New Mexico (Fries 1981). Interestingly, 1 American parastrelle was discovered entangled in a spider web, where it died, but apparently was not consumed by the spider (LaDuc 1993).

Copulation occurs in the fall and winter, and fertilization occurs in the spring (Hoffmeister 1986; Krutzsch 1975). Parturition occurs from late May to early June after a gestation period of approximately 40 days (Hayward and Cross 1979). One female collected in Val Verde County on April 16 had 2 embryos with crown-rump lengths of 7 mm; another female collected in Culberson County (Guadalupe Mountains National Park) on June 8 had 2 embryos that were 10 mm in length (Genoways et al. 1979); and 17 pregnant females with embryos 15–16 mm in length were collected in Presidio County (Big Bend Ranch State Park) on June 9. Of the 17 pregnant fe-males collected, 12 were carrying twins, and 5 had a single embryo (Yancey 1997).

Small maternity colonies may be established in buildings or rock crevices, although it is unknown when such colo-nies begin forming (Cross 1965). One of us (Loren K. Ammerman) has cap-tured lactating females as early as May 20 in Brewster County. Newborn bats weigh slightly less than 1 gram at birth but grow quickly, and juveniles become volant at about 1 month of age (Hayward and Cross 1979). In Big Bend National Park, volant juveniles have been cap-tured as early as the third week of June (Easterla 1973; Higginbotham and Am-merman 2002), and in Dickens County, Robert J. Baker (Texas Tech University) collected 3 volant young on July 8.

Possible predators include lizards and birds of prey, although predation is probably not a major cause of mortal-ity for this species (Hayward and Cross 1979; Jones and Gillespie 2009). Only 1 American parastrelle has been reported to the DSHS, and that individual was not infected with rabies. In California and Arizona, *P. hesperus* were found to have a unique strain of rabies virus not found in other bats; this particular strain has not been implicated in any humans cases of rabies (Franka et al. 2006).

Status. The IUCN 2011 status of the American parastrelle is "least concern." This species is one of the commonest bats of the desert Southwest, and populations appear to be stable in Texas.

Remarks. Another commonly used name for *Parastrellus hesperus* is the canyon bat. Until recently, the 2 species of North American pipistrelles (eastern and western) were in the genus *Pipistrellus*. However, based on genetic and morphological evidence, it was determined that they are not each others' closest relative, nor are they close relatives of the pipistrelles present in the Eastern Hemisphere (Hoofer and Van Den Bussche 2003). This discovery led to changes in the taxonomy of this group and to the formal description of a new genus, *Parastrellus* (Hoofer et al. 2006). Thus, the bat formerly known as the western pipistrelle (*Pipistrellus hesperus*) has been placed in this new genus as *Parastrellus hesperus* (American parastrelle). In addition, the genus *Perimyotis* (Menu 1984) was resurrected to apply to the eastern pipistrelle (*Pipistrellus subflavus*), now known as the American perimyotis (*Perimyotis subflavus*).

Specimens Examined. Go to www.batsoftexas.com for more detailed information about the total of 887 specimen records of *P. hesperus* from Texas. Additional records: *Brewster County* (30) (Bradley et al. 1999).

References. 42, 47, 54, 62, 86, 92, 93, 102, 106, 140, 142, 147, 166, 168, 172, 178, 179, 259, 289, 296, 318, 323, 336, 339, 355, 359, 381, 405, 423, 428, 440, 443, 450, 466, 475, 485, 489, 511, 513, 525, 528, 531, 541, 564, 570, 574, 575, 583, 584, 625, 626, 633, 645, 667, 681, 686, 720, 727, 767, 768, 797, 823, 830, 833, 837, 840, 850, 872, 884, 947, 953, 976, 988, 999, 1001, 1002, 1024, 1026, 1028, 1029, 1071, 1091, 1109, 1110, 1112, 1164, 1200, 1246, 1248, 1254

Perimyotis subflavus (F. Cuvier, 1832) American Perimyotis

Etymology. The generic name, *Perimyotis*, comes from the Greek word *peri*, meaning "around," and *Myotis*, which refers to another genus of vespertilionid bat (Hoofer et al. 2006). The specific name is derived from Latin words that mean "yellow belly" or "somewhat yellow" (Fujita and Kunz 1984; Stangl et al. 1993).

Subspecies. Two subspecies occur in Texas, according to the most recent review of the taxonomy of the species (Davis 1959). The most wide-ranging subspecies, *P s. subflavus* (F. Cuvier, 1832), occurs in 10 physiographic regions of Texas. The other, *P. s. clarus* Baker, 1954, which is slightly larger and paler than *P. s. subflavus*, occurs in Val Verde County and adjacent localities in Coahuila, Mexico. These 2 subspecies are ecologically separable, and the geographic area that separates the 2, including much of the Rio Grande Valley, appears to be uninhabited by the species.

Description. This is a small (forearm = 31–35 mm), yellowish-brown or reddish bat that is characterized by the unique "tricolored" pelage, in which the base of each individual hair is dark, the middle

American perimyotis (*Perimyotis subflavus*)

band is lighter, and the tips are dark or dusky. This feature gives this species a grizzled appearance that readily distinguishes *P. subflavus* from other similar-sized species. The calcar is not keeled, and the leading edge of the wing membranes is paler (pinkish) than the rest of the membrane. Ears and facial skin are pinkish brown.

The American perimyotis can be confused with the American parastrelle (*Parastrellus hesperus*), from which it is distinguished by its unique, tricolored fur, unkeeled calcar, and paler wing membranes and ears. Also, the tragus in *P. subflavus* is not curved, as in *Parastrellus hesperus*. Dental formula: I 2/3, C 1/1, Pm 2/2, M 3/3 × 2 = 34. Females are slightly larger than males. Average external measurements are as follows:

total length, 77 mm; tail, 35 mm; hind foot, 9 mm; ear, 12 mm; forearm 33 mm. Weight: 3–6 g.

Distribution. The American perimyotis has been documented from 10 physiographic regions of Texas but is most common in the eastern portion of the state, where it commonly has been collected along bottomland streams and forest flyways. However, they also have been taken from higher elevations (2,100 m) in the Chisos Mountains (Ammerman 2005). This species also is abundant in caves along the southern and eastern edges of the Balcones Escarpment. It is known to hibernate in caves and box culverts within its summer range and is a year-round resident of Texas (Sandel et al. 2001). Recent records from Brewster and Presidio coun-

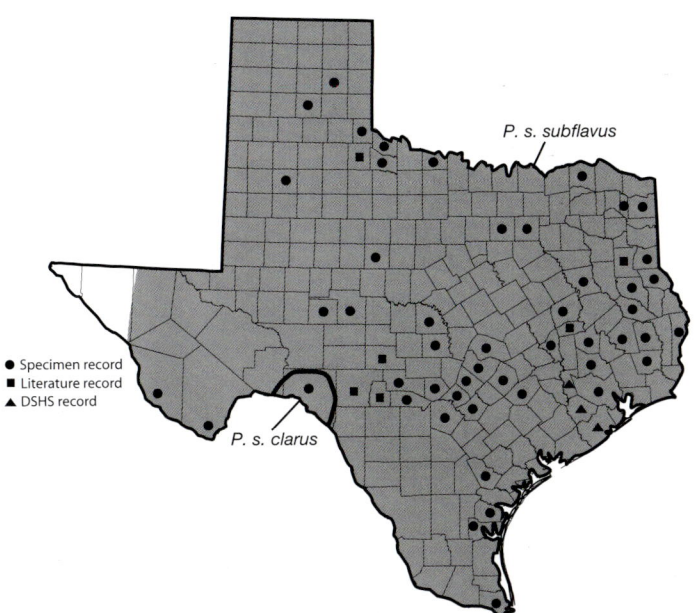

P. s. subflavus

● Specimen record
■ Literature record
▲ DSHS record

P. s. clarus

ties in Texas (Ammerman 2005; Yancey et al. 1995b), from southeastern New Mexico (Valdez et al. 2009a; White et al. 2006), and from as far west as Wyoming, Colorado, and Nevada (Adams 2003; Armstrong et al. 2006; Geluso et al. 2005; Ports and Bradley 1996) suggest a northward and westward expansion of its range.

Life History. The American perimyotis uses caves, mine tunnels, dams, box culverts, and rock crevices as hibernacula. The bats prefer the warmest locations within such sites (Rabinowitz 1981; Raesly and Gates 1987). Most hibernating bats roost alone, although clusters of 2–3 bats occasionally have been observed (Briggler and Prather 2003; Jones and Suttkus 1973; Sandel et al. 2001). They appear to be one of the first species to enter hibernacula in the fall and are one of the last to emerge in spring (LaVal and LaVal 1980). In Arkansas, *P. subflavus* are more frequently found in larger, longer caves. Larger caves

provide a wide range of temperature profiles within a season, while maintaining a consistent temperature throughout the year (Briggler and Prather 2003). Concrete box culverts are used as hibernacula along Interstate Highway 45 in southeastern Texas. As many as 254 American perimyotis have been found in 1 culvert. The bats can be difficult to arouse and are sometimes covered with water droplets and spider webs, suggesting a fairly deep state of torpor (Walker et al. 1996). Bats arrive in these winter roosts in September and are gone by April (Sandel et al. 2001).

In summer, American perimyotis roost under bridges and in caves, mines, buildings, tree cavities, and the live and dead foliage of deciduous trees; individuals have occasionally been documented using squirrel nests and Spanish moss (Carter and Menzel 2007). Males typically roost alone, whereas females roost alone or in small (fewer than 50 individuals) maternity colonies (Perry and Thill

2007c). They have been documented roosting under bridges in Arkansas, Illinois, and Louisiana in the summer (Feldhamer et al. 2003; Ferrara and Leberg 2005; Gardner and McDaniel 1978). Flat slab bridges are the only bridge type not occupied in Illinois (Feldhamer et al. 2003). In Louisiana, the bats roost in the narrowest spaces available in the darkest portion of the bridge. It has been suggested that such sites are selected to minimize their visibility and accessibility to predators. Roost sites also are in the warmest portion of the bridge (Ferrara and Leberg 2005).

When roosting in tree foliage, the dead foliage of oaks is preferred as a roost site (Leput 2004; Perry and Thill 2007c; Veilleux et al. 2003). It was previously thought that American perimyotis did not utilize coniferous trees as roosts, but they recently have been documented using clusters of dead needles in shortleaf pines (*Pinus echinata*) as maternity roosts in Arkansas (Perry and Thill 2007c). Roosts are located primarily in upland habitat in Indiana, but this habitat type also has a greater availability of preferred roost trees (oaks) than other habitat types (Veilleux et al. 2003). This tendency is similar in South Carolina, where all of the roosts are located in bottomland hardwood forest where there is a high abundance of oak trees (Carter et al. 1999). In Arkansas, roosts are located in forested sites at least 50 years old with a mature overstory and a complex vertical structure with abundant midstory trees (Perry and Thill 2007c). This preference for more mature stands with a complex structure also was found in North Carolina and South Carolina (Leput 2004; O'Keefe et al. 2009).

It has been widely demonstrated that many species of foliage-roosting bats prefer larger trees in which to roost. However, this is not always found to be true for the American perimyotis. In Arkansas, females select trees that are larger than those randomly available in mature forest; however, males demonstrate no such preference (Perry and Thill 2007c). In North Carolina males typically roost close to streams (O'Keefe et al. 2009). Reproductive females also roost closer to a permanent water source and farther from the forest edge when compared to nonreproductive females in Indiana. Roosts of both genders are located farther from roads than random in Arkansas (Perry et al. 2008). Home range size for 1 adult female in Georgia was 389.2 hectares (Krishon et al. 1997).

As reproductive females wean their young and transition to postlactating, they increase their use of tree species other than oaks (Veilleux et al. 2004). Interestingly, male American perimyotis most frequently use hickory, maple (*Acer* spp.), and birch (*Betula* spp.) trees as roosts in North Carolina even though oak trees are abundant (O'Keefe et al. 2009). The similarity of roost selection between males and nonreproductive females in a wider variety of tree species suggests that the tight leaf clusters in oaks impart some selective advantage to reproductive females, perhaps by providing better concealment and more protection from the cooling effects of the elements (Veilleux et al. 2004).

Roost fidelity of American perimyotis appears to be higher in tree roosts than for other foliage-roosting bats but lower in building roosts when compared to other species of bats that roost in buildings. In South Carolina and Indiana, reproductive females used the same foliage roost for an average of 3.3 and 3.9 consecutive days, respectively, with a maximum of 17 days (Leput 2004). Maternity colonies located in leaf clusters contained 3–13 individuals (adults and juveniles), with an average of 6.9. Although bats occupied specific leaf clusters for fairly short periods of time, they demonstrated strong fidelity to a small roosting area (0.4–2.3 hectares) during a single summer season and across multiple years (Veilleux et al. 2003). Male American perimyotis in Arkansas are known to use the same foliage roost for up to 33 consecutive days (Perry and Thill 2007c). While in the roost, bats spend most of their time resting (77%), followed by times of alertness (16%), grooming (7%), and crawling (1%; Winchell and Kunz 1996).

Roost switching to different sites within a building and between buildings seems to be common. One maternity colony in Massachusetts was observed to roost in 4 distinct locations within a barn; the colony sometimes used several roost sites within a single 24-hour period (Winchell and Kunz 1996). In Indiana, maternity colonies in 6 houses have been observed for 2–4 years (for a total of 20 bat years). The colony remained in the same roost for the entire summer season for only 2 of those years. Females would often move their young to an alternate roost for various lengths of time before returning to the one where they originally had been observed. The largest maternity colony comprised 29 adult females and their offspring (Whitaker 1998). Roost switching also is common in colonies located under bunkers in southern Louisiana. During a 6-year period, 41.2% of females and 68.7% of males roosted in more than 1 building (Jones and Pagels 1968; Jones and Suttkus 1973).

Little is known of their food habits in Texas, but elsewhere they are known to have a broad diet of at least 10 different orders of insects (Griffith and Gates 1985; Lacki et al. 1996). Different insects dominate their diet in various locations. In Georgia, moths are the most important food item (Carter et al. 1998); in West Virginia, flies are most commonly eaten, although 4 other orders also are frequently consumed (Carter et al. 2003a); and in Indiana, true bugs, beetles, and flies are consumed with nearly equal frequency (Whitaker 1972, 2004). Their diet is highly diverse in Maryland, where beetles, true bugs, flying ants, moths, and flies are present in more than 40% of the fecal samples examined (Griffith and Gates 1985). One study in Georgia found gender-based differences in their diet; females consume mostly true bugs and beetles, and males forage on flies and true bugs (Menzel et al. 2002a).

Shortly after sunset, *P. subflavus* emerge to feed and can be recognized by their slow, fluttering flight, which resembles that of a large moth (Fujita and Kunz 1984). Insects are caught in considerable quantities in a short period, and within 30 minutes the bats are gorged. They probably feed at intervals throughout the night and digest their meals at a night roost between feeding bouts. Several studies have found that they preferentially feed in riparian areas (Ellis et al. 2002; Ford et al. 2005, 2006a; Menzel et al. 2005). In fact, activity of American perimyotis increases downstream from wastewater treatment–plant effluent, suggesting that this species is tolerant of and may even benefit from anthropogenic input into such systems (Kalcounis-Rueppell et al. 2007).

American perimyotis typically mate in the fall before entering hibernation, although they also have been observed mating in January and February in Louisiana (Jones and Suttkus 1973). Some springtime mating also occurs at the time of ovulation (Guthrie 1933). The young may be born any time from late May to early July after a gestation period of about 44 days. Litter size is usually 2, and young bats are able to fly within a month of birth (Kurta and Kunz 1987; Whitaker 1998). The weight of the pups is about 50% of the postpartum weight of the female (as opposed to 20–40% in other vespertilionid bats), which makes the maternal effort of pregnant *P. subflavus* among the highest reported for bats. In New England, growth and development of the pups are delayed by low roost temperature, higher-than-average rainfall, and low insect availability (Hoying and Kunz 1998).

Mortality of *P. subflavus* is apparently greatest during their first year of life, probably due to heavy losses during winter, and the highest rate of survival is for those that are approximately 3.5 years of age. Males typically live longer than females (Davis 1966b). The greatest longevity record of an American perimyotis is 15 years (Fujita and Kunz 1984). Of 10 American perimyotis reported to the DSHS between 1984 and 1987, only 1 tested positive for the rabies virus. An additional 40 individuals were submitted between 1996 and 2000, and all proved to be negative for rabies.

Status. The IUCN 2011 status of the American perimyotis is "least concern." There is no concern at this time for the conservation status of this species in Texas as it is common in the eastern part of the state and seems to be expanding its range westward. In the eastern United States, there is some concern over potential declines in populations due to fatalities caused by wind turbines—up to 25% of dead bats found at wind farms were *P. subflavus*, most of which were males (Arnett et al. 2008; Kunz et al. 2007).

Remarks. This species is also known as the tricolored bat. Recent taxonomic changes of the former eastern pipistrelle (*Pipistrellus subflavus*) are summarized in the species account for *Parastrellus hesperus* (Hoofer and Van Den Bussche 2003; Hoofer et al. 2006).

Specimens Examined. Go to www.batsoftexas.com for more detailed information about the total of 167 specimen records of *P. subflavus* from Texas. Additional records: *Kimble County* (8) (Goetze et al. 1996), *Presidio County* (1) (Geluso et al. 2005), *Madison, Freestone,* and *Walker County* (Walker et al. 1996), *Rusk County* (1) (Johnson et al. 2005).

References. 8, 37, 47, 54, 56, 59, 73, 74,
86, 91, 92, 102, 106, 136, 143, 145, 144,
185, 205, 212, 213, 215, 216, 225, 226,
262, 292, 306, 313, 318, 319, 336, 340,
341, 359, 388, 395, 405, 409, 418, 433,
435, 443, 459, 469, 478, 487, 504, 509,
511, 513, 524, 571, 583, 584, 589, 611, 623,
630, 631, 635, 636, 651, 665, 676, 696,
706, 710, 714, 729, 742, 760, 767, 768,
781, 790, 798, 803, 804, 805, 823, 830,
833, 859, 881, 918, 924, 953, 959, 962,
976, 999, 1001, 1012, 1024, 1025, 1026,
1028, 1030, 1040, 1044, 1091, 1112, 1145,
1153, 1160, 1161, 1162, 1164, 1175, 1194,
1196, 1197, 1214, 1236, 1237, 1250, 1252,
1257

Eptesicus fuscus (Palisot de Beauvois, 1796)
Big Brown Bat

Etymology. The generic name, *Eptesicus,* comes from the Greek words *petesthai,* meaning "fly," and *oikos,* meaning "house," which can be translated as "house flyer." The specific name, *fuscus,* comes from the Latin word for "dusky" (Stangl et al. 1993).

Subspecies. Two subspecies of big brown bat are known from Texas and occur in distinct geographic areas (Burnett 1983; Manning et al. 1989). *Eptesicus f. fuscus* (Palisot de Beauvois, 1796), is known from the South Central Plains (Pineywoods) of East Texas, where it is apparently restricted to pine-oak and longleaf pine vegetation zones (Schmidly et al. 1977). *Eptesicus f. pallidus* Young, 1908, which has a lighter pelage than *E. f. fuscus,* has been collected from wooded, montane areas of the Chisos, Chinati, Davis, Eagle,

and Guadalupe mountains and in the Sierra Vieja range of the Trans-Pecos of western Texas. Records of this subspecies also are available throughout winter from caves and canyons of the High Plains and the Rolling Plains regions of northwestern Texas. Jones and Manning (1990) have commented that individuals in the Panhandle (Garza County) appear somewhat intermediate between the 2 recognized subspecies—they were larger and gave birth to 2 young like *fuscus,* but their color was pale like *pallidus.* At a contact zone between the 2 subspecies in Colorado, no consistent differences in body size, color, litter size, or overwintering behavior were found that correlated with divergent eastern and western mitochondrial DNA lineages (Neubaum et al. 2007).

Description. This is a medium to large bat (forearm = 42–51 mm) with large, broad wings and a broad nose. The pelage varies from pale brown to chestnut above and slightly paler below. The fur is long and lax and has an oily appearance. The ears are relatively small, leathery, and black; the tragus is broad and curved. Finally, the calcar is keeled, and, as with many other vespertilionids, males are often slightly smaller than females.

The large size and strong, steady flight of the big brown bat distinguishes it on the wing. *Eptesicus fuscus* is very vocal and often produces an audible chatter during flight. Dental formula: I 2/3, C 1/1, Pm 1/2, M 3/3 × 2 = 32. Average external measurements are as follows: total length, 111 mm; tail, 43 mm; hind foot, 11 mm; ear, 17 mm; forearm 48 mm. Weight: 13–20 g.

Big brown bat (*Eptesicus fuscus*)

Distribution. The big brown bat is distributed throughout the United States except for the extreme southern tip of Florida. It has a somewhat disjunct distribution in Texas, having been recorded primarily from the eastern, northern, and western parts of the state (Manning et al. 1989). Few specimens are available from the central and southern parts of the state, and there is an unverified literature record from Bexar County. Although most commonly encountered in summer, this bat apparently is a year-round resident of Texas.

In Big Bend National Park, male and female big brown bats are known to segregate elevationally in summer. During this season, females are more common in the lowlands, where conditions are more favorable for raising young, and males occupy the higher elevations of the Chisos Mountains (Easterla 1973).

Life History. In summer, big brown bats roost in hollow trees, rock crevices, bridges, and tunnels but seem to prefer buildings when they are available. A study conducted in Alberta, Canada, found that the benefits gained from roosting in a building as opposed to a rock crevice included lower energy use, near absence of predation, earlier parturition, and faster growth of juveniles (Lausen and Barclay 2006). Buildings preferred as maternity roosts in Colorado and Pennsylvania are warmer, taller, and more accessible (had more

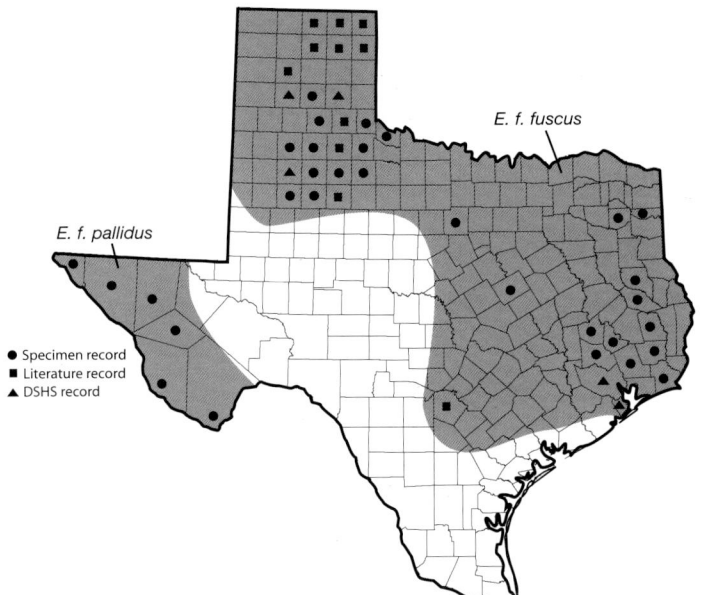

MAP 19. Distribution of the two subspecies of the big brown bat, *Eptesicus fuscus.*

E. f. fuscus

E. f. pallidus

● Specimen record
■ Literature record
▲ DSHS record

or larger openings) to bats than buildings unoccupied by bats (Neubaum et al. 2007; Williams and Brittingham 1997). Individuals show high roost fidelity to artificial structures and will return if attempts to exclude them are not thorough (Brigham 1991; Brigham and Fenton 1986). They seem relatively intolerant of high temperatures at the roost site. When the temperature rises above 33–35°C (91–95°F), the bats move to a cooler location within the roost or abandon the roost entirely (Davis et al. 1968; Ellison et al. 2007a).

Before buildings became widely available, big brown bats roosted predominately in tree hollows and rock crevices, as well as under exfoliating bark. Bats are much less loyal to these ephemeral roosting sites when compared to relatively long-lasting buildings (Brigham 1991). Although bats switched roosts every 2–3 days on average, they were loyal to small roosting areas of forest in Saskatchewan, Canada, over a 3-year period (Brigham

1991; Lausen and Barclay 2002; Willis and Brigham 2004). *Eptesicus fuscus* has been observed roosting in a variety of deciduous and coniferous trees, the most frequently reported of which include trembling aspen and ponderosa pine (Betts 1996; Christian 1956; Kalcounis and Brigham 1998; Kurta 1994; Perry and Thill 2008b; Vonhof 1996). Trees preferred as roosts are usually taller than surrounding trees and are located on flat areas (as opposed to slopes) in relatively open forest (Betts 1996; Loeb and Waldrop 2008; Perry and Thill 2008b; Rancourt et al. 2007; Vonhof and Barclay 1996). They also have been reported using pine snags as roosts in forested areas in Arkansas (Perry et al. 2007b).

Maternity colonies vary in size from 2 (mother and pup) to 700 individuals, with colonies of 25–75 most typical; the largest colonies are usually found in buildings (Cope et al. 1961; Kurta and Baker 1990; Mills et al. 1975; Rancourt et al. 2007; Timpone et al. 2006;

Whitaker and Gummer 1992). Genetic evidence indicates that each maternity colony comprises 5–15 maternal lineages, which suggests that females frequently move among colonies (Vonhof et al. 2008) . Moreover, roosting associations within maternity colonies were not influenced by the relatedness of individuals (Metheny et al. 2008). Adult males most frequently roost solitarily in the summer but also are found in small, all-male colonies or maternity colonies (Hamilton and Barclay 1972; Kurta and Baker 1990; Perry and Thill 2008b). Their presence in maternity colonies increases later in the summer after the pups are weaned (Davis et al. 1968; Mills et al. 1975).

In winter, these bats hibernate in caves, mine tunnels, rock crevices, storm sewers, and buildings. The single most important feature of a hibernaculum appears to be the maintenance of the ambient temperature above freezing; if the temperature falls below freezing, the bats will rouse and move to a different, more favorable location (Barbour and Davis 1969; Whitaker and Gummer 1992). When hibernating in caves, they generally choose sites near the entrance where the temperature is low and the relative humidity is less than 100 percent (Beer and Richards 1956). *Eptesicus fuscus* hibernates singly or in small groups, and males enter hibernation earlier than females do (Phillips 1966). It has been suggested that the availability of heated buildings for hibernation has allowed the big brown bat to extend its range northward and expand its population (Whitaker and Gummer 2000; Winhold et al. 2008).

Big brown bats feed on a variety of medium-sized (6–12 mm long) night-flying insects, particularly beetles, but curiously they do not appear to prey upon moths to an appreciable degree (Agosta 2002). The digestive tracts of *E. f. pallidus* from Arizona and Mexico were found to contain termites, true bugs, leafhoppers, June beetles, leaf beetles, and flying ants (Ross 1967). In Indiana and Illinois, the insects that big brown bats consume most frequently are agricultural pest species, primarily those associated with corn (Whitaker 1995). Moreover, *E. fuscus* has been categorized as a beetle strategist in New Mexico (Black 1974), and beetles constitute 50% or more of the entire volume of food this species consumes in Indiana, Illinois, and Idaho (Lacki et al. 2007; Mumford and Whitaker 1982; Whitaker 2004). Juveniles tend to consume softer prey items than adults (Hamilton and Barclay 1998).

Although their roosts are often in urban areas, big brown bats forage more frequently in rural areas perhaps because insect density is lower in urban settings (Everette et al. 2001; Geggie and Fenton 1985; Menzel et al. 2001b). In Indiana, big brown bats readily cross large continuous blocks of urban habitat to reach agricultural and wooded areas for foraging (Duchamp et al. 2004). Wastewater treatment–plant effluent seems to have a negative impact on the foraging activities of these bats in North Carolina (Kalcounis-Rueppell et al. 2007). Home-range size was 2,906 hectares for female bats in Georgia (Menzel et al. 2001b). In Alberta, Canada, males were found to forage over larger areas than females do, but no estimation of home-range size was given (Wilkinson and Barclay 1997).

As is common in other vespertilionid bats, *Eptesicus* mates in the fall and stores sperm until ovulation in the spring. Gestation lasts about 60 days (Kurta and Baker 1990). The time of parturition varies with latitude, but in Texas most births occur from late May to June. Interestingly, big brown bats in the eastern part of the United States (subspecies *fuscus*) usually produce 2 young per litter, whereas in the Rocky Mountains and westward (subspecies *pallidus*) litters contain only 1 pup (Barbour and Davis 1969). Since Texas spans both of these ranges, it is probable that bats in the Trans-Pecos have a litter size of 1, whereas those in the South Central Plains (Pineywoods) typically produce twins. Indeed, 6 gravid females from the Trans-Pecos each carried a single fetus, in keeping with their assignment to *E. f. pallidus* (Manning et al. 1989). Fetal counts for bats from northwestern Texas are 2, consistent with their assignment to *E. f. fuscus* (Jones and Manning 1990). In this subspecies, multiple paternity was documented in 12 of 26 sets of twins from maternity colonies in Indiana and Illinois (Vonhof et al. 2006). Thus, females commonly mate with and store sperm from at least 2 different males in the fall.

At birth, young *E. fuscus* weigh an average of 3 g and grow quickly, gaining as much as 0.5 g per day (Kunz 1974b). Young bats can make quite noisy chirps when they are separated from their mothers. Mothers are known to retrieve their fallen young from the floor of maternity roosts but do not carry them during foraging flights (Davis et al. 1968). Mortality before weaning was 7% at a maternity colony in Maryland (Christian 1956). Lactating females have been collected in mid-July in Texas; lactation lasts approximately 34 days (Kurta et al. 1990). At 4 weeks of age the young bats begin foraging for themselves and reach adult size approximately 2 months after birth (Kunz 1974b).

Predators include Barn Owls, horned owls, kestrels, rat snakes, bull snakes, rattlesnakes, weasels, rats, and cats (Barbour and Davis 1969; Black 1976; Kurta and Baker 1990; Lausen and Barclay 2006). Avoiding these hazards, big brown bats are long lived, and 1 banded individual has been recorded as having lived for 19 years (Hitchcock 1965); ages in excess of 10 years are not uncommon (Schowalter et al. 1978). Thirty-three big brown bats have been reported to the DSHS (in the years 1984–1987 and 1996–2000), of which only 2 proved rabid. However, nationwide in 2006, *E. fuscus* accounted for 64% of all rabies-positive bats (Blanton et al. 2007).

Status. The IUCN 2011 status of the big brown bat is "least concern." There is no known threat to this species at this time, as it is common in most of its range in Texas.

Specimens Examined. Go to www.batsof texas.com for more detailed information about the total of 318 specimen records of *E. fuscus* from Texas. Additional records: *Garza County* (18) (Yancey and Jones 1996b), *Hall County* (44) (Howell et al. 2009), *Hansford County* (1) (Dalquest et al. 1990), *Kent County* (147) (Yancey and Jones 1996b), *Lipscomb*

County (7) (Dalquest et al. 1990), *Ochiltree County* (4) (Dalquest et al. 1990), *Potter County* (6) (Yancey et al. 1998).

References. 19, 39, 54, 62, 86, 92, 102, 106, 122, 132, 136, 139, 140, 141, 142, 147, 148, 159, 166, 168, 183, 184, 186, 188, 198, 205, 228, 232, 284, 306, 313, 315, 316, 318, 319, 322, 323, 324, 332, 336, 339, 342, 355, 361, 381, 397, 398, 404, 405, 409, 443, 463, 475, 478, 494, 511, 515, 516, 517, 524, 564, 571, 574, 575, 588, 617, 625, 626, 633, 635, 639, 651, 652, 691, 705, 707, 718, 723, 724, 725, 727, 752, 760, 767, 769, 774, 781, 802, 811, 822, 823, 830, 833, 840, 851, 862, 897, 921, 923, 934, 964, 976, 999, 1001, 1024, 1025, 1026, 1028, 1029, 1030, 1034, 1040, 1044, 1072, 1091, 1101, 1102, 1112, 1120, 1121, 1136, 1138, 1137, 1153, 1164, 1169, 1170, 1171, 1172, 1179, 1195, 1197, 1200, 1202, 1203, 1207, 1222, 1225, 1227, 1230, 1231, 1236, 1238, 1248, 1250, 1253, 1254

Nycticeius humeralis (Rafinesque, 1818) Evening Bat

Etymology. The generic name comes from the Greek word *nyktios,* meaning "of the night." The specific name is derived from the Latin word *umerus,* for "shoulder" (Stangl et al. 1993).

Subspecies. Texas specimens are referable to the subspecies *N. h. humeralis* (Rafinesque, 1818), according to the latest taxonomic assessment of the species in the state (Schmidly and Hendricks 1984).

Description. This is a small (forearm = 33–39 mm), rather nondescript bat whose pelage is dark brown dorsally and paler below. The wings are short and narrow, and the ears are small. The wing and tail membranes are dark and leathery in texture, as are the ears. The calcar is not keeled, and juvenile specimens are darker than adults.

Nycticeius humeralis resembles many species of *Myotis* but may easily be distinguished by its short, blunt tragus—as opposed to the long and sharp-pointed tragus of myotises. Dental formula: I 1/3, C 1/1, Pm 1/2, M 3/3 × 2 = 30. As with many vespertilionids, females are slightly larger than males. Average external measurements are as follows: total length, 87 mm; tail, 32 mm; hind foot, 8 mm; ear, 12 mm; forearm, 35 mm. Weight: 5–10 g. They can weigh up to 15 g in the fall, when fat deposition is at its maximum.

Distribution. This species occurs throughout most of the eastern United States, west to Nebraska, and southward into northeastern Mexico. In Texas it is found in the eastern part of the state, but scattered records from farther west may suggest a range extension in Texas, as has presumably occurred in Nebraska and Kansas (Dowler et al. 1999; Geluso et al. 2008; Phelps et al. 2008). The evening bat is commonly encountered in 7 ecological regions in the eastern half of Texas—the South Central Plains (Pineywoods), East Central Texas Plains, Cross Timbers, Blackland Prairies, Edwards Plateau, Southern Texas Plains, and Gulf Coastal Plains. It is a forest inhabitant and is commonly found along watercourses throughout the year (Schmidly et al. 1977).

Throughout much of its range in the United States, *N. humeralis* is migratory and favors southern climes in winter,

Evening bat (*Nycticeius humeralis*)

although it is unknown how far north or south this bat migrates (Boyles et al. 2003; Humphrey and Cope 1968; Saugey et al. 1988). In Texas, the evening bat has been collected in abundance from late March through September and only occasionally in winter. However, the winter habits of this bat are poorly known, and no winter roosting sites in Texas have been reported.

Life History. Evening bats occupy a large variety of roosts and have been found in the cavities and crevices of live or dead trees, in woodpecker holes, behind loose bark of trees, in tree foliage and Spanish moss, in buildings, in leaf litter, between rocks, and even in abandoned underground burrows (Bowles et al. 1996; Boyles et al. 2005; Chapman and Chapman 1990; Hein et al. 2009; Menzel et al. 2001a; Perry and Thill 2008a). They are a rather cosmopolitan tree-roosting species and readily use a large number of tree species as roosts; those

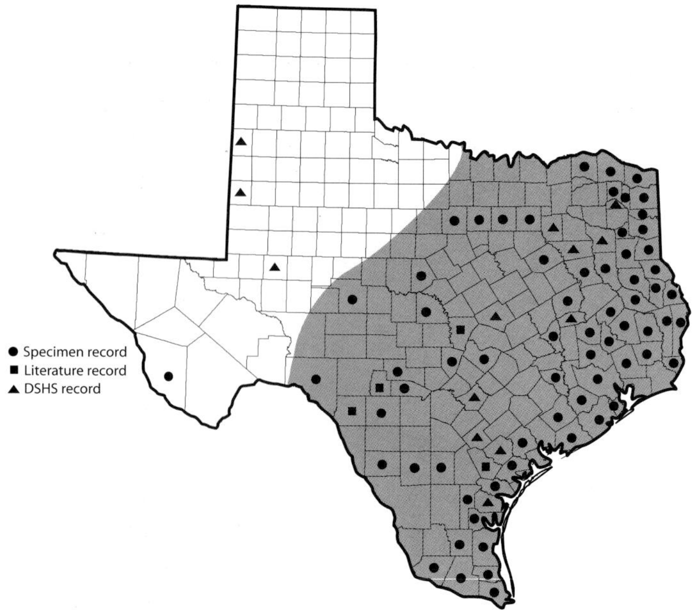

MAP 20. Distribution of the evening bat, *Nycticeius humeralis*.

● Specimen record
■ Literature record
▲ DSHS record

most commonly used include longleaf pine (*Pinus palustris*), loblolly pine, shortleaf pine, bald cypress (*Taxodium distichum*), red maple (*Acer rubrum*), oaks, and mesquite (*Prosopis glandulosa*). The size of trees they use as roosts also is highly variable; bats may occupy larger, older trees in some areas and smaller understory trees in others (Hein et al. 2009; Menzel et al. 2001a; Menzel et al. 2000; Perry and Thill 2008a). As is typical of many tree-cavity dwellers, roost fidelity is low, and bats move to a nearby tree (about 100 m distant) every 2–4 days even in winter (Bowles et al. 1996; Boyles and Robbins 2006; Duchamp et al. 2004; Menzel et al. 2001a). However, in buildings, nursery colonies are usually occupied throughout the summer months (Watkins 1969; Watkins and Shump 1981; Wilkinson 1992a).

Nycticeius demonstrates equal plasticity in regard to the habitat where roosts are located. For example, in South Carolina and Georgia, roosts are mostly located in mature longleaf pine habitat, while some occur in riparian hardwood stands or intensely managed loblolly pine stands (Hein et al. 2009; Menzel et al. 2001a; Miles et al. 2006). However, in one study in Missouri, every roost examined (63 total) occurred in forests subjected to prescribed burning (Boyles and Aubrey 2006). In another Missouri study, roosts were located exclusively in mature hardwood forest (Boyles and Robbins 2006). In Arkansas, roosts were found close to water in thinned, mature pine-hardwood stands (Perry et al. 2008). Their heights ranged from 5 to 24 m above the ground (Hutchinson 2001; Perry and Thill 2008a). Some studies have found that evening bats select these sites based on roost characteristics, whereas others have found that roosts are selected in the context of the landscape (Hein et al. 2009; Miles et al. 2006; Perry et al. 2008).

Characteristics of roost trees and

habitat also differ according to the season and between genders. In summer, roosts are more frequently located in snags in the late stages of decay, but during winter a higher proportion are in live trees (Boyles and Robbins 2006). Males are typically solitary roosters, whereas, during summer, females establish nursery colonies in trees or buildings, which can contain up to 950 individuals, although colonies in trees tend to have fewer than 100 individuals (Bowles et al. 1996; Hein et al. 2009; Menzel et al. 2001a; Perry and Thill 2008a; Watkins 1969; Wilkinson 1992a). In Arkansas, male evening bats roost in small hardwoods in unthinned (i.e., more cluttered, shaded, and cooler) tree stands, allowing them to enter deeper torpor more frequently to reduce energy expenditure (Perry and Thill 2008a). Females inhabited buildings 20 km distant from the study area (Saugey et al. 1989).

Buildings more typically house nursery colonies in the summer, although small groups of bats or solitary individuals sometimes overwinter in artificial structures (Bain and Humphrey 1986; French and Bunyard 2002; Saugey et al. 1989; Watkins and Shump 1981; Wilkinson 1992a). An attic roost in an apartment building in College Station, Brazos County, contained several hundred adult females and young bats in late June. The temperature in the attic was extremely warm (about 45°C/113°F), and the bats were spaced evenly over the ceiling, as well as in the wall spaces. The bats entered and left the attic along an open crack between the roof and the wall. They remained at this location until late August, at which time they left the roost and dispersed. Nearby, another colony of adult male and female bats was subsequently located in a hollow tree in the wooded area adjacent to the apartments (Schmidly 1991). Obviously, more research is needed to understand the complex roosting preferences of this species.

Evening bats make 1 to several foraging excursions each evening, the number of which is influenced by the reproductive stage of the females. Pregnant and postlactating bats usually forage only once per night, but lactating bats leave the roost on short foraging bouts 2 or more times per night (Clem 1993; Duchamp et al. 2004; Wilkinson 1992a). The time of first emergence is shortly before to just after sunset (Bowles et al. 1996; Clem 1993; Miller 2003). These bats prey primarily on beetles but also take other small, night-flying insects such as true bugs, flying ants, pomace and crane flies, and moths, although moths appear to be relatively unimportant in their diet (Carter et al. 1998, 2004; Geluso et al. 2008; Mumford and Whitaker 1982; Whitaker 1972, 2004). In Illinois and Indiana the single most important food item (23.5% and 14.2 % of the total volume, respectively) was the spotted cucumber beetle (*Diabroctica undecimpunctata*), the larvae of which is the southern corn rootworm, a significant agricultural pest (Feldhamer et al. 1995; Whitaker and Clem 1992). The evening bat's role as an agent of agricultural pest control has not been examined in Texas, but it is expected that it also feeds on the spotted cucumber beetle and its relatives in areas where corn and cotton are grown.

In addition to cornfields, evening bats frequently forage in open marshy areas, along the edges of forests, above the canopy, and around and below the crowns of trees in less cluttered habitats, such as pine savannah (Bowles et al. 1996; Clem 1993; Cochran 1999; Duchamp et al. 2004; Ford et al. 2006a; Menzel et al. 2005). In South Carolina and Georgia, they forage primarily in pine forests (Carter et al. 2004; Krishon et al. 1997). Demonstrating some foraging-site fidelity, they return to productive spots on consecutive nights at a site outside of Indianapolis, Indiana (Duchamp et al. 2004). Their home range is relatively small in Georgia (15.11 hectares; Krishon et al. 1997), and in Indiana they forage less than 2.75 km from their roost (Clem 1993).

Copulation is thought to take place in the fall (Watkins 1972), although *Nycticeius* has been observed copulating during winter months in Florida (Bain and Humphrey 1986). Scrotal males with mature sperm in their epididymides have been reported as early as June in Florida; interestingly, these also were young of the year (Bain and Humphrey 1986). Early sexual maturity of males also has been reported in Arkansas, where scrotal males have been taken in August (Saugey et al. 1988). Adult males are rarely encountered in nursery colonies (Humphrey and Cope 1970).

One to 3 offspring are born to each female, with an average of 2 (Bain and Humphrey 1986; Watkins 1969; Watkins and Shump 1981). Young are typi-
cally born from late May to mid-June, although a pregnant female has been captured on July 8 in Missouri (Bain and Humphrey 1986; Saugey et al. 1988; Saugey et al. 1989; Watkins 1969; Watkins and Shump 1981). In Presidio County, Texas, a juvenile female was captured on April 20. This suggests a very early parturition date (late March/early April) for this individual, as young bats do not become volant until about 20 days of age (Dowler et al. 1999).

The young are born pink and hairless and weigh approximately 2.25 g (Saugey et al. 1989; Watkins 1972). Females have been known to carry their young while feeding in captivity; in the wild it appears that they carry their young only when moving to a new roost (Gates 1941). Communal nursing of young has been documented in this species in a nursery colony occupying an attic in northern Missouri. Females nurse their own young during the first 2 weeks subsequent to birth, after which time about 20% of nursing periods are between females and nondescendant pups (Watkins and Shump 1981; Wilkinson 1992b). The young reach adult size at about 1 month of age.

Known predators of evening bats include black rat snakes, Long-eared Owls, raccoons, and domestic and feral cats (Cahn and Kemp 1930; Sparks et al. 2003; Watkins 1972). Of 221 evening bats reported to the DSHS between 1984 and 1987, 6 tested positive for rabies. Between 1996 and 2000, 410 additional evening bats were submitted, and only 3 were positive. That corresponds to an overall infection rate of 1.4%.

Status. The IUCN 2011 status of the evening bat is "least concern." This species is common in East Texas, and recent records west of its historical range suggest a westward expansion. Populations appear to be stable as long as forested habitat to meet foraging and roosting needs is conserved. However, some evidence of population declines of this species has been documented in Indiana, where big brown bats (*Eptesicus fuscus*) may outcompete it for roosts in buildings (Whitaker et al. 2006).

Specimens Examined. Go to www.batsof texas.com for more detailed information about the total of 649 specimen records of *N. humeralis* from Texas. Additional records: *Lamar County* (6) (Edwards and Johnson 2007).

References. 30, 54, 60, 86, 87, 92, 102, 106, 115, 136, 144, 145, 170, 171, 172, 173, 174, 175, 202, 214, 215, 225, 226, 248, 251, 306, 318, 333, 336, 358, 359, 361, 388, 389, 405, 410, 433, 443, 445, 461, 462, 467, 486, 488, 490, 511, 524, 536, 595, 596, 602, 635, 676, 760, 767, 768, 781, 798, 800, 801, 816, 817, 833, 848, 851, 920, 924, 926, 931, 970, 1017, 1019, 1025, 1026, 1028, 1029, 1030, 1040, 1044, 1079, 1091, 1112, 1153, 1164, 1180, 1181, 1194, 1197, 1198, 1199, 1218, 1219, 1220, 1250

Euderma maculatum (J. A. Allen, 1891)
Spotted Bat

Etymology. The generic name, *Euderma*, comes from the Greek words *eu*, meaning "true," and *derma*, meaning "skin." The specific name comes from the Latin word *macula*, for "stain" or "spot" (Stangl et al. 1993).

Subspecies. *Euderma maculatum* (J. A. Allen, 1891) is monotypic, and no subspecies are recognized. Some geographic variation in size has been documented across its range, although this was based on a relatively small (67) sample size. Individuals were largest in the southern part of their range and smallest in the west (Best 1988).

Description. This is a relatively large (forearm = 44–55 mm), strikingly distinctive bat. It has jet black fur dorsally with a conspicuous white spot present on each shoulder and on the rump. Also, a patch of white fur is located at the base of each ear. The ears are huge—longer than in any other North American bat. Membranes of ears, wings, and tail in living specimens are uniquely pinkish-red. Spotted bats are strong, swift fliers and may be recognized at night by their distinctive, high-pitched calls (Easterla 1973). Dental formula: I 2/3, C 1/1, Pm 2/2, M 3/3 × 2 = 34. Average external measurements are as follows: total length, 118 mm; tail, 50 mm; hind foot, 11 mm; ear, 43 mm; forearm, 51 mm. Weight: 15 g.

Distribution. This species occurs in semi-arid regions of the southwestern United States and Mexico, north to British Columbia, Canada, and east to south-central Montana. In spite of a fairly wide range in the United States, *E. maculatum* is one of the least known of American bats because it is infrequently captured due to its habit of flying and foraging high above the ground. It has been captured with regularity only in California, Arizona, New Mexico, southern Utah, and southern Colorado. In Texas this

Spotted bat (*Euderma maculatum*)

species has been captured only in Big Bend National Park in Brewster County, where it was first recorded in 1967 (Easterla 1970a). Additional specimens have since been obtained from several localities within Big Bend National Park, but no specimens have been captured outside the park. Spotted bats have been collected there in May, June, July, and August. The Big Bend area represents the southeasternmost locale where this bat has been recorded in the United States.

Its distribution is apparently unaffected by plant associations or elevation. Spotted bats have been captured in habitats that range from spruce-fir forests that are 3,230 m in elevation, to ponderosa pine forests at somewhat lower elevations, to juniper-sagebrush steppe and semiarid, desert scrub at still lower elevations (Easterla 1973; Geluso 2000; Perry et al. 1997; Reynolds 1981; Rodhouse et al. 2005). The single habitat component seemingly consistent with the capture of spotted bats is the presence nearby of broken canyon country or cliffs, which are their preferred roosting sites (Easterla 1973; Pierson and Rainey 1998; Poche 1981; Priday and Luce 1999).

Life History. Spotted bats favor the crevices found in vertical cliff faces as roosting sites but also have been documented roosting in caves, mines, and buildings (Easterla 1973; Geluso 2000; Mickey 1961; Miller 1903; Poche 1975; Poche and Ruffner 1975; Rodeck 1961; Sherwin and Gannon 2005). Studies in British Columbia indicate that individuals consistently return to the same cliff face to roost (Leonard and Fenton 1983; Wai-Ping and Fenton 1989). This species roosts solitarily or in small groups of fewer than 10 individuals (Easterla 1973; Mead and Mikesic 2001; Poche 1975). The winter habits of this species are poorly known, and there are only a handful of records from November through February (Ruffner et al. 1979; Sherwin and Gannon 2005; Szewczak et al. 1998).

Once thought to be relatively late fliers, spotted bats actually leave the roost within about 1 hour of sunset and do not return until early morning (Leonard and Fenton 1983; Mead and Mikesic 2001; Poche 1981; Priday and Luce 1999; Storz 1995; Szewczak et al. 1998; Wai-Ping and Fenton 1989; Woodsworth et al. 1981). Their characterization as a late flier was due to the long distance they travel between the roost and foraging areas at some locations; thus, they were not captured until later in the evening (Geluso 2000). In British Columbia, they forage within 10 km of their roost site (Wai-Ping and Fenton 1989). However, in the Grand Canyon, Arizona, they commute up to 77 km each night between their roost in the canyon and the meadows where they forage up on the canyon rim. In this case, these bats use a night roost in groves of trees after feeding and before commuting back to their day roost (Rabe et al. 1998b). Otherwise, this species flies continuously all night, and its activity is fairly constant throughout the night (Leonard and Fenton 1983; Storz 1995; Wai-Ping and Fenton 1989).

Euderma usually forage over open meadows and other clearings within ponderosa pine forests, but not in the forest itself, as well as in pinyon-juniper habitat (Navo et al. 1992). They have been found feeding over a golf course in Washington and over agricultural fields in British Columbia (Gitzen et al. 2001; Leonard and Fenton 1983). They do not typically forage over streams or other riparian areas, as do many other species of bats, but use streams as flight corridors where they may opportunistically feed while traveling between foraging grounds (Leonard and Fenton 1983; Pierson and Rainey 1998; Storz 1995; Szewczak et al. 1998; Wai-Ping and Fenton 1989; Woodsworth et al. 1981). They are fast, high fliers that travel about 19 km per hour while foraging and up to 50 km per hour while commuting from their roost to their foraging grounds (Rabe et al. 1998b; Wai-Ping and Fenton 1989). They forage in large elliptical orbits 10–30 m above the ground and circumnavigate the same group of foraging areas several times per night (Navo et al. 1992;

Rodhouse et al. 2005; Storz 1995). They return to feed along the same circuit night after night (Wai-Ping and Fenton 1989; Woodsworth et al. 1981).

Although spotted bats are not territorial, they do not tolerate conspecifics in proximity to them (< 50 m) while foraging (Leonard and Fenton 1983; Storz 1995; Woodsworth et al. 1981). They exhibit strong agonistic behavior toward other spotted bats by chasing those that are too close or by attacking speakers that are playing spotted bat calls (Obrist 1995). However, several spotted bats will use the same foraging area sequentially during the night (Leonard and Fenton 1983). More than 1 spotted bat will forage in adjoining areas simultaneously, although they maintain a distance of at least 50 m (Storz 1995).

Spotted bats are unquestionably moth specialists, as moths constitute nearly 100% of their diet. Moths from 3 families (Noctuidae, Lasiocampidae, and Geometridae) account for 88–100% of the moths they consume in Arizona. They also consume an occasional beetle or true bug; insects other than moths make up less than 5% of their diet (Easterla 1971, 1973; Painter et al. 2009; Ross 1961; Wai-Ping and Fenton 1989). One spotted bat was observed landing on the ground to capture grasshoppers in Utah, but this seems to be anomalous behavior (Poche and Bailie 1974).

The echolocation calls of *Euderma* are of low frequency (9–12 kHz) and are easily heard by humans but are below the hearing range of their prey. In fact, moths can likely detect the calls of *Euderma* only from distances of less than 1 m, if at all (Fullard and Dawson 1997). Obviously this would greatly enhance *Euderma*'s hunting ability. Because calls of lower frequency by definition have a longer wavelength, which limits the lower size threshold of a target's detectability, spotted bats can detect only objects that are 10 mm in diameter or larger, which is consistent with the size of prey they are known to capture (Leonard and Fenton 1984).

Data on reproduction are sparse. It is unknown when this species mates, but males with scrotal testes have been taken in June in Arizona and in August in Utah (Easterla 1965; Poche 1975; Poche and Ruffner 1975). Six males collected from March to August (no location was given) were found to have mature spermatozoa (Poche 1981). One pregnant female was captured in Big Bend National Park in June, and a pregnant female has been taken in both Nevada and Arizona in the same month (Easterla 1973; Geluso 2000; Poche 1975). Lactating females have been captured in Big Bend, Texas, and in Wyoming, Nevada, New Mexico, and Arizona from June through August (Berna 1990; Easterla 1973; Findley and Jones 1965; Geluso 2000; Perry et al. 1997; Priday and Luce 1999). Thus, parturition probably occurs from late May to mid-June at Big Bend National Park. The scant evidence available indicates that only 1 young is born annually (Easterla 1973; Findley and Jones 1965; Geluso 2000). The young are altricial and do not show the characteristic spotted color pattern of adults (Watkins 1977).

The only confirmed predator of spotted bats is the American Kestrel, although other raptors have been observed in pursuit of spotted bats (Black 1976; Easterla 1973). The mummified remains of a spotted bat found in a cave in northern Arizona were found to be 9,180 years old based on radiocarbon dating (Mead and Mikesic 2001). No spotted bats have been submitted to the DSHS for rabies testing.

Status. The IUCN 2011 status of the spotted bat is "least concern." The U.S. Fish and Wildlife Service has taken the position that much more knowledge is needed before any type of endangered or threatened listing should be considered. In a large-scale echolocation survey throughout their known range in North America, they were encountered less often than other species, which caused the authors to reject the hypothesis that they were common but rarely captured (Fenton et al. 1987). The species' restricted range in Texas accounts for its listing by Texas Parks and Wildlife as threatened. Two male spotted bats were captured in Big Bend National Park in 1997 and 1998, confirming its more recent presence in the park (Higginbotham and Ammerman 2002).

Specimens Examined. Go to www.batsoftexas.com for more detailed information about the total of 5 specimen records of *E. maculatum* from Texas. Additional records: *Brewster County* (2) (Higginbotham and Ammerman 2002).

References. 31, 106, 128, 129, 140, 141, 272, 288, 336, 373, 376, 378, 381, 383, 385, 416, 423, 425, 452, 465, 479, 511, 518, 564, 575, 625, 626, 740, 741, 767, 792, 814, 821, 833, 858, 885, 900, 925, 939, 946, 947, 948, 949, 954, 958, 982, 992, 993, 998, 999, 1002, 1024, 1026, 1028, 1056, 1091, 1099, 1109, 1153, 1164, 1174, 1200, 1244

Corynorhinus rafinesquii (Lesson, 1827) Rafinesque's Big-Eared Bat

Etymology. The generic name comes from the Greek words *coryn* and *rhinos*, which mean "club nosed." The specific name refers to French biologist and collector C. S. Rafinesque, who traveled and studied wildlife in the United States in the early 1800s (Jones 1977).

Subspecies. Texas specimens are referable to the subspecies *C. r. macrotis* Le Conte, 1831, according to the latest taxonomic revision of the species (Handley 1959).

Description. This is a medium-sized bat (forearm = 40–46 mm) with very large ears that measure more than 2.5 cm in length. The ears are held erect during flight and can droop back over the shoulders when at rest; in this position they resemble a ram's horns. A large, distinctive facial gland is present on each side of the snout. The pelage is gray above and nearly white below. The ventral pelage is conspicuously bicolored—the bases of individual hairs are black, and the tips are much lighter. Long toe hairs extend beyond the tips of the claws. Dental formula: I 2/3, C 1/1, Pm 2/3, M 3/3 × 2 = 36. Average external measurements are as follows: total length, 95 mm; tail, 45 mm; hind foot, 11 mm; ear, 32 mm, forearm, 42 mm. Weight: 7–13 g.

Corynorhinus rafinesquii is similar in appearance to Townsend's big-eared bat (*Corynorhinus townsendii*). However, the 2 species are geographically separated in

Rafinesque's big-eared bat (*Corynorhinus rafinesquii*)

Texas, and they may be distinguished as discussed in the account for *C. townsendii*.

Distribution. Rafinesque's big-eared bat is distributed throughout the forested areas of the southeastern United States and reaches the westernmost boundary of its range in extreme eastern Texas. Here, this bat is found in small numbers at scattered localities in the forested South Central Plains region (Pineywoods), where it has been captured from May through December (Mirowsky et al. 2004; Schmidly 1983). Two winter roosts have been found in East Texas, but it is not known whether this species hibernates there or migrates elsewhere during this season (Mirowsky et al. 2004).

Life History. Rafinesque's big-eared bats roost in hollow trees, crevices behind loose tree bark, caves, culverts, water wells, cisterns, bridges, barns, and abandoned buildings. Much recent work on this species' roost preferences in the southeastern United States indicates that it prefers hollows in water tupelo and black gum trees in bottomland hardwood forest (Carver and Ashley 2008; Cochran 1999; Gooding and Langford 2004; Lance et al. 2001). It occasionally uses hollow magnolia (*Magnolia grandiflora*), American beech (*Fagus grandifolia*), sycamore (*Platanus occidentalis*), and oak trees (Clark 1990; Mirowsky et al. 2004; Trousdale and Beckett 2005). The openings of most cavities used as roosts are triangular in shape and located at the base of the tree. Some cavities have openings located farther up the tree trunk. Trees used as roosts tend to be larger than those randomly available in the area, with a diameter at

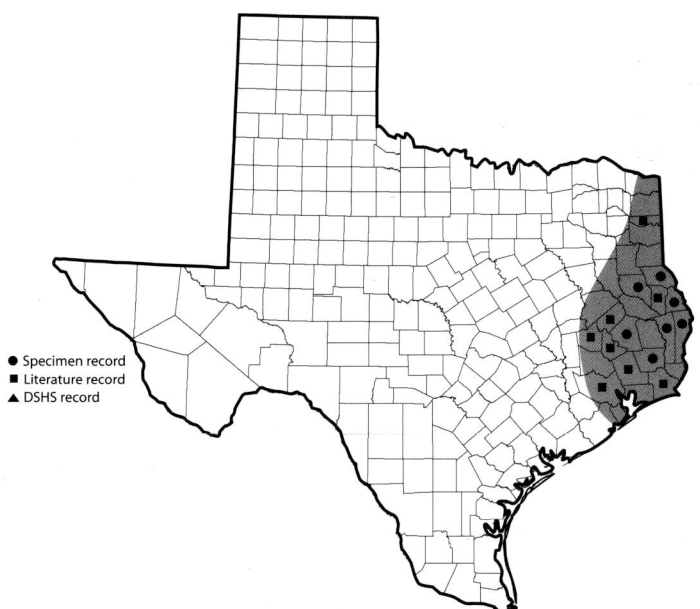

MAP 21. Distribution of Rafinesque's big-eared bat, *Corynorhinus rafinesquii.*

● Specimen record
■ Literature record
▲ DSHS record

breast height usually in excess of 100 cm (Cochran 1999). There is some evidence that this species selects roosts that do not receive direct sunlight (Carver and Ashley 2008; Clark et al. 1997b).

These bats are frequently found roosting under bridges; they prefer concrete bridges with girder construction. They have never been found under wooden bridges and rarely roost under flat slab concrete bridges (Bennett et al. 2008; Lance et al. 2001). Unlike many species of bats that use bridges as day roosts, *C. rafinesquii* roost in the open spaces between beams, not in small crevices (Ferrara and Leberg 2005; Keeley and Tuttle 1999). In Louisiana, bridges located in mature deciduous forest are occupied more frequently than are those in other habitat types. The bats change roosts frequently and spend their time equally in bridge and black gum tree roosts (Lance et al. 2001). They also roost in the warmest, darkest part of the bridge, perhaps to minimize their visibil-

ity and accessibility to predators (Ferrara and Leberg 2005).

Abandoned buildings also are frequently used as roosts, particularly by maternity colonies (Clark 1990; England et al. 1990; Horner and Maxey 1998; Jones and Suttkus 1975; Saugey et al. 1993). Such colonies may contain more than 100 individuals; however, 10–50 bats represent a more typical colony size, and bats are often found roosting singly (Cochran 1999; Ferrara and Leberg 2005; Menzel et al. 2001c; Mirowsky et al. 2004; Trousdale and Beckett 2004). They have been found sharing roost sites with southeastern myotis (*Myotis austroriparius*), American perimyotis (*Perimyotis subflavus*), and big brown bats (*Eptesicus fuscus*), but they cluster separately from these other species.

In the northern part of its range, *C. rafinesquii* remains throughout the winter and hibernates in caves and mines (Hurst and Lacki 1999; Jones

1977; Saugey et al. 1993), but the winter habits of this bat in the southern United States have not been well documented. They have been found using abandoned water wells throughout the winter in southern Arkansas (England et al. 1990), and a group was found hibernating in a cistern in western Tennessee (Goodpaster and Hoffmeister 1952).

Rafinesque's big-eared bats emerge late in the evening to forage and, like other big-eared bats, are very strong and agile fliers. Most of their foraging activity occurs during the first 4 hours after sunset and last 2 hours before sunrise (Menzel et al. 2001c). Their home range size is relatively small, from 24 to 260 hectares. They tend to forage in upland pine and hardwood habitats in South Carolina and Kentucky even though their roosts are usually located in mature bottomland forest (Hurst and Lacki 1999; Lacki et al. 1996; Menzel et al. 2001c). Small moths (31–57 mm wingspan) constituted more than 90% of the diet of these bats in Kentucky; the remainder included beetles and 5 other orders of insects in small amounts (Hurst and Lacki 1997).

Breeding probably occurs in fall and winter, and a single young is produced after the nursery colonies form in spring (Gardner and McDaniel 1978; Jones 1977). Parturition occurs from late May to early June in the northern part of the range and is slightly earlier in the South (Jones 1977; Trousdale and Beckett 2004). Pregnant individuals have been taken in May and June in Arkansas (Cochran 1999). The young are born naked and with their eyes closed; by 3 weeks of age their eyes have opened, permanent dentition is present, and the young bats are volant. Adult size is reached in approximately 4 weeks, and the pups are weaned at 2 months (Schmidly 1983). Lactating females have been documented in southern Mississippi as late as July 25 (Trousdale and Beckett 2004). In Texas, maternity colonies in manufactured structures were abandoned by October (Mirowsky et al. 2004).

This species is long lived, and the maximum recorded age is 10 years (Paradiso and Greenhall 1967). Circumstantial evidence suggests rat snakes may prey on these bats (Clark 1990). Only 2 specimens of *C. rafinesquii* were reported to the DSHS in 9 years, and 1 of these tested positive for rabies.

Status. The IUCN 2011 status of Rafinesque's big-eared bat is "least concern." At the federal level, it is a "species of conservation concern," and in Texas it is classified as "threatened." It bears special watching because of its scarcity, lack of knowledge about population levels, and the considerable potential that exists for degradation of roosting and feeding sites by commercial logging practices in the bats' preferred habitat. There is a real need to determine the effects of modern timber-management practices on this species. The replacement of older girder concrete bridges with slab-style bridges raises concerns as well because this decreases the availability of roost sites (Lance et al. 2001). Moreover, these bats appear to be extremely sensitive to human disturbance at their roost sites (Lacki 2000). Because of their penchant for using anthropomorphic structures as maternity roosts, increased human intrusion at these sites could negatively impact bat populations.

Remarks. Rafinesque's big-eared bat previously was classified in the genus *Plecotus*. However, analyses based on morphological and genetic differences within the genus *Plecotus* resulted in the recognition of a unique genus (*Corynorhinus*) that applies to 2 species of big-eared bats in North America (Bogdanowicz et al. 1998; Frost and Timm 1992; Hoofer and Van Den Bussche 2001; Tumlison and Douglas 1992).

Specimens Examined. Go to www.batsoftexas.com for more detailed information about the total of 46 specimen records of *C. rafinesquii* from Texas. Additional records: *Hardin, Jasper, Jefferson, Liberty, San Augustine, San Jacinto,* and *Trinity counties* (Mirowsky et al. 2004), *Marion* and *Harrison counties* (Schmidly 1983), *Walker County* (Thies 1994).

References. 14, 45, 98, 106, 115, 126, 161, 219, 245, 246, 247, 251, 336, 392, 401, 418, 451, 459, 492, 493, 511, 518, 524, 532, 582, 586, 599, 600, 632, 660, 712, 714, 721, 760, 767, 806, 812, 826, 833, 856, 902, 1018, 1025, 1026, 1028, 1030, 1035, 1040, 1044, 1112, 1116, 1125, 1126, 1129, 1251

Corynorhinus townsendii (Cooper, 1837)
Townsend's Big-Eared Bat

Etymology. The generic name comes from the Greek words *coryn* and *rhinos*, which mean "club nosed." The specific name refers to naturalist Charles H. Townsend (Kunz and Martin 1982).

Subspecies. Texas specimens are referable to the subspecies *C. t. pallescens* (G. S. Miller, 1897) and *C. t. australis* (Handley, 1955). Previously, it was thought that only *C. t. pallescens* occurred in the state, but recent genetic analysis of specimens from the Chihuahuan Desert ecoregion has classified them as *C. t. australis* (Piaggio and Perkins 2005). The subspecies *C. t. pallescens* has been confirmed (genetically) to occur in western Oklahoma (Smith et al. 2008). Based on their proximity, this suggests that the subspecies that occurs in the Central Great Plains and Southwestern Tablelands ecoregions also is *C. t. pallescens*. However, additional work is needed to define the distributional limits of each of these subspecies in Texas.

Description. This is a medium-sized bat (forearm = 39–46 mm) whose most distinctive features are its extremely large ears, which typically measure more than 2.5 cm in length, and the presence of a large and distinctive facial gland on either side of the snout. The function of these glands is unclear, although they may secrete pheromones important in mating rituals. The pelage in *C. townsendii* is pale to dark brown dorsally and ventrally. Dental formula: I 2/3, C 1/1, Pm 2/3, M 3/3 × 2 = 36. Females are typically slightly larger than males. Average external measurements are as follows: total length, 98 mm; tail, 46 mm; hind foot, 11 mm; ear, 34 mm; forearm, 42 mm. Weight: 7–12 g.

Townsend's big-eared bat is similar in appearance to the pallid bat (*Antrozous pallidus*) and to Rafinesque's big-eared bat (*Corynorhinus rafinesquii*). *Antrozous*

Townsend's big-eared bat (*Corynorhinus townsendii*)

pallidus is much lighter in coloration—yellowish above with white below—and lacks the 2 nose lumps. *Corynorhinus rafinesquii*, which also has the 2 nose glands, is found only in the extreme eastern portion of Texas and is not sympatric with *C. townsendii* in Texas. The dorsal pelage of *C. townsendii* is darker at the base than at the tips but is not as sharply contrasted as in *C. rafinesquii*. In addition, the ventral fur of *C. rafinesquii* is tipped with white, which sharply contrasts with blackish bases. Careful inspection of the middle upper incisors reveals a secondary cusp present in *C. rafinesquii* that is uncommonly encountered in *C. townsendii* (Jones 1977; Kunz and Martin 1982).

Distribution. Townsend's big-eared bat ranges across the entire western United States (with disjunct populations in the Ozarks and the Appalachians) and has been documented in Texas from 6 ecological regions—the Central Great Plains, Southwestern Tablelands, High Plains, Edwards Plateau, Chihuahuan Desert, and Arizona/New Mexico Mountains. Its distribution is not restricted by vegetative associations, and specimens have been captured in habitats ranging from desert scrub to pinyon-juniper woodlands. However, the presence of rocky, broken country is consistent with the capture of these bats. This is perhaps the most characteristic bat of caves and mine tunnels in the Trans-Pecos (Schmidly 1977).

A year-round resident of Texas, *C. townsendii* hibernates in caves across its range. It is one of the few species of Trans-Pecos bats that regularly may be found in winter, and records also exist

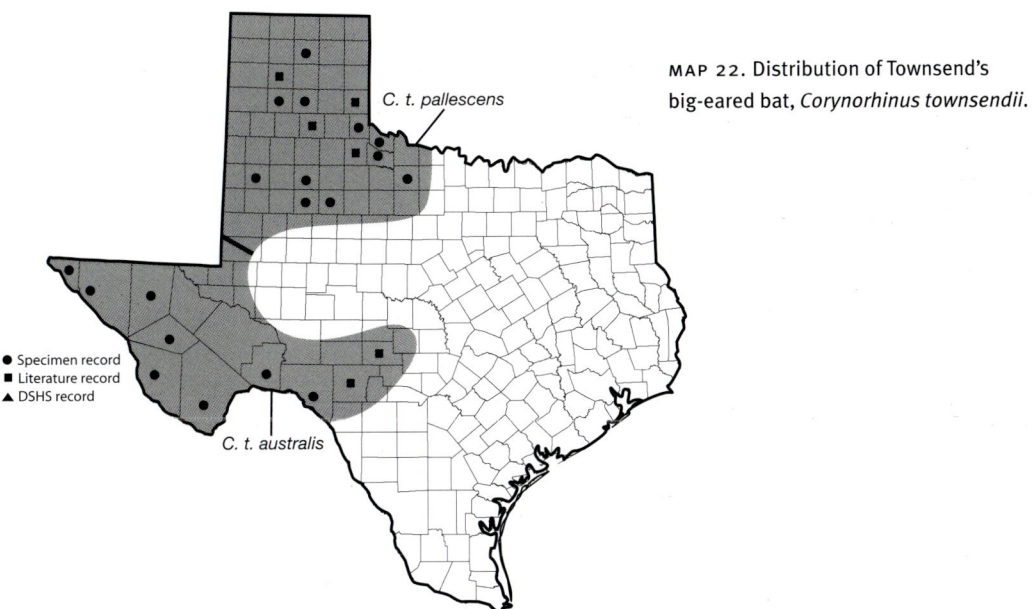

C. t. pallescens

C. t. australis

● Specimen record
■ Literature record
▲ DSHS record

from the High Plains, Southwestern Tablelands, and Central Great Plains during this season.

Life History. Although *C. townsendii* roosts primarily in caves and mines, it occasionally occupies buildings. It also has been found roosting in tree hollows in northern California (Gellman and Zielinski 1996). Unlike many other species of bats, Townsend's big-eared bats do not use rock crevices and cracks as roosting sites but instead hang pendulum-like by 1 or both feet from the roost ceiling (Dalquest 1947a; Genter 1986). These bats inhabit gypsum caves found throughout northern Texas and are common in caves and mine tunnels of the Trans-Pecos. In a cave or mine these bats select roost sites with dim light near the zone of total darkness (Kunz and Martin 1982), although they occasionally have been found hanging in fairly well-lit areas of caves and abandoned buildings (Twente 1955a).

During the winter months *C. townsen-dii* hibernates throughout its range in caves and mines. They do so in the coldest parts of caves but do not tolerate temperatures below 1.5°C (34.7°F) for extended periods (Clark et al. 2002; Genter 1986; Nagorsen et al. 1993). The bats can be found hibernating singly close to cave entrances early in the winter. As temperatures drop below freezing, they move farther into the caves, where they hibernate in tight clusters, which may help stabilize their body temperature against external changes in temperature (Humphrey and Kunz 1976). While the animals are torpid, their large ears are coiled from the tips down and laid back against their necks; in this position they resemble a ram's horns. Although they hibernate during winter, these bats remain fairly active and move within and between hibernacula (Alcorn 1944; Clark et al. 1997a, 2002). Sexual differences in winter activity and hibernacula site selection suggest that male bats select warmer hibernation sites and are more easily aroused and active compared with females (Barbour and Davis 1969). The largest number of hibernating bats reported from 1 cave is 485 in Oklahoma and 1,000 in Kentucky, but hibernacula more typically contain fewer than 100 individuals (Clark et al. 1997a; Rippy and Harvey 1965; Sherwin et al. 2000a; Szewczak et al. 1998). The bats lose more than half their body weight during hibernation, and most mortality is thought to occur during this time (Kunz and Martin 1982). The annual survival of wintering individuals over a 10-year period in Washington was 54–76% (Ellison 2008).

In Oregon, these bats use a series of interim roosts during their move from hibernacula to foraging areas, which can take up to 2 months and cover about 24 km (Dobkin et al. 1995). However, in other parts of their range, they use the same cave as both a hibernaculum and a maternity roost (Clark et al. 1997a; Sherwin et al. 2000b, 2003). Males and females occupy separate roosting sites during the summer. During this season, males appear to lead a solitary lifestyle while females and young form maternity colonies that contain 15–550 individuals. Bachelor male groups are smaller (1–7 individuals; Sherwin et al. 2000b), although 1 large bachelor roost containing more than 500 individuals has been documented in Kentucky (Lacki et al. 1994). Larger, more complex mines and caves with multiple openings and multiple levels seem to be preferred to small, simple workings as maternity roosts. These sites are more likely to provide a suite of microhabitats that allow the bats to use the site both as a maternity roost and a hibernaculum. However, mines and caves with one low (< 1.5 m) opening are preferred by bachelor groups and hibernating bats (Sherwin et al. 2000b, 2003). Vegetative structure outside the cave was not predictive of cave use in eastern Oklahoma (Wethington et al. 1997).

In Utah and Nevada, maternity colonies in caves are more stable than those in mines. Those found in mines relocate 2–3 times during the maternity period, but those in caves do not move. Bachelor colonies in mines are the most labile, although individuals demonstrate more fidelity to a group of mines than to any individual mine (Sherwin et al. 2000a, 2000b, 2003). Mines have been found to be of particular importance as roost sites in Durango, Mexico. There bats prefer mines with small openings and long entrances (> 50 m), with temperatures below 10°C (50°F) in the winter and greater than 16°C (61°F) the remainder of the year (López-González and Torres-Morales 2004). Sherwin et al. (2003) commented that the patterns of roost selection and fidelity are complex and vary locally for this species.

These bats emerge late in the evening to forage, usually 1–2 hours after sunset, and are highly maneuverable fliers (Bagley and Jacobs 1985; Cockrum and Cross 1964; Kunz and Martin 1982). Despite the report of 1 individual that traveled more than 150 km during 1 night of foraging, Townsend's big-eared bats are relatively sedentary and have small (90–120 hectares) home ranges (Adam et al.

1994; Bagley 1984; Piaggio and Perkins 2005; Wethington et al. 1996; Wilhide et al. 1998b). They seem to prefer foraging along the edge of vegetation in riparian zones and other ecotones (Caire et al. 1984; Clark et al. 1993; Dobkin et al. 1995; Kunz and Martin 1982). They have been seen foraging 10–30 m above the ground in the upper third of the canopy, where they follow the contour of the vegetation. Although it has been suggested that they feed by gleaning insects from the surface of vegetation, based on their wing morphology, they have not been observed to hover or pause as one would expect if they were using this feeding strategy. They also avoid open areas, and if it is necessary to cross such an area, they drop to within 1 m of the ground and fly straight and fast across it (Fellers and Pierson 2002). Their diet comprises small moths almost exclusively, although small numbers of flies, lacewings, dung beetles, and sawflies also are ingested (Lacki et al. 1996; Ross 1967; Sample and Whitmore 1993; Whitaker et al. 1977).

Reproduction in these bats has been studied extensively in California (Pearson et al. 1952), and, although the timing in Texas populations may be somewhat different, general reproductive habits and behaviors are likely similar. Most breeding occurs at winter roosts from November to February, although some females may be inseminated prior to their arrival at the winter roosts. The sex organs of the males do not fully develop until their second year of life, which means that young male bats are not reproductively active in their first

year. Young females, however, may breed as early as 4 months of age. Males have been observed copulating with torpid females, which is one reason that young females may become pregnant at such an early age. The large nose glands are probably of importance in precopulatory behavior. Males approach the females while making a twittering sound and vigorously rub their snout over the female's face, neck, forearms, and ventral surface, producing a strong and noticeable odor (Barbour and Davis 1969; Pearson et al. 1952).

A firm vaginal plug does not form in females following copulation, which means that repeated inseminations can occur during the winter. This situation is unlike that of several other vespertilionids, in whom repeated copulations are prevented by the presence of such a plug. Sperm is stored by the female until ovulation occurs in late winter or early spring. Fertilization of the ovum takes place shortly afterward. Climatic factors and temperature may affect the length of gestation, which ranges from 56 to 100 days.

A single young is born between late May and early June. The baby bat weighs approximately 2.4 g at birth and is pink, naked, and completely helpless. At 4 days the newborn bat begins to display hair growth, and by 1 month of age is volant and nearly adult size. At 2 months the juveniles are weaned, and the nursery colonies begin to disperse (Humphrey and Kunz 1976; Lacki et al. 1994; Pearson et al. 1952). In Oklahoma, it was found that females alter their foraging activity according to the stage of their reproductive cycle. Before parturition and after the young become volant, the females leave the cave after dusk and remain out for the entire night. During lactation, they exhibit a trimodal foraging pattern, leaving to forage for short periods and revisiting the cave to nurse their young (Clark et al. 2002).

The estimated lifespan of these bats is 16 years (Paradiso and Greenhall 1967). Known predators include roof rats and eastern woodrats, which is somewhat unusual since rodents are not generally implicated as bat predators (Clark et al. 1990; Fellers 2000). No *C. townsendii* have been submitted by the public to the DSHS for rabies testing. However, 4 free-flying individuals captured at Big Bend Ranch State Park were submitted for testing, and all were negative (Yancey et al. 1997).

Status. The IUCN 2011 status of Townsend's big-eared bat is "least concern." Federal and state governments classify *C. townsendii* as a "species of concern." Populations appear to be decreasing in southern California (Miner and Stokes 2005) but seem stable at a major hibernaculum in South Dakota (Choate and Anderson 1997). This species is among the rarest bats in Texas, and the practice of blasting old mine tunnels to shut them off permanently could destroy significant numbers of these bats. Reports of their sensitivity to disturbance are variable; some accounts suggest they are intolerant of human intrusion and will quickly abandon a roost site that has been disturbed, whereas others indicate they are more tolerant of human activities (Clark et al. 1996a, 1997a; López-

González and Torres-Morales 2004; Sherwin et al. 2000b). This definitely is a species that bears watching in the future.

Two subspecies that do not occur in Texas, the Virginia big-eared bat (*C. t. virginianus*) and the Ozark big-eared bat (*C. t. ingens*), are on the federal endangered subspecies list.

Remarks. Recent genetic studies found distinct lineages of *Corynorhinus townsendii* in the subspecies that occur in eastern (*C. t. ingens*) and western (*C. t. pallescens*) Oklahoma (Smith et al. 2008; Weyandt et al. 2005). Lineages can be traced to specific caves, some of which are located less than 20 km apart, which suggests a limited dispersal of females and a strong tendency for juveniles to return to their maternity roost (philopatry). This is further supported by morphological differences that have been documented between populations that occur in the Texas Panhandle and western Oklahoma and Kansas; those that are found south of the Red River are relatively smaller than those to the north (Smith and Tumlison 2004). Gene flow appears to be mediated by male dispersal. There is a need for similar genetic studies of these bats in Texas.

Specimens Examined. Go to www.batsof texas.com for more detailed information about the total of 232 specimen records of *C. townsendii* from Texas. Additional records: *Briscoe County* (1) (Roberts et al. 1997a), *Brewster County* (2) (Yancey et al. 2006), *Crosby County* (2-TTU 11973, 34548), *Potter County* (1) (Yancey et al. 1998).

References. 14, 18, 35, 40, 45, 60, 62, 77, 84, 85, 86, 92, 98, 102, 106, 115, 136, 140,

Pallid bat (*Antrozous pallidus*)

145, 146, 147, 154, 168, 197, 205, 227, 235, 236, 237, 238, 239, 259, 306, 308, 313, 318, 332, 336, 339, 351, 355, 381, 396, 405, 411, 412, 413, 423, 443, 464, 475, 477, 478, 511, 518, 524, 564, 574, 575, 579, 598, 617, 625, 626, 633, 635, 686, 713, 714, 727, 755, 767, 823, 824, 830, 833, 856, 857, 902, 908, 935, 976, 986, 988, 990, 999, 1008, 1009, 1024, 1026, 1028, 1029, 1040, 1044, 1057, 1058, 1059, 1075, 1076, 1109, 1110, 1112, 1129, 1136, 1137, 1138, 1147, 1153, 1164, 1167, 1188, 1189, 1192, 1200, 1207, 1217, 1248, 1253, 1254, 1255

Antrozous pallidus (Le Conte, 1856)
Pallid Bat

Etymology. The generic name is derived from the Greek words *antron* and *zoos*, which translate as "cave living." The spe-cific name is the Latin word for "pale," referring to its light color (Stangl et al. 1993).

Subspecies. In the past, all Texas speci-mens of *A. pallidus* have been assigned to the subspecies *A. p. pallidus* (Le Conte, 1856) (Martin and Schmidly 1982). However, it has been determined that pallid bats from the vicinity of the Red River and the Panhandle of north-central Texas are referable to another subspecies, *A. p. bunkeri* Hibbard, 1934, which is a somewhat larger and darker race with a limited range that extends into south-western Kansas and western Oklahoma (Manning et al. 1988). The subspecies *A. p. pallidus* occupies the remainder of the pallid bat's range in Texas. Intergrada-tion between the 2 subspecies may occur

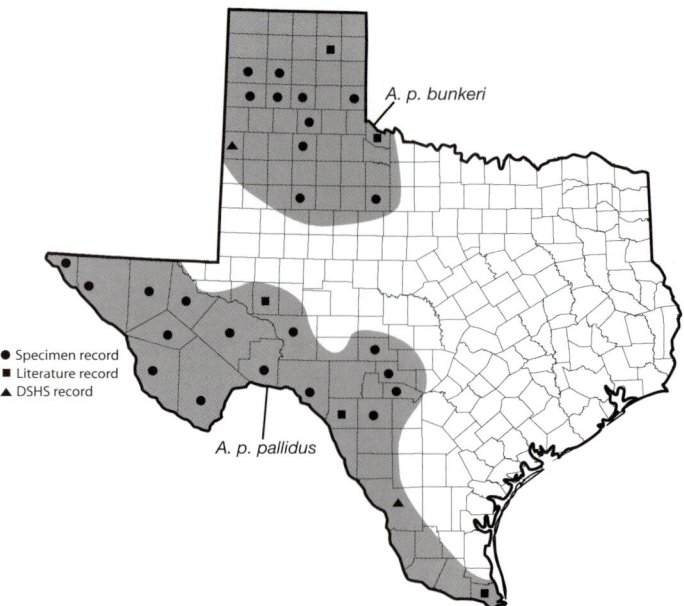

MAP 23. Distribution of the two subspecies of the pallid bat, *Antrozous pallidus.*

A. p. bunkeri

- ● Specimen record
- ■ Literature record
- ▲ DSHS record

A. p. pallidus

in the western Canadian River Valley (Manning et al. 1988). Despite their morphological differences, no genetic distinction between the 2 subspecies was documented based on maternally inherited mitochondrial DNA (Weyandt and Van Den Bussche 2007).

Description. This is a large bat (forearm = 47–57 mm) with broad wings and large ears that are approximately 2.5 cm in length and slightly over 1.25 cm in width. The snout is large and bare and has a horseshoe-shaped ridge over the piglike nose. There are pale purple wartlike bumps present on the face that become obviously darker in April and May and produce a musky secretion. The significance of these pararhinal glands has not been established. The pelage in *A. pallidus* is light yellow above, sometimes washed with brown or gray, and pale cream to white below. The bases of the hairs are pale. Membranes are brownish and mostly naked. Dental formula: I 1/2, C 1/1, Pm 1/2, M 3/3 × 2 =

28. Females are often slightly larger than males. Average external measurements are as follows: total length, 106 mm; tail, 44 mm; hind foot, 11 mm; ear, 28 mm; forearm, 51 mm. Weight: 12–20 g.

The pallid bat is a distinctive bat that is not easily confused with other species. Big-eared bats of the genus *Corynorhinus* (*C. townsendii* and *C. rafinesquii*) are distinguished by the presence of nose lumps and darker pelage, and the spotted bat (*Euderma maculatum*) is easily identified by its unique markings. Only *A. pallidus* has the combination of large ears and pale yellowish color.

Distribution. The pallid bat is known in Texas from the Chihuahuan Desert, Arizona/New Mexico Mountains, Edwards Plateau, Southern Texas Plains, Central Great Plains, Southwestern Tablelands, and High Plains ecoregions. One of the most abundant bats of the Trans-Pecos, the pallid bat inhabits mountainous areas, intermontane basins, and lowland desert scrub habitats at elevations that range from 600 to 2,000 m. It is considerably less abundant toward the eastern margin of its range on the Edwards Plateau.

Little is known about the migratory habits of this species. Pallid bats have been collected from late March through November in Texas, but they have not been obtained during the winter. Very few winter records exist for this bat anywhere in its range; hibernating bats have been reported from Kansas, Nevada, New Mexico, and California (Alcorn 1944; Geluso 2007; O'Farrell and Bradley 1970; Orr 1954; Poche 1981; Twente 1955a). *Antrozous pallidus* is not known to perform long migrations and is thought to hibernate throughout much of its summer range in mines or caves.

Life History. Pallid bats inhabit rocky outcrops, where they commonly roost in crevices, caves, and mine tunnels, but they may also roost in the attics of houses, under the eaves of barns, behind signs, under bridges, in the hollows of live trees and snags, and in abandoned adobe buildings (Baker et al. 2008; Davis and Cockrum 1963a; Orr 1954; Rabe et al. 1998a; Tatarian 1999). Colonies are usually small and contain from 12 to 100 adults (Chapman et al. 1994). Males reside in bachelor colonies apart from females while the young are raised (Beck and Rudd 1960; Dalquest 1947a; Davis and Cockrum 1963a; O'Shea and Vaughan 1977).

During spring and autumn, when nights are cooler, they roost behind thin slabs of rock and in vertical crevices that are passively heated in late afternoon. In summer, they seek refuge from the heat in deep, horizontal crevices that main-tain more stable, moderate temperatures (Vaughan and O'Shea 1976). Pallid bats are metabolically best adapted for roosting at temperatures near 30°C (86°F; Trune and Slobodchikoff 1976). Demonstrating low roost fidelity, they change roosts every 1–2 days in Oregon. Lewis (1996) proposed that this is a strategy to decrease ectoparasite loads by disrupting the parasites' reproduc-tive cycle. Bats found at the same sites as *A. pallidus* include the big brown bat (*Eptesicus fuscus*), Townsend's big-eared bat (*Corynorhinus townsendii*), Brazilian free-tailed bat (*Tadarida brasiliensis*), American parastrelle (*Parastrellus hespe-rus*), Yuma myotis (*Myotis yumanensis*), fringed myotis (*Myotis thysanodes*), and California myotis (*Myotis californicus*). With the exception of the Brazilian free-tailed bat (*Tadarida brasiliensis*), which is often closely associated with *A. pallidus,* these bats typically segregate from *A. pallidus* at the roosting site (Orr 1954). The winter habits of *A. pallidus* are poorly known; it is thought that these bats hibernate on their summer range, although hibernacula have not been discovered.

Pallid bats employ a foraging strategy somewhat unusual among bats. They generally feed on insects greater than 17 mm in size, including both night-flying and flightless ground insects. The prevalence of nocturnal flightless insects in their diet indicates that they alight on the ground to capture such prey. In other words, to some extent they are terrestrial foragers. When they do prey upon flying insects, they use a novel be-havior—they force their quarry against a surface (such as the ground) with their wings before capturing it (Johnston and Fenton 2001). The membranes of this bat often show evidence of injuries, as one would expect for a bat that forages close to the ground and gleans insects from vegetation that is sometimes spiny (Davis 1968). Holes in membranes heal rather quickly, however, and leave only a pale, unpigmented blotch. A 14-mm hole heals completely in 22–33 days (Davis and Doster 1972).

Pallid bats generally forage close to their day roost. In northern California, postlactating females have a foraging area of almost 600 hectares, whereas lactating pallid bats forage in a much smaller area (160 hectares; Baker et al. 2008). Pallid bats do not use echolocation to locate prey but instead localize their victim by listening for the sounds it produces, particularly walking sounds. They do so with amazing accuracy— from a distance of 5 m they can land within 7.6 cm of a moving cricket (Fuzessery et al. 1993). Interestingly, they ignore the calls of crickets (Bell 1982). Pallid bats have been observed searching for flightless prey in their natural habitat. When doing so, they appear to fly at random over an area at 1 meter or less above the ground. It scouts by flying back and forth until it locates the prey by sound. Then, abruptly it drops to the ground, searches briefly, grabs a victim in its mouth, and takes off—not without some difficulty (Bell 1982; Ross 1967).

Captured prey may be eaten on the ground or carried to a night roost or feeding station for consumption (Black 1974). Small, soft-bodied prey may be eaten on the wing (Johnston and Fenton 2001). Locations of night roosts can be identified by the presence of guano, wings, and other hard parts of insects that are discarded and accumulate on the ground under the roost. Small trees, rocks crevices and overhangs, bridges, and abandoned buildings are used as night roosts. In the Mojave Desert, desert willows (*Chilopsis linearis*) were the most commonly used tree species, where bats spent up to 30 minutes processing prey that had been captured nearby (Hirshfeld et al. 1977). In addition to processing prey in isolation, bats will cluster in night roosts and enter short periods of torpor before departing for another feeding bout (Hermanson and O'Shea 1983). In Oregon, pallid bats demonstrate relatively high night-roost fidelity within and between years. Roostmates that share a night roost may or may not roost together during the day (Lewis 1994).

Their diet appears to be substantially different from the availability of insects in their local environment, suggesting they are selective foragers (Johnston and Fenton 2001; Lenhart et al. 2010). A summary of their food habits states that specific prey in California included Jerusalem crickets, June beetles, grasshoppers, ground beetles, and scorpions (Orr 1954). A detailed analysis of prey consumed at 4 locations in Arizona revealed that Jerusalem crickets (*Stenopelmatus* spp.) are the most important component of their diet (Ross 1961). Although pallid bats have often been portrayed as frequently attacking and eating scorpions, the remains of only 1 scorpion was found in this sample, leading the author to speculate that scorpions are probably encountered by bats searching for Jerusalem crickets and flightless beetles. A comprehensive food list for the pallid bat based on a survey of all of the published literature on the subject revealed 54 different types of prey, of which large, night-flying insects (20–70 mm) and flightless insects (20–50 mm), chiefly ground-dwelling arthropods, were the most prevalent (Ross 1967).

The stomach contents of 9 individuals from Big Bend National Park contained moths (22.2%), crickets (11.1%), ground beetles (11.1%), ant lions (8.9%), froghoppers and leafhoppers (1.7%), and unidentified insects (45.0%; Easterla 1973). A study conducted in the Indio Mountains of Hudspeth County, Texas, based on culled parts of prey found under a night roost, found that grasshoppers and crickets were the most preyed-upon group (44.1%), followed by beetles (26.8%), sun spiders (16.2%), moths (3.5%), walking sticks (3%), and true bugs (2.7%). Other groups represented less than 1% of the total. Unlike in the study in Arizona, Jerusalem crickets were not found in the diet of *Antrozous* because they likely do not occur in the Indio Mountains. Instead, pallid bats most commonly took toad hoppers (*Phrynotettix* spp.) at this location (Lenhart et al. 2010).

For an insectivorous bat, their diet is quite variable both geographically and seasonally. In addition to arthropods, pallid bats also are known to eat lizards, rodents, and possibly other bats (Bell 1982; Engler 1943; Hermanson and O'Shea 1983; Johnston and Fenton 2001; Lenhart et al. 2010). They also are known to feed among flowering agave plants (Barbour and Davis 1969). An early study using carbon-isotope data suggested that pallid bats obtained carbon from cactus and agave, but it was unclear whether the carbon was directly from the plants or from insects that had fed on the plants (Herrera M. et al. 1993). However, recent observations clearly indicate that pallid bats consume nectar from flowering cardón cactus (*Pachycereus pringlei*) in Baja California (Frick et al. 2009). They found that the head and the torso of 52% of the pallid bats captured were covered with pollen. In Big Bend National Park, one of us (Loren K. Ammerman) noticed a high proportion of the pallid bats captured from May to July were covered with agave pollen, which suggests they might consume the nectar of agaves in Texas.

Interestingly, in Baja California there is potential for competitive exclusion between pallid bats and nectar-feeding lesser long-nosed bats (*Leptonycteris yerbabuenae*), which is an important pollinator of cacti and agave in the desert southwest. Pallid bats visit cactus early in the evening at locations where lesser long-nosed bats (*Leptonycteris yerbabuenae*) occur, but they utilize this food resource throughout the night at locations where they are not competing with this well-adapted nectar-feeding species. Pallid bats appear to be legitimate pollinators of cardón cactus and could be effective pollinators of other cactus and agave species as well (Frick et al. 2009).

In California, mating occurs in late

October, followed by parturition in late May to early June (Orr 1954). Most births have been recorded in late June in Kansas (Twente 1955a). In Texas, lactating females have been captured in May, June, and July (Easterla 1973; Higginbotham and Ammerman 2002; Manning et al. 1987b; Yancey et al. 1996). Parturition occurs in mid-June in southern Arizona (Davis 1969) and probably occurs from early May to mid-June in Texas.

Females may carry 1–4 embryos (Manning et al. 1987a), but the birth of twins is normal for mature females. Yearling females usually bear only 1 pup, and young males are not sexually active their first fall (Davis 1969). The length of gestation is 53–71 days. Newborn bats weigh about 3 g at birth and seem to develop more slowly than other species (Bassett 1984). The eyes open at 8–10 days, hair is evident at 10 days, and the young are volant by 6 weeks of age (Orr 1954). Cool springs tend to delay parturition and development in the northern part of their range (Lewis 1993).

In Crockett County, Texas, 2 young A. pallidus were captured whose stomachs were found to contain both insect remains and milk, suggesting that young pallid bats continue to nurse after they become volant (Manning et al. 1987b). Volant juveniles have been taken from July to September in Big Bend National Park (Higginbotham and Ammerman 2002; Yancey et al. 2006). Juveniles are allowed to occupy the central location in clusters of roosting pallid bats. This facilitates conservation of energy in young bats, which enables them to store as much fat as possible for their first winter

(Trune and Slobodchikoff 1978). In spite of this, juveniles weigh less than adults when they enter hibernation, indicating they have less stored body fat (Davis 1969). This could result in higher mortality of juvenile bats their first winter.

Five pallid bats were reported to the DSHS (from 1984 to 1987 and from 1996 to 2000), and all tested negative for the rabies virus.

Status. The IUCN 2011 status of the pallid bat is "least concern." This species is one of the most abundant bats in the western part of Texas. In fact, pallid bats are the second most common species captured in Big Bend National Park; only Brazilian free-tailed bats (*Tadarida brasiliensis*) are more abundant (Higginbotham and Ammerman 2002). However, there is some evidence of declining populations of this species in southern California, where it appears to be intolerant of urban development (Miner and Stokes 2005). At this time, there is no need to worry about the conservation status of this species in Texas.

Specimens Examined. Go to www.batsof texas.com for more detailed information about the total of 851 specimen records of A. pallidus from Texas.

References. 12, 40, 47, 54, 62, 86, 90, 92, 102, 106, 115, 117, 121, 123, 136, 140, 142, 145, 147, 159, 166, 168, 224, 225, 226, 231, 308, 315, 318, 320, 323, 325, 326, 328, 329, 330, 332, 333, 336, 355, 381, 386, 402, 405, 423, 443, 447, 454, 466, 475, 511, 528, 541, 550, 559, 564, 570, 573, 575, 579, 624, 625, 626, 633, 635, 636, 641, 647, 714, 727, 733, 739, 744, 745, 747, 765, 766, 767, 768, 771, 773, 790, 808, 823, 824, 830, 833, 847, 848,

872, 883, 890, 891, 897, 941, 947, 953, 957, 963, 974, 976, 988, 998, 999, 1024, 1026, 1028, 1029, 1042, 1063, 1091, 1110, 1111, 1112, 1127, 1128, 1136, 1138, 1137, 1150, 1153, 1158, 1164, 1177, 1191, 1200, 1207, 1248, 1254

Family Molossidae

Worldwide, 100 species belong to the family Molossidae, or free-tailed bats. They primarily can be found in tropical and subtropical regions of the Old World and from the southern United States and Mexico south through South America. These bats are medium to large in size, insectivorous, and characterized by having a tail that extends beyond the free edge of the uropatagium. They have long, tactile hairs on their feet and short, velvety fur. Unlike most species of bats, members of the family Molossidae demonstrate sexual size dimorphism, with males slightly larger than females. Members of this family do not use the reproductive strategy of delayed implantation and fertilization.

Molossids are swift, strong fliers and often fly great distances between roosting and feeding sites. Many species also make extensive migrations between their winter and summer ranges. Because of their narrow wings, free-tailed bats have limited maneuverability, and taking off from the ground is difficult. They often roost high in buildings, cliffs, and caves. They require a free fall for takeoff to enable them to achieve sufficient momentum to sustain level flight. Because of their wing morphology, they also require larger, unobstructed bodies of water from which to drink.

In North America free-tailed bats occur from Canada to Mexico, but they are most common in the southern and southwestern regions of the United States. Of the 7 species that occur in the United States, 4 have been documented from Texas.

Tadarida brasiliensis (I. Geoffroy Saint-Hilaire, 1824)
Brazilian Free-Tailed Bat

Etymology. There is no consensus on the derivation of the generic name *Tadarida*. It can be translated in Latin as "withered toad" but also translates in Greek as "long ones" (Tuttle 2003). It is conceivable that the author of the name thought the species looked like a withered toad; however, the Greek derivation could refer to this species' long, narrow wings. The specific name refers to Brazil, where the type specimen was collected.

Subspecies. Historically, 2 subspecies of Brazilian free-tailed bats were recognized in Texas, *T. b. mexicana* and *T. b. cynocephala* (Owen et al. 1990; Schmidly 1991, 2004; Schmidly et al. 1977). The 2 bats are very similar morphologically, and their primary differences seem to be behavioral. *Tadarida b. cynocephala* (Le Conte, 1831) is a nonmigratory resident of the eastern quarter of the state, and *T. b. mexicana* (Saussure, 1860) is a highly migratory subspecies found throughout the remainder of the state. Morphologically, the 2 are distinguished by several differences in skull characteristics (e.g., greatest length of skull, zygomatic breadth, and breadth of cranium), all of which average larger in *T. b. cynocephala*.

Most populations of the migratory subspecies, *mexicana*, have normally completed their move into Mexico prior to the onset of breeding, whereas

Brazilian free-tailed bat (*Tadarida brasiliensis*)

cynocephala remains in the United States during the breeding season. This movement pattern indicates that the 2 races are reproductively isolated and possibly even separate species. However, overwintering populations of *mexicana* have been discovered in an area of contact between the 2 subspecies in southeastern Texas. A colony of *mexicana* was known to overwinter at the old animal pavilion on the Texas A&M University campus in College Station (Brazos County; Spenrath and LaVal 1974), which is only 160 km from colonies of *cynocephala* in extreme eastern Texas. A morphological analysis of cranial measurements from free-tailed bats captured near Navasota (Grimes County) found

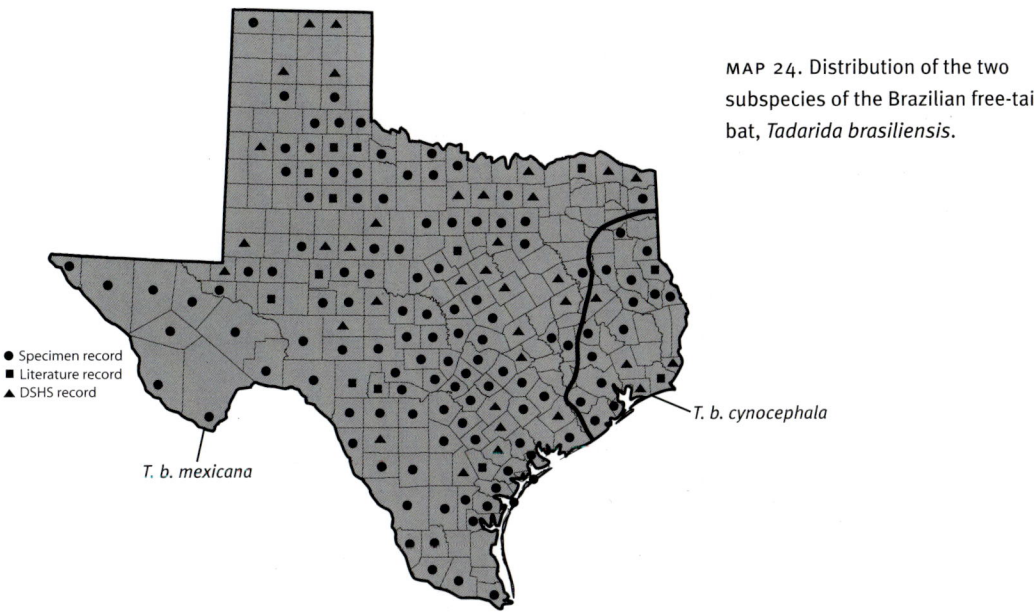

MAP 24. Distribution of the two subspecies of the Brazilian free-tailed bat, *Tadarida brasiliensis*.

● Specimen record
■ Literature record
▲ DSHS record

T. b. cynocephala

T. b. mexicana

these bats to be intermediate between *cynocephala* and *mexicana* (Schmidly et al. 1977). Thus, it appears the 2 subspecies are not reproductively isolated and that they likely interbreed in this part of Texas.

Although ethological differences exist between the recognized subspecies of *T. brasiliensis*, recent molecular analyses have found no significant genetic structuring to support the recognition of subspecies throughout their range in North America (McCracken and Gassel 1997; McCracken et al. 1994; Russell and McCracken 2006; Russell et al. 2005). Nonetheless, we continue to recognize 2 subspecies in Texas based on their behavioral and morphological differences.

Description. This is a medium-sized bat (forearm = 36–46 mm) with broad ears that do not join at the midline of the head, vertical wrinkles on the lips along the muzzle, and dark brown to dark gray pelage. The individual hairs of the pelage are almost uniform in coloration from base to tip, with a small, lighter-colored region at the base. Individuals with irregular patches of white fur are found occasionally. Long bristles are present on the feet, and, as with all molossids, the terminal third of the tail projects beyond the free edge of the interfemoral membrane. All other free-tailed bats in Texas have ears that are obviously joined. Dental formula: I 1/2 or 1/3, C 1/1, Pm 2/2, M 3/3 × 2 = 30 or 32. Average external measurements are as follows: total length, 93 mm; tail, 33 mm; hind foot, 8 mm; ear, 16 mm; forearm, 43 mm. Weight: 11–15 g.

Distribution. Brazilian free-tailed bats are the commonest species of bat in Texas and are found statewide. These bats appear every year in Texas in multimillion numbers to inhabit a few select caves (known as "guano caves") located in the Balcones Escarpment and the adjacent Edwards Plateau. Those caves with the largest populations of *Tadarida* in Texas include, in descending order, Bracken,

Goodrich, Rucker, Frio, Ney, Fern, Devil's Sink Hole, and James River caves (McCracken 2003; Wahl 1993). Over most of Texas their presence is seasonal, although they are nonmigratory and year-round residents of the eastern part of the state (Carter 1962; Scales and Wilkins 2007; Spenrath and LaVal 1974). Individual bats rarely are captured in the winter in the remainder of the state (Goetze et al. 2003; Roberts et al. 1997b; Yancey 1997).

Life History. This is among the best known of North American bats because of its widespread occurrence and abundance. Historically, the total population of these bats that inhabited Texas caves during the summer was estimated at 95–104 million (Davis et al. 1962). This number was obtained by estimating their numbers in many Texas caves with several techniques, including counting them as they exited the cave, extrapolating colony size from roosting densities, and using mark-and-recapture techniques (McCracken 2003). The largest numbers of bats occurred at Bracken, Goodrich, Rucker, and Frio caves, each of which contained an estimated 10–20 million bats during the summer population peak periods (Davis et al. 1962).

Recently, new thermal-imaging techniques have been used to assess the abundance of bats in these caves (Betke et al. 2008; McCracken 2003). Six major colonies, including those at Bracken, Davis, James River, Frio, and Ney caves in Texas and at Carlsbad Caverns in New Mexico, were censused from 2000 to 2006; this technique yielded an estimate of 4 million bats inhabiting these 6 caves.

Historic estimates (in 1957) for these same colonies totaled 54 million. Two factors have been proposed to explain the large discrepancy between the 2 studies. First, the decline documented since the late 1950s could be real and severe. Second, methods used to count bats in the 1950s had high error rates, which resulted in large overestimates. Although the second explanation seems plausible, researchers evaluated this hypothesis and found that volunteers generally underestimated colony size based on visual assessment of video alone. Thus, it is possible that much of the difference between historic and current population estimates reflects a population decline, although the authors suggest a combination of the 2 factors as the most likely scenario (Betke et al. 2008).

If the large decrease in population size is real, the decline could be due to pesticide exposure, particularly DDT and its metabolites. Bats are sensitive to DDT, and populations could experience major mortality by ingesting pesticides in the insects they consume (Clark 1981; Geluso et al. 1976). Residue from organochlorine insecticides has been detected in bats at several caves in Texas and elsewhere in the southwest (Clark 2001; Clark et al. 1975; Cockrum 1970; Thies and Thies 1997). It has been suggested that these residues have caused major population declines at these caves. However, only data collected from Carlsbad Caverns, New Mexico, provide conclusive evidence in support of this hypothesis.

Bats collected across several decades from Carlsbad Caverns were analyzed for accumulated DDT residues. The toxicity of the accumulated DDT was about 2.7 times higher in 1956 and 4.8 times higher in 1965 than in 1973. In 1988 it was about half of the 1973 level (Clark 2001). This timing makes sense in light of the fact that DDT was first used as an agricultural pesticide in 1939 and was banned in 1961. The authors concluded that DDT probably played a major role in the decline of the bat population at Carlsbad Caverns.

Population trends of Brazilian free-tailed bats from 4 caves and 2 bridges near New Braunfels, Texas, were analyzed from 1995 to 2005 using NEXRAD Doppler radar images. Although the population fluctuated during this time, there was no evidence of a general downward population trend (Horn and Kunz 2008). Based on the information about DDT presented earlier, it is possible that the massive decrease in bat populations seen in this area since 1957 occurred before 1995 and that the population has remained stable since then.

In Texas, the Brazilian free-tailed bat seems to be primarily a cave dweller, and its use of buildings as roosts is likely a relatively recent, possibly expanding practice. In East Texas, this species most likely used tree hollows as roosting sites before buildings existed (Wilkins 1989). Only a small fraction of the numbers of bats found in caves is ever found in the total of all roosts in buildings. Every town in the Brazilian free-tailed bats' range in Texas is likely to have at least 15 roosts per 5,000 humans, but the occupation of buildings is especially common

in eastern Texas (Davis et al. 1962). Most roosts in buildings house fewer than 100 bats at a time, but a few buildings traditionally house many hundreds each year. Overwintering in buildings occurs infrequently in the southern Gulf Coast prairies of Texas.

No particular style, size, age, state of repair, or use by people exempts a building from use by Brazilian free-tailed bats. The critical feature is whether the building has any accessible small cracks or niches into which bats can retreat into semidarkness during the day (Wilkins 1989). Such openings usually are found even in the most modern, compact structures. One architectural type common in South Texas, the Spanish-style building with a clay tile roof, is among the most vulnerable to invasion by Brazilian free-tailed bats. The bats roost under the tiles and seldom can be driven out permanently either by killing those present or by chemical treatment of the surface of the roost. The simplest, most effective method is to close the entrance to the roost. With clay tile roofs this is almost impossible unless the tile is replaced by some other kind of roofing material.

Carter (1962) summarized the features of buildings typically chosen by *T. b. cynocephala* for roost sites. Among these characteristics are "(1) the building is not permanently occupied by man or there is an unoccupied space between the roost and the area occupied by man; (2) the buildings are not heated during the winter or there is an unheated area between the space used for hibernating and the heated area, such as an unoccupied second floor in a 2-story building; (3) there is a fairly open area, such as an attic, which the bats usually occupy during the summer; and (4) there is an area of limited space, such as that between interior walls or between floors, that is protected from a moderately severe drop in outside temperature." Carter went on to state that churches and old 2-story buildings provide admirable roosts for these bats.

In Texas, the abundance of Brazilian free-tailed bats in roosts in buildings follows an annual pattern of 1 peak in spring and another in fall, with general midsummer and midwinter lows or periods of complete absence (Sherman 1937; Spenrath and LaVal 1974). This pattern complements that of bat abundance in the guano caves. In Texas, Brazilian free-tailed bats in buildings during spring and fall usually are itinerant between tropical latitudes and the midlatitude guano caves of Texas, Oklahoma, Kansas, and New Mexico. Sufficient interchange of banded Brazilian free-tailed bats has occurred among colonies in the guano caves of Texas and between those in Texas and colonies in neighboring states to demonstrate that individual bats are not compelled to return each year to the cave of their birth. Rather, Brazilian free-tailed bats exhibit an ability to range over great distances and find the widely separated, often well-hidden entrances to the few traditional guano caves and roosts in buildings (Glass 1982).

Bridges in Texas provide important stopover roosts for migrating bats and in the spring are used as mating sites (Keeley and Keeley 2004; Keeley and

Tuttle 1999). Colony size increases at bridges in south-central Texas (Travis and Williamson counties) in early March, 2 months before such an increase occurs at maternity roosts in caves. In autumn, bridges are used later into the fall migration period than are caves. Activity at bridges has been detected into October, when no activity was recorded at caves in the same area. This may be because the crevices used as roosting sites in bridges warm up more quickly during cooler months than do caves (Horn and Kunz 2008).

Although it seems that the noise and pollution produced by traffic, as well as higher ambient-light levels found under bridges, would negatively impact bat colonies living there, this does not appear to be the case. One detailed study that compared bats roosting in caves with those roosting under bridges in south-central Texas (Medina and Uvalde counties) found that bats utilizing bridges were healthier than those in caves. Not only were they in better condition, but they also had fewer parasites, lower levels of stress hormone, and bigger, faster-growing offspring than did their cavern-dwelling counterparts (Allen 2009).

Caves utilized by large colonies of these bats are often referred to as "guano caves" due to the tremendous amount of excrement that accumulates. Bat guano is an excellent fertilizer, and bat caves have been mined for guano since the 19th century, continuing even to the present (Constantine 1970; López-González and Best 2006). The decomposing feces release ammonia, which can accumulate to dangerous levels—well above the levels tolerated by humans (> 500 ppm). *Tadarida*, however, can tolerate ammonia levels in excess of 5,000 ppm with no ill effects (Studier and Fresquez 1969).

Few, if any, Brazilian free-tailed bats ever overwinter in the Texas guano caves. Populations of the subspecies *mexicana* that occupy these caves are inefficient hibernators, which is the reason they spend the depth of winter, from early December to late February, at lower latitudes, probably in Mexico, Central America, or even South America (Davis et al. 1962). Migratory movements of close to 1,800 km have been recorded (Glass 1982; Villa-R. and Cockrum 1962). Most individuals may accomplish the annual movement between Texas and Mexico in a few direct, long-distance flights between guano caves. During migration, these bats also use rock fissures, crevices in structures such as bridges and signs, bat houses, and cliff swallow nests as roosting sites (Buchanan 1958; Fraze and Wilkins 1990; Keeley and Tuttle 1999; Pitts and Scharninghausen 1986; Ritzi et al. 1998; Scales and Wilkins 2007; Sgro and Wilkins 2003). Although the migration of this species has been well documented, we still do not know where the majority of the individuals of these large summer colonies spend the winter (López-González and Best 2006). In East Texas, where these bats commonly inhabit old buildings and similar structures, they are nonmigratory and are year-round residents (Spenrath and LaVal 1974; Wilkins 1989).

Bracken Cave, located near New Braunfels, along the southeastern edge of the Hill Country, is home to the world's largest bat colony. The cave serves as a nursery colony for adult female free-tailed bats and their young and houses perhaps as many as 20 million bats during the summer (Davis 1962, but see Betke et al. 2008). Likewise, the largest urban colony of bats in North America is located at the Congress Avenue Bridge in Austin. This colony, which comprises about 2.5 million free-tailed bats in the summer, has become one of the major tourist attractions in that city. So common and well known are these bats in Texas that they were designated as the official flying mammal of Texas in 1995.

In summer, Brazilian free-tailed bats choose the warmest areas of the roost site, an adaptation thought to contribute to the quick development of the young. These bats are often found in association with other bat species. In the west these include the ghost-faced bat (*Mormoops megalophylla*), cave myotis (*Myotis velifer*), and Townsend's big-eared bat (*Corynorhinus townsendii*; Eads et al. 1957b). In the east, *T. b. cynocephala* occasionally is found in the same roost with the evening bat (*Nycticeius humeralis*), the big brown bat (*Eptesicus fuscus*), and the southeastern myotis (*Myotis austroriparius*; Jennings 1958). The different species almost always segregate and use different areas of the roost site.

Brazilian free-tailed bats appear on the wing several minutes before dark. The famous bat flights at Carlsbad Caverns, New Mexico, are made up almost entirely of this species. The late William B. Davis (pers. comm.) watched these bats emerge from the attic of a house one evening. They fell from the exit, dropped nearly to the ground, then zoomed upward, and, flying high, disappeared from view, each bat following the general direction of the one in front of it. When foraging, the bats fly rather high (15 m or more as a rule) except when sweeping over a body of water to drink. Their flight is rapid and aggressive, reminding one of swifts, and the long, angular, and narrow wings, plus relatively large size, make them easy to identify.

The major event in the life of the free-tailed bats summering in Texas is the birth and development of their young. Females become sexually mature in their first year of life and may become pregnant as yearlings (Short 1961b). Females ovulate in late March and produce a single young after a gestation period of 77–86 days. Occasionally the females may carry 2 embryos. The left horn of the uterus usually does not carry an embryo (Rasweiler and Badwaik 2000). Lactation begins after delivery of the young, and 2 mammae are located laterally, each with 1 functional pectoral teat.

Well over 90% of the returning females produce young each year. Most mating in the Texas population is accomplished each spring before the bats arrive at the Texas caves, although a recent study has documented mating in a population roosting under a bridge outside of Austin, Texas, from March 23 to April 4 (Keeley and Keeley 2004). Male bats predominate at the caves for a

brief period in early spring, but they are quickly outnumbered by females as the populations build steadily with the approach of parturition. By mid-June, adult females outnumber adult males by more than 3 to 1 (Davis et al. 1962).

Males do not become sexually mature until 18–22 months of age (Short 1961b). They are sexually active from February to early April, just prior to the bats' arrival in Texas caves. By the time they arrive there, sperm production is waning. Their sex glands decrease steadily in size in spring and reach a resting stage by early May. The small proportion of the male population that shows no sexual activity comprises principally the youngest age class. In these, the testes, prostate, and hedonic glands are smaller than the resting-stage sizes of the same glands in adult males (Sherman 1937). In late fall, the few adult males remaining in Texas again show some increase in size of the testes and the prostate, but sperm are absent. Peak production of sperm thus either occurs during winter, while the males are in lower latitudes, or in early spring, when the bats first arrive at transient roosts, such as bridges, in Texas. Since the highly disproportionate ratio of female to male Brazilian free-tailed bats in Texas cannot easily be explained as resulting from higher mortality among males, it may be that most males do not summer in Texas. Apparently they generally remain in the tropical and subtropical portion of their range and play no part in the sociology of bearing and rearing the young (Davis et al. 1962).

Males use either an aggressive or a

passive copulation strategy with approximately equal frequency. During aggressive copulation, the male separates a female from a roosting cluster and physically restrains her by biting her neck. The female usually resists strongly, and both produce audible squeaks, which represent mating "songs," at least in males (Bohn et al. 2008, 2009). During passive copulation, the male slowly approaches a female while she roosts in a dense cluster. The female does not resist and sometimes grooms herself or appears to sleep during copulation. No audible calls are emitted, and no neck biting occurs (Keeley and Keeley 2004).

Although a vaginal or copulatory plug is formed after copulation, it is easily removed by males that subsequently mate with a female, so the exact function of the plug is unclear. The gular (throat) gland of the males secretes a musky, oily fluid that they rub on the walls of the roost during the mating season. Males do not appear to defend territories in the wild, as has been documented in captivity (Bohn et al. 2008, 2009; French and Lollar 1998; Keeley and Keeley 2004; Schwartz et al. 2007).

Parturition occurs within a much narrower time span than is exhibited by vespertilionid bats. Peak birthing occurs from early to mid-June, during which time 70% of the young bats are born within a span of 10 days. More than 90% of the newborn bats make their appearance within 15 days (Schmidly 2004). The baby bats weigh only about 2.5 g at birth and are blind, naked, and pink. The newborn young are deposited together, naked and flightless, on specific areas of the ceiling in continuous colonies and are not carried by their mothers during the nocturnal feeding flights (Davis 1970). Nursery colonies may contain clusters of more than a million baby bats roosting in densities of 5,000 pups per square meter.

The almost simultaneous arrival of the babies creates marked crowding in the cave colony clusters. In the past it was thought that adult females made no attempt to locate their own young within these masses but nursed the first 2 young they encountered upon their nightly return to the roost (Davis et al. 1962). More recent studies have shown that females do indeed recognize their own young about 83% of the time, which is a remarkable feat given the confusion within such huge swarms of bats. However, some communal nursing still occurs. The mothers locate their pups using a combination of locational, vocal, and olfactory information (McCracken 1984).

Even in the warm nursery colonies, neonates of *T. brasiliensis* mature more slowly than the young of most vespertilionid species. However, preweaning mortality is lower in *Tadarida* than in vespertilionids. Young free-tailed bats are more precocial than the young of other bat species and sometimes are able to climb back up to the roosting site if they fall (Hermanson and Wilkins 1986). By 6 weeks of age the young free-tails have attained adultlike dentition, pelage, and body weight and have attempted their first flights (Pagels and Jones 1974). At Frio Cave in Uvalde County, juvenile bats were found to emerge earlier in the evening and return earlier at dawn than the adult bats (Lee and McCracken 2001).

The sudden increase in the numbers of flying bats resulting from mass achievement of fledgling status among the babies creates additional congestion in the caves. The overcrowding is relieved by the rapid disappearance of the adults as the fledglings appear. These adults presumably migrate southward out of Texas; the missing adults have not been found elsewhere in Texas at this time. After late July, the fledglings predominate in the diurnal feeding flights from the caves, and they tend to reside at the cave of their birth until the onset of cool weather in October and November drives them southward out of Texas.

Tadarida brasiliensis forages over long distances each night. The bats disperse at least 56 km in all directions from Carlsbad Caverns, New Mexico, and forage in an area that encompasses nearly 4,000 km^2 (Best and Geluso 2003). In south-central Texas, they disperse in the direction of dense corn and cotton fields, on average 15 km distant, where pest insects are highly abundant (Horn and Kunz 2008). These bats also feed anywhere from ground level to 900 m in altitude, and echolocation calls have been recorded at up to 1,118 m. The majority of feeding buzzes are detected either at ground level or 400–500 m above ground level. The peak feeding activity at the higher altitude coincides with the altitude of the atmospheric boundary layer that is a major aeroecological corridor for the nocturnal dispersal of noctuid moths (McCracken et al. 2008).

Samples of free-tailed bat droppings collected at San Antonio (Bexar County) contained remains of the following insects: moths (nearly 90% of the total number of insects eaten), ground beetles, leaf chafers, weevils, leaf beetles, flying ants, water boatmen, green blowflies, and leafhoppers. A separate food habits study showed these bats take small prey from 2–10 mm in length and listed the following food items and proportions for 88 specimens: moths (34%), flying ants (26.2%), June beetles and leaf beetles (16.8%), leafhoppers (15%), and true bugs (6.4%). It also was noted that *T. brasiliensis* feeds in groups of 10–13 individuals and often preys on densely swarming insects, a foraging method called filter feeding (Ross 1967). Where this species occurs together with the pocketed free-tailed bat (*Nyctinomops femorosaccus*) in Big Bend National Park, there is substantial dietary overlap. Both species feed on the widest variety of prey in June and narrow their diet to predominantly moths in July (Matthews et al. 2010).

The diet of lactating females is variable throughout the night; they feed mostly on beetles and cinch bugs in the evening and switch to moths in the morning (Whitaker et al. 1996). They also double the amount of food they eat between pregnancy and lactation and triple that amount during the first half (4 weeks) of lactation (Kunz et al. 1995). Although moths and beetles are the most common insects taken, free-tailed bats feed on a large variety of prey, including 12 orders and 38 families of insects. Moreover, the relative proportions of different types of insects preyed upon varies seasonally and is proportional to insect availability in the foraging areas (Lee and McCracken 2005; McWilliams 2005).

The enormous summer populations of these bats consume an equally enormous number of insects, many of which are serious pests that cause significant damage to crops in Texas each year. Two species of migratory noctuid moths, corn earworms or cotton bollworms (*Helicoverpa zea*) and fall armyworms (*Spodoptera frugiperda*), are particularly damaging. These also are favored prey of free-tailed bats (Lee and McCracken 2005).

Based on the size and energetic requirements of these bats, it has been suggested that 100 million Brazilian free-tailed bats consume approximately 4 billion corn earworm–sized insects each night (Kunz et al. 1995). In other terms, 1 million bats will prevent the development of about 5 million moth larvae per night simply by foraging for adult moths. This service, which the bats provide for free, has substantial economic value for agriculture in Texas. In an 8-county region in south-central Texas (the Winter Garden region), it is estimated that the services bats provide save farmers approximately 12% of the total value of the cotton crops produced there, or roughly $741,000 per year (Cleveland et al. 2006).

Mortality of adults is low, and longevity is probably great: Their average lifespan is more than 11 years. Predators include snakes, raccoons, opossums, skunks, and a variety of raptors (Black

1976; Wilkins 1989). At Frio Cave in Uvalde County, Red-tailed Hawks and Peregrine Falcons were observed preying on Brazilian free-tailed bats during the evening emergence and the dawn return, respectively. Hawks were successful on nearly 80% of attempts made in the evening and 25% made in the morning. Falcons were less efficient, with only 42% success in the evening and 25% in the morning. However, falcons captured more bats per raptor (2.5–5) than did hawks (0.5–2.8) during a hunting session (Lee and Kuo 2001).

Great-horned Owls are another common avian predator of free-tailed bats in Texas. Observations of Great-horned Owls preying on free-tailed bats were made during the evening emergence at Clarity Tunnel in Caprock Canyons State Park in Floyd County. Owls were successful at capturing bats up to 80% (4 out of 5 attempts) of the time. This suggests that, in northwestern Texas, bats could be an important source of food for Great-horned Owls during certain times of the year (Roberts et al. 1997b).

Brazilian free-tailed bats are susceptible to rabies, although it appears that this species may be able to survive rabies infections (Baer and Bales 1967; Messenger et al. 2003). Populations of *Tadarida brasiliensis* with high levels (up to 80%) of rabies-neutralizing antibodies have been found at Lava Cave, New Mexico, and in Texas caves, suggesting previous exposure to the virus (Constantine 1988; Steece and Altenbach 1989). The possibility of the bats' acquired immunity to rabies by exposure could explain the rarity of major outbreaks of rabies in these dense populations. Nonetheless, some individual bats are susceptible to rabies and become paralyzed and die soon after exposure.

As the most widespread and abundant bat in Texas, Brazilian free-tailed bats also are more frequently submitted to the DSHS for rabies testing than any other species in the state. The prevalence of rabies in bats submitted for testing in Texas varies annually, but the average is about 17.8% (443/2,492) over 9 years (see Table 7). However, small, localized groups of bats sometimes have a much higher prevalence. For example, in 1 group of dead bats found in Mineral Wells, Texas, 74% (17/23) were positive (Clark et al. 1996b). Nonetheless, the total number of confirmed rabies cases in free-tailed bats is miniscule in relation to the population of these bats as a whole.

Healthy bats that are capable of flight are rarely infected with rabies (< 0.5%, Brass 1994). If a bat incapable of flight is encountered, it could be rabid. For this reason, precautions, including the use of leather gloves, should be taken to avoid being bitten. If reasonable caution is exercised, one's chances of being bitten by a bat are slim. However, if a bite occurs, medical advice should be sought immediately.

Status. The IUCN 2011 status of Brazilian free-tailed bats is "least concern." This is probably the most common species of bat in Texas. The major threat to it would be the destruction and disturbance of cave sites either in the United States or Mexico, where the bats roost in great concentrations.

Several roosting sites of these bats (e.g., Bracken Cave, James River Bat Cave in Mason County, and the Fredericksburg railroad tunnel in Gillespie County) are protected by private conservation organizations or Texas Parks and Wildlife. The other cave sites are on private property. In Mexico, cave destruction has been on the increase under the mistaken belief that it is a way to control vampire bats. Moreover, DDT and other toxic pesticides are still used in Mexico. So, although this species appears to be in good shape in Texas, it may be vulnerable where it overwinters in Mexico.

A new potential threat to this species in Texas is wind turbines, which have been constructed for harvesting sustainable wind energy (Kunz et al. 2007). Studies to quantify bat fatalities caused by wind turbines have not been published for Texas, but one such study in nearby Oklahoma found that 86% of bat fatalities were *Tadarida* (Arnett et al. 2008). Therefore, it is important that the impact of wind facilities on bat populations in Texas be monitored and that steps be taken to minimize bat fatalities at these important power-generating sites.

Remarks. The impressive evening emergence of this species can be viewed at numerous sites. James River Cave, southwest of Mason, near State Highway 290 in Mason County, and Old Tunnel Wildlife Management Area, south of Fredericksburg in Kendall County, are 2 popular viewing locations in Texas that do not required prearranged tours. Bracken Cave, near New Braunfels

in Comal County; Devil's Sinkhole
State Natural Area, near Rocksprings
in Edwards County; Clarity Tunnel in
Caprock Canyons State Park, 161 km
southeast of Amarillo in Briscoe County;
and Stuart Bat Cave in Kickapoo Caverns
State Park, north of Brackettville on the
Kinney/Edwards county line, also provide views of large evening emergences,
but only prearranged tours are offered.
In addition, Congress Avenue Bridge in
Austin, Travis County, is 1 of the largest
urban bat colonies in the world and is
easily accessible to visitors who wish to
view the evening emergence.

Specimens Examined. Go to www.batsof
texas.com for more detailed informa-
tion about the total of 2,610 specimen
records of *T. brasiliensis* from Texas.
Additional records: *Bee County* (6)
(Chapman and Chapman 1990), *Donley
County* (5) (Stangl et al. 1989), *Erath
County* (7) (Goetze et al. 2003), *Jefferson
County* (1) (Yancey and Jones 1996b),
Kent County (92) (Yancey et al. 1995a),
Lamar County (1) (Edwards and Johnson
2007), *Motley County* (11) (Howell et al.
2009), *Shelby County* (1) (Johnson et al.
2005).

References. 34, 36, 44, 48, 50, 51, 54, 59,
62, 74, 77, 79, 80, 81, 86, 88, 92, 106,
119, 130, 131, 140, 141, 142, 145, 144, 146,
147, 159, 162, 163, 166, 168, 182, 183, 194,
199, 201, 205, 210, 225, 226, 240, 241,
242, 243, 244, 249, 250, 257, 258, 263,
264, 266, 271, 273, 276, 277, 278, 281,
306, 313, 314, 318, 323, 327, 331, 336, 339,
355, 370, 371, 372, 381, 389, 405, 423,
441, 442, 443, 444, 446, 471, 475, 478,
480, 481, 483, 489, 491, 511, 524, 539,
541, 542, 544, 543, 545, 546, 547, 548,
549, 553, 552, 551, 555, 554, 556, 557, 558,
564, 574, 575, 585, 588, 594, 610, 611,
616, 617, 618, 623, 625, 626, 633, 635,
646, 658, 660, 668, 685, 696, 704, 727,
735, 736, 737, 754, 760, 767, 768, 777, 781,
782, 784, 785, 786, 787, 788, 789, 790,
791, 809, 819, 823, 831, 830, 832, 833,
838, 848, 892, 899, 926, 927, 944, 965,
966, 971, 975, 984, 988, 989, 991, 998,
999, 1005, 1006, 1020, 1025, 1026, 1028,
1029, 1030, 1038, 1039, 1040, 1044,
1045, 1048, 1049, 1052, 1060, 1061,
1062, 1071, 1083, 1085, 1086, 1087, 1088,
1091, 1094, 1095, 1098, 1102, 1103, 1107,
1108, 1110, 1112, 1117, 1120, 1133, 1164,
1165, 1173, 1178, 1187, 1200, 1208, 1226,
1246, 1247, 1248, 1250, 1249, 1254

Nyctinomops femorosaccus (Merriam, 1884) Pocketed Free-Tailed Bat

Etymology. The generic name, *Nyctinomops,* comes from 3 Greek words, *nyktios,* or "night"; *nomos,* which means the
"custom of grazing"; and *ops,* or "face,"
which can be translated as "resembling a
night feeder" (Kumirai and Jones 1990).
The species name comes from the Latin
word *femur,* meaning "thigh," and the
Greek word *sakkos,* or "sack," in reference to the shallow pocket formed by a
fold of skin near the knee (Stangl et al.
1993).

Subspecies. *Nyctinomops femorosaccus*
(Merriam, 1884) is a monotypic species,
and no subspecies are recognized.

Description. This is a medium-sized
bat (forearm = 44–50 mm) with long,
narrow wings, vertical wrinkles on the
lips along the muzzle, and broad ears
that are joined basally at the midline of

Pocketed free-tailed bat (*Nyctinomops femorosaccus*)

the head. The pelage is dark brown to gray above and below, and the bases of the individual hairs are nearly white. Several long rump hairs just above the uropatagium extend beyond the normal length of the pelage and possibly serve as sensory "whiskers" when this bat scoots backward into its crevice roost. The common and scientific names of this bat refer to a shallow fold of skin on the underside of the leg where the wing membrane attaches to form a pocketlike area near the knee. Dental formula: I 1/2, C 1/1, Pm 2/2, M 3/3 × 2 = 30. Average external measurements are as follows: total length, 113 mm; tail, 42 mm; hind foot, 10 mm; ear, 20 mm; forearm, 47 mm. Weight: 10–18 g.

The Brazilian free-tailed bat (*Tadarida brasiliensis*) is similar in size and overall appearance to the pocketed free-tailed bat, but the ears meet only on the forehead and are not joined and raised above the forehead. *Tadarida* also has monocolored, not bicolored, hair and lacks the long hairs on the rump. Other free-tailed bats (*Nyctinomops macrotis* and *Eumops perotis*) can be distinguished by their larger size. The presence of the "pocket" in this species cannot be used as a diagnostic character because other free-tailed bats also possess one.

Distribution. This species is found throughout the Mexican plateau. The northern edge of its range extends into southern California, New Mexico, Arizona, and West Texas. It typically occurs in desert areas with rugged canyons, rock outcrops, and high cliffs. This rare bat was first recorded in Texas from Big Bend National Park in 1967 (Easterla 1968). Since that time, additional speci-

● Specimen record
■ Literature record
▲ DSHS record

mens from Brewster County (Easterla 1970b, 1973), as well as Presidio (Higginbotham et al. 2002) and Terrell counties (Ammerman et al. 2002) have been recorded. The specimens from Terrell County along the Rio Grande represent the easternmost record of this species. It has been documented as far north as Carlsbad Caverns, New Mexico, from a single specimen (Geluso and Geluso 2004). Studies conducted since its initial discovery in Big Bend have revealed that this bat is a well-established, year-round resident of the area (Higginbotham and Ammerman 2002).

Life History. This species is an inhabitant of semiarid desert lands. They roost in caves and rock crevices of rugged canyon country, although they also have been reported in buildings. They are colonial roosters, and their colonies generally are thought to be small and contain fewer than 100 individuals (Easterla 1973). However, in Big Bend National Park, up to 700 individuals were observed exiting

1 rock crevice high in a canyon wall on September 23. Michael Dixon and Jana Higginbotham Baldwin (pers. comm.) observed bats using this roost in January and February as well, and although they did not emerge, at least 12 individuals were observed in the roost, and chattering noises could be heard.

The time of emergence from the roost in Big Bend National Park was before sunset in fall and spring; emergences were not observed during the summer months (Higginbotham et al. 2002). Other researchers have found that this species leaves the roost well after dark (Kumirai and Jones 1990). Captures in mist nets set over open water in the summer peaked between 4 and 6 hours after sunset (Matthews et al. 2010), suggesting that *N. femorosaccus* either emerge later in the summer than in the spring and fall or forage immediately after leaving the roost and visit water resources later at night. They are swift, powerful flyers, but the distance this species covers during a typical night is not known.

On March 17 the following observations were made at a San Diego County, California, colony of 50–60 of these bats found at a daytime crevice roost in the face of a cliff: The first bats left the colony at 6:15 PM; others followed in twos and threes for another half hour. They were not observed to go down to drink from the stream below. The bats dropped from 1 to 1.5 m before taking wing. Their flight appeared to be a rapid, complete wing beat. When first taking flight, they uttered a shrill, sharp, high-pitched, chattering call, which was repeated while they were in full flight. They also squeaked a great deal while in the roost. Their odor was described as similar to that of the Mexican free-tailed bat but not quite so strong (Krutzsch 1944).

When this bat is foraging, its echolocation call is sufficiently low in frequency (17–18 kHz) to be audible to some people. This species can be quite vocal, and noisy social calls often can be heard during the evening emergence as individuals circle near the opening of the roost. It appears that a few scouts leave the roost first and then return to coax their roostmates out of the roost. Their loud squeaks and squawks can easily be heard emanating from the roost site during the day.

This fast-flying bat pursues insects on the wing. Easterla (1973; Easterla and Whitaker 1972) examined the stomachs of 13 individuals from Big Bend National Park and reported the following con-

tents: moths (36.9%), crickets (3.8%), flying ants (18.8%), stinkbugs (2.3%), froghoppers and leafhoppers (6.9%), lacewings (1.2%), and unidentified insects (30%). A more recent study of the diet of this species has found that dietary composition changes seasonally, with a higher proportion of moths eaten in June and July (48.3% and 64.8%, respectively), but more true bugs in September and March (61.7% and 56.9%, respectively). Beetles also were important prey in September. Other prey items included leafhoppers, crickets, and ants (Matthews et al. 2010).

Females give birth to a single young between late June and early July. Pregnant bats, each containing a single embryo, have been captured from June 10 to July 12 in Big Bend National Park, and lactating females from June 20 to August 17. Volant juveniles have been captured from August 29 to November 27 (Easterla 1973; Higginbotham and Ammerman 2002). Limited evidence suggests that males and females roost together. One banded individual was recaptured at the same site in Big Bend National Park 2 years and 8 months after its original capture, but no other information about its longevity is available (Matthews and Ammerman 2003). Obviously, more work is necessary to understand the basic natural history of this species.

Status. The IUCN 2011 status of the pocketed free-tailed bat is "least concern." In Texas, the species is restricted in distribution but locally common. At 5 out of 17 sites surveyed in Big Bend

National Park, this species was second in abundance only to *Tadarida brasiliensis*, and it accounted for 13% of the total bat captures (Higginbotham and Ammerman 2002).

Remarks. In her comprehensive study of the Molossidae, Freeman (1981b) proposed the use of the generic name *Nyctinomops* for the big free-tailed bat and pocketed free-tailed bat. The Brazilian free-tailed bat (*Tadarida brasiliensis*) was found to be more closely related to Old World species and remains classified in the genus *Tadarida*. *Nyctinomops femorosaccus* is most closely related to a widespread species of free-tailed bat common in Central and South America, *N. laticaudatus* (Dolman 2009).

Specimens Examined. Go to www.batsof texas.com for more detailed information about the total of 25 specimen records of *N. femorosaccus* from Texas. Additional records: *Brewster County* (256) (Higginbotham and Ammerman 2002), *Presidio County* (1) (Higginbotham et al. 2002).

References. 16, 60, 106, 127, 269, 336, 352, 375, 377, 381, 386, 423, 444, 472, 496, 511, 564, 566, 574, 575, 677, 678, 767, 776, 777, 833, 988, 999, 1026, 1028, 1030, 1049, 1091, 1133, 1164, 1178, 1200

Nyctinomops macrotis (Gray, 1839)
Big Free-Tailed Bat

Etymology. The genus, *Nyctinomops*, comes from 3 Greek words: *nyktios,* or "night"; *nomos,* which means the "custom of grazing"; and *ops,* or "face." This can be translated as "resembling a night feeder" (Milner et al. 1990). The specific name also is Greek, from the words *mak-* *ros* and *ous,* meaning "long ear" (Stangl et al. 1993).

Subspecies. *Nyctinomops macrotis* (Gray, 1839) is a monotypic species, and no subspecies are recognized.

Description. This is the largest species (forearm = 58–64 mm) in the genus *Nyctinomops,* and adults weigh from 22 to 37 g (they are heavier in the fall due to fat deposition). It is characterized by large, broad ears that are joined basally at the midline of the head and extend beyond the tip of the snout when laid forward. Deep vertical wrinkles are present on the lips along the muzzle. The pelage is reddish brown to dark brown or gray and is somewhat glossy. Individual hairs are bicolored; the basal portions are white. Dental formula: I 1/2, C 1/1, Pm 2/2, M 3/3 × 2 = 30. Average external measurements are as follows: total length, 132 mm; tail, 53 mm; hind foot, 11 mm; ear, 25 mm; forearm, 61 mm. Weight: 27 g.

The large size of this bat serves to distinguish it from both the Brazilian free-tailed bat (*Tadarida brasiliensis*) and the pocketed free-tailed bat (*Nyctinomops femorosaccus*). The forearm length of these 2 species is always less than 52 mm, whereas that of *N. macrotis* is always greater than 55 mm. The bicolored hair condition further distinguishes *N. macrotis* from *Tadarida brasiliensis*.

Distribution. These bats have been recorded primarily from the Chihuahuan Desert and the Arizona/New Mexico Mountains ecoregions of Texas, where they apparently are inhabitants of rugged, rocky country in both lowland and

Big free-tailed bat (*Nyctinomops macrotis*)

highland habitats. Big free-tailed bats have also been recorded from the High Plains, Southwestern Tablelands, Edwards Plateau, East Central Texas Plains, and Gulf Coastal Plains regions. This species has been collected in the state from March to November.

With the exception of the single specimen from San Patricio County, which was found hanging on a screen door at the Welder Wildlife Refuge on December 23, 1959 (Raun 1961), there are no winter records of this species in Texas. In summer, segregation of the sexes apparently occurs, as evinced by the fact that few adult males have been taken in the Trans-Pecos (Easterla 1973; Higginbotham and Ammerman 2002). Presumably males remain in more southerly locations in Mexico.

Life History. This bat is uncommon and poorly known in Texas. It is believed to roost primarily in cracks and crevices in high canyon walls of the rocky cliff country typical of western Texas, but it also has been reported in buildings. A specimen from Brazos County was obtained when it flew down a chimney and into the owner's house. A day roost is known in Fern Canyon, Chihuahua, Mexico, which is very near Big Bend National Park (Easterla 1972a).

The only known nursery colony of these bats in the United States was discovered in the Chisos Mountains of Brewster County in Big Bend National Park on May 7, 1937. The roost was discovered in a horizontal crevice in a cliff near the head of Pine Canyon in the Chisos Mountains. The number of adults

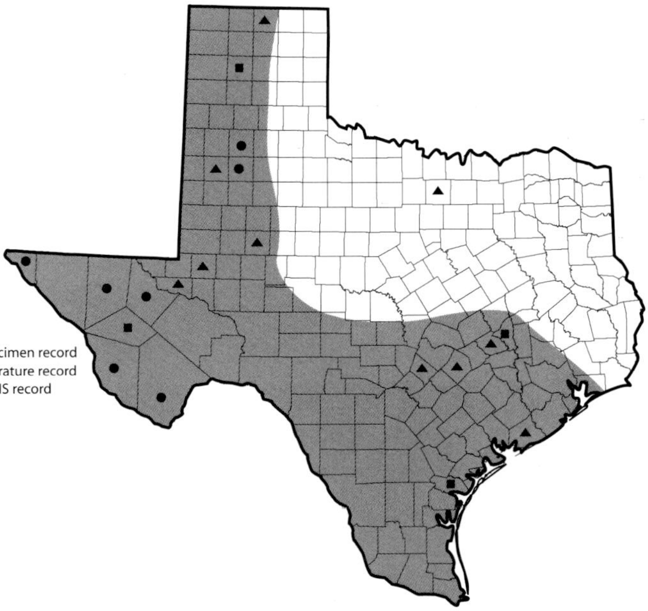

MAP 26. Distribution of the big free-tailed bat, *Nyctinomops macrotis*.

● Specimen record
■ Literature record
▲ DSHS record

using the site was estimated to be about 150, and all of the individuals collected were adult females, most of which were pregnant. The colony was revisited on October 19, 1938, at which time 4 more specimens were collected, again all females (Borell 1939). On October 27, 1958, some 20 years later, William B. Davis and Richard D. Porter visited the colony. Their notes, written the next day, are as follows: "We hiked up Pine Canyon as far as the falls (a trickle of water over a cliff about 100 feet [30 m] high). The canyon is narrow and steep-sided and has a few large yellow pines, but most of them are dead. To the right of the falls the cliff is overhanging, and it has several more-or-less horizontal crevices paralleling the top. One of them, about 50 feet [15 m] above the talus and some 100 feet [30 m] north of the falls, contained the colony. We could clearly hear the bats chattering, much like the muted coo of doves." A considerable quantity of guano on the talus at the

cliff base marked the place above which the bats were roosting. None of the bats voluntarily left the roost while Davis and Porter were there. Subsequent investigators have been unable to relocate this colony (Easterla 1973).

Borell (1939) found that the bats left the roost on May 7 at 8:20 PM (CST), when it was almost dark. The bats left in small groups during a period of 15 minutes. The swish of their wings was plainly audible, and their flight was rapid. It was so dark when they emerged that Borell could not determine whether they flew up or down the canyon. It is possible that they are seldom seen and rarely collected because they leave their daytime roost so late.

Another maternity colony is thought to be in McKittrick Canyon in the Guadalupe Mountains National Park. In June 1968 and August 1970 Richard LaVal (1973) netted 14 *N. macrotis* at a pool 8 km inside the canyon, where steep walls rise nearly 540 m above the narrow canyon floor. In this section of the canyon the bats were heard vocalizing far above the floor. All of the individuals captured were females. Eight of the 12 taken in June contained a single large embryo each. One of the 2 females captured in August was lactating.

The winter habits of this bat are unknown, although they may possibly overwinter in the Trans-Pecos. It was found that individuals kept in a refrigerator at 5°C (41°F) for 24 hours entered a deep torpor, from which they emerged within 15 minutes after their removal (LaVal 1973). Moreover, although molossids are not known for their ability to hibernate, a study on a European species of molossid suggests that some can indeed hibernate but not as efficiently as species of vespertilionids (Arlettaz et al. 2000). Another bit of evidence that suggests hibernation is that adult females taken in October were very fat and weighed about 20% more than nonpregnant females captured in June. The location of hibernacula for individuals that remain in the state for the winter is unknown.

It is possible that big free-tailed bats migrate out of Texas in the winter, as do some other members of the family Molossidae that occur in the state. The increase in fat reserves could be for migration and not hibernation. The highly skewed sex ratio encountered in populations in Texas also suggests seasonal migration to facilitate mating. In one study in Big Bend National Park, only 1 adult male was captured out of 411 individuals (Easterla 1973). Over a 13-year period in the park (1996–2008), 10 males (3 adults) out of 185 individu-als were captured (Ammerman, unpublished data). However, in the state of Hidalgo, Mexico, 9 out of 11 individuals captured on March 20 were males (Polaco et al. 1992). Therefore, mating may occur at more southerly latitudes, where more males are present. Finally, these bats are strong fliers and prone to wander somewhat in fall; they often turn up far from their normal range during this season (Di Salvo et al. 1992; Nagorsen et al. 1993; Pitts et al. 1996). This also points toward the possibility of migration. Unfortunately, no data exist to clarify this question, and the winter habits of big free-tailed bats remain a mystery.

In Arizona, females moved an average of 15 km from their roosts and were detected up to 36 km away from them while they foraged. Small maternity colonies of 4 or fewer females were found in cracks and crevices in the upper portions of vertical cliffs that faced south or southeast. This could be beneficial to a maternity colony as these sites would receive greater sun exposure and create a warmer microclimate within the roost, thus enhancing the growth of the pups. The average area 1 of these bats used while foraging over a 6-night period was estimated to be 29,590 hectares. This is much larger than home ranges reported for other insectivorous bat species. To cover this area, it was conservatively estimated that the average flight speed of this bat was 19 km per hour, with a maximum speed of 61 km per hour. Bats are active in several habitat types, including pinyon-juniper, ponderosa pine, and desert scrub (Corbett et al. 2008).

Big free-tailed bats are powerful fliers that leave their day roosts late in the evening to forage (Borell 1939; Higginbotham and Ammerman 2002). The stomachs of 49 bats from Big Bend National Park were found to contain moths (86.1%), crickets (6.7%), flying ants (4.1%), stinkbugs (1.3%), froghoppers and leafhoppers (0.1%), and unidentified insects (1.7%). In the stomachs that contained crickets and longhorn grasshoppers, these items usually made up less than 25% of the contents, but in a few they constituted as much as 50%. One stomach contained only small flying ants, and 1 contained only large ants. The researchers speculated that while in flight the bats capture ground-dwelling insects (crickets, longhorn grasshoppers, and large ants) by picking them from the walls of the cliffs (Easterla 1973; Easterla and Whitaker 1972).

The diet of individuals captured from May to September in Big Bend National Park comprised mostly moths (87.5%) and beetles (4.6%). Other items included froghoppers and leafhoppers (4.1%), crickets (1.1%), and other insects (2.4%). Although other sympatric molossids also feed on moths, it was found that *N. macrotis* eats significantly more moths and fewer true bugs and beetles than do Brazilian free-tailed bats (*Tadarida brasiliensis*) or pocketed free-tailed bats (*Nyctinomops femorosaccus;* Debelica et al. 2006). In New Mexico, the diet of *N. macrotis* appears to be more diverse, with leafhoppers (27.6%)

and wasps (19.5%) the most frequently consumed insects. Moths constituted only 17.2% of the diet (Sparks and Valdez 2003).

Little is known about reproduction and development of the young of this bat, although mating is presumed to occur in the early spring. Seemingly, each gravid female gives birth to a single offspring in late June to early July. Thirty-six pregnant females were captured in Big Bend National Park between June 10 and July 7. Each contained a single embryo (Easterla 1973). One pregnant female was captured in Big Bend Ranch State Park in Presidio County on May 29, and 8 lactating females were taken on July 6 (Higginbotham et al. 2002). A total of 73 lactating females have been captured over a 12-year period (1996–2008) at Big Bend National Park from June 28 to September 27 (Ammerman unpublished data). At the same location, volant juveniles were captured from August 5 to October 8. Development is rather rapid because by October the young are nearly full grown and difficult to distinguish from adults.

Like the pocketed free-tailed bat (*Nyctinomops femorosaccus*), the echolocation calls of this bat are audible to some people. They also emit high-intensity sounds when handled. Seven big free-tailed bats have been reported to the DSHS, and all proved to be nonrabid.

Status. The IUCN 2011 status of the big free-tailed bat is "least concern." However, considerable additional data are needed to establish the facts necessary to arrive at a meaningful and biologically defensible position as to its status. As with the other large molossid species in Texas, the western mastiff bat (*Eumops perotis*), open-surface water from which to drink could be of critical importance to this species (Corbett et al. 2008).

Specimens Examined. Go to www.batsof texas.com for more detailed information about the total of 63 specimen records of *N. macrotis* from Texas. Additional records: *Brewster County* (84) (Higginbotham and Ammerman 2002), *Jeff Davis County* (2) (Bradley et al. 1999; Higginbotham et al. 2002), *Presidio County* (13) (Higginbotham et al. 2002; Jones and Lockwood 2008; Yancey 1997).

References. 21, 49, 62, 72, 76, 106, 167, 168, 178, 225, 253, 271, 272, 286, 336, 346, 349, 355, 379, 381, 386, 423, 444, 475, 511, 564, 566, 575, 625, 626, 629, 635, 727, 767, 823, 833, 834, 853, 857, 897, 945, 950, 969, 976, 988, 999, 1021, 1024, 1026, 1028, 1029, 1049, 1080, 1091, 1112, 1133, 1153, 1155, 1164, 1178, 1200, 1201, 1246

Eumops perotis (Schinz, 1821)
Western Mastiff Bat

Etymology. The generic name, *Eumops*, comes from a combination of the Greek word *eu*, or "true," and the Malay word for "bat," *mops*. The specific name comes from a combination of the Latin prefix *per-* (meaning "through" or "complete") and the Greek word *ous*, for "ear" (Best et al. 1996; Stangl et al. 1993).

Subspecies. Texas specimens are referable to the subspecies *E. p. californicus* (Merriam, 1890), according to the most recent taxonomic review of the species (Eger 1977).

Description. This is the largest (forearm = 72–83 mm) bat in the United States, with adults weighing up to 84 g. Besides its large size, this species is characterized by large, broad ears that join basally at the midline of the head. The ears are not carried erect but slant forward and extend beyond the nose, nearly concealing the eyes. The lips are smooth along the muzzle and are not wrinkled vertically. The pelage is dark gray to pale chocolate brown dorsally and slightly paler below; individual hairs are bicolored and are nearly white at the base. Membranes are dark brown and leathery. Unlike many other species in Texas, male *E. perotis* are larger than females. Dental formula: I 1/2, C 1/1, Pm 2/2, M 3/3 × 2 = 30. Average external measurements are as follows: total length, 171 mm; tail, 57 mm; hind foot, 16 mm; ear, 31 mm; forearm, 78 mm. Weight: 70 g.

The large size of this bat and the absence of vertical wrinkles on the lips along the muzzle serve to distinguish this species from the Brazilian free-tailed bat (*Tadarida brasiliensis*) and both species of *Nyctinomops* in Texas.

Distribution. This is a bat of the arid Southwest, where it inhabits rugged, rocky canyon country. In Texas, it has been recorded from localities close to the Rio Grande in Presidio, Brewster, Terrell, and Val Verde counties. Western mastiff bats have been observed or captured in all months of the year and are probably permanent residents of the Trans-Pecos.

Life History. Away from human habitations, this bat generally seeks diurnal refuge in crevices in rocks that form ver-

Western mastiff bat (*Eumops perotis*)

tical or nearly vertical cliffs. Occupied roosts sites have obvious urine stains on the rock below the crevice entrance. The roost entrances typically are horizontally oriented, have moderately large openings, and face downward so they can be entered from below. Most authors agree that these bats choose a roost below which there is an unobstructed drop of several meters so that emerging bats can drop and gain sufficient momentum to become airborne (Hoffmeister 1986; Howell 1920; Krutzsch 1955; Little 1920; Vaughan 1959). This requirement is due to their long, narrow wings, which, while excellent for flying long distances at high speeds, provide insufficient lift for static takeoff. Captive bats are unable

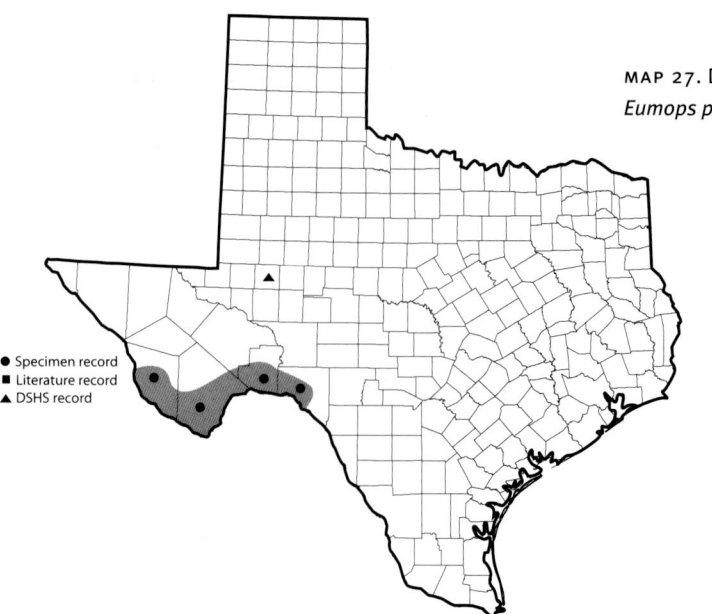

MAP 27. Distribution of the western mastiff bat, *Eumops perotis*.

● Specimen record
■ Literature record
▲ DSHS record

to take off from the ground or from flat surfaces and also are unable to maintain flight after launching themselves from the tops of tables. Bats tossed 4.5 m high in the air, however, are able to become airborne, but those thrown half that distance cannot (Vaughan 1959).

A colony of mastiff bats was observed in Capote Canyon, which is located near Candelaria in Presidio County. This site is situated on the western face of the Sierra Vieja at an elevation of 1,130 m. The bats were roosting in a crevice formed by exfoliation of the nearly vertical rimrock. Openings were present on both the lower and the upper sides of the slab and were unobstructed by vegetation. *Eumops* were observed at this site in January, February, May, June, July, August, September, and November. It was estimated that at least 71 mastiff bats used this roost; they were usually quite vocal and audible at a distance of 180 m from the roost. No other species were seen

around the *Eumops* roost, although the American parastrelle (*Parastrellus hesperus*), big brown bat (*Eptesicus fuscus*), and Brazilian free-tailed bat (*Tadarida brasiliensis*) used other crevices within 90 m of the *Eumops* colony (Ohlendorf 1972).

In California, *Eumops* were found sharing a roost with pocketed free-tailed bats (*Nyctinomops femorosaccus*). However, they used a different part of the roost and were not found intermingling with the pocketed free-tailed bats (*Nyctinomops femorosaccus*; Krutzsch 1945). Also in California, a colony was found that regularly moved among 3 different, nearby crevice roosts. One acted as the main roost, where the young remained throughout their development. Once the young were able to fly, the entire colony would shift from one roost to another without provocation (Krutzsch 1955).

Colony size varies from 2 or 3 individuals to several dozen. Twenty individuals is a large colony of these bats, although colonies of up to 70 are known. On January 30, 71 individuals were counted as they left the Capote Canyon roost about sunset. The first 4 bats emerged about 6:45 PM (CST), and within 10 minutes, 30 more of them had taken wing. During the next 15 minutes, 19 more emerged; 12 more took off during the next 10 minutes; 4 more in the next 15 minutes; and 2 more in the last 10 minutes. Thus, the exodus of the 71 bats was strung out over a period of 50 minutes. Just before launching themselves into flight and then during flight, the bats uttered a series of loud, shrill, chattering calls that could be heard for a considerable distance (Ohlendorf 1972).

Ohlendorf's observations clearly suggest that *Eumops* is a year-round resident of the Trans-Pecos. Mastiff bats are active throughout winter; they enter a daily torpor and then arouse at night to forage rather than entering the extended hibernation displayed by other bats. An adult male was captured on December 17 in Big Bend National Park approximately 1.5 hours after sunset. The temperature was 8°C (46°F), and other *Eumops* were heard foraging in the area (Ammerman, unpublished data). Barbour and Davis (1969) report that *Eumops* do not emerge from their roost when temperatures are below 5°C (41°F).

These bats leave their day roosts late in the evening to forage. The stomachs of 18 bats collected in Big Bend National Park contained moths (79.9%), crickets (16.5%), grasshoppers (2.8%), and unidentified insects (0.7%; Easterla

and Whitaker 1972). In Arizona, they feed exclusively on the abdomens of large hawk moths up to 60 mm long (Ross 1967). Bees, dragonflies, leafbugs, beetles, and cicadas also have been reported in their diet.

These bats are not believed to use night roosts but instead soar at great altitudes (up to 1,000 m) all night long so that they can feed over wide areas (up to 25 km from the roost) over a 6–8 hour period (Vaughan 1959). Insects carried aloft by thermal currents probably furnish an important portion of their diet. The presence of flightless insects, such as crickets, in their diet is interesting as these bats are unable to take off from the ground and, therefore, cannot alight to capture such prey. These prey items could be picked from canyon walls as the bats forage. The lower jaw is thin for a bat of this size, which is consistent with a diet of soft-bodied prey (Freeman 1981a, 1981b).

Eumops uses low echolocation frequencies to detect its large-bodied prey. These loud "cheeps" are emitted at regular intervals of 2–3 seconds and are audible to humans from distances of 200–300 m (Ohlendorf 1972). Known predators of western mastiff bats include Peregrine Falcons, American Kestrels, and Red-tailed Hawks (Best et al. 1996).

Observations indicate that males and females of this species remain together throughout the year, even during the period when young are produced. This is a rare situation among species of bats in North America. Females generally outnumber males by about 3 to 1 in Big Bend National Park (Higginbotham and Ammerman 2002). Male western mastiff bats have a skin gland in the throat region that enlarges in early spring (Hoffmeister 1986). The gland secretes a thick, oily secretion with a strong odor that may be important for attracting females or marking territories during the breeding season (Cockrum 1960). Unlike other species of *Eumops*, *E. perotis* lacks a baculum (Best et al. 1996).

As a nonmigrating, nonhibernating mammal, western mastiff bats mate in the spring. Testes are descended in early spring (March), and females are thought to ovulate at this time (Cockrum 1960). The gestation period is approximately 80–90 days. Normally only 1 young is produced per pregnancy, but occasionally a female may give birth to twins. The period of parturition is most commonly from June to early July, although young have been born as early as April and as late as August. Pregnant females have been captured in Big Bend National Park from June 10 to July 21, and lactating females from August 4 to August 18. A nursery colony may contain young ranging from newborn individuals to ones that are several weeks old. Volant juveniles have been taken from August 5 to September 27 (Easterla 1973; Higginbotham and Ammerman 2002).

One *Eumops perotis* from Midland County was submitted for rabies testing in 2009 and found to be positive. Genetic analysis of the rabies virus indicated that it was infected with a rabies variant commonly associated with hoary bats (*Lasiurus cinereus;* Tipps et al. in press).

Status. The IUCN 2011 status of the western mastiff bat is "least concern." This is another rare and poorly known species whose situation bears watching. The lack of information on the animal, as well as its apparent decline in other parts of its range (e.g., California), indicate that more study is needed about its basic biology and population dynamics in Texas before a meaningful and biologically defensible position as to its status can be made.

Because of their limited maneuverability, *Eumops* requires open, unobstructed waterways for drinking. They are not typically found at small water holes or at sites where vegetation encroachment has occurred. Because of this requirement, plant species, such as salt cedar and mesquite, that invade riparian zones in the Trans-Pecos pose a potential threat to this bat by further constraining the available water resources.

Specimens Examined. Go to www.batsof texas.com for more detailed information about the total of 38 specimen records of *E. perotis* from Texas. Additional records: *Brewster County* (88) (Higginbotham and Ammerman 2002), *Midland County* (1) (Tipps et al. in press), *Presidio County* (2) (Higginbotham et al. 2002; Jones and Lockwood 2008).

References. 2, 86, 92, 106, 255, 271, 336, 372, 381, 386, 387, 390, 423, 443, 444, 511, 564, 566, 575, 587, 629, 678, 680, 751, 830, 833, 853, 886, 998, 999, 1010, 1024, 1026, 1028, 1091, 1112, 1124, 1133, 1154, 1155, 1164, 1176, 1178, 1200

Hypothetical Species

As evinced by the records of museum specimens summarized in this and the first edition of *The Bats of Texas*, the distributions of bats in Texas are quite dynamic; eastern species are moving westward, western species are moving eastward, and subtropical species are moving to the north. Five species that have been documented relatively close to the Texas border and for which suitable habitat occurs in the state might eventually be documented in the state either as resident populations or wandering migrants. Their possible occurrence should be considered when conducting surveys (netting or acoustic) in the state. Here, we briefly describe each of these species and explain how to distinguish them from similar species, where they currently occur, and where we believe they could possibly be found in Texas in the future.

Myotis auriculus (Baker and Stains, 1955)
Southwestern Myotis

This species has been reported from the Fronteriza Mountains in Coahuila, Mexico, less than 50 km from the Chisos Mountains in Texas (Easterla and Baccus 1973). It also has been found as far east as Santa Fe County, New Mexico, about 320 km west of the Texas border (Findley et al. 1975). It occurs in desert scrub/grassland and mesquite forest to pine-fir forest (366–2,266 m; Warner 1982). The southwestern myotis is a medium-sized myotis (forearm = 36–40 mm, 4–6 g) without a keeled calcar. The distinguishing features of this species are large ears (> 18 mm) and tragus, with brownish, not blackish, membranes. The face is poorly furred, and a large patch of pink skin shows around the eyes. The fringed myotis (*Myotis thysanodes*) is similar but larger (forearm = 40–45 mm) and has an obvious fringe of hairs on the edge of the uropatagium (*M. auriculus* may have an inconspicuous fringe visible under a microscope). Other smaller myotis species with which it might be confused can be distinguished by their keeled calcars, black membranes, or shorter ears.

Myotis lucifugus (LeConte, 1831)
Little Brown Bat

The little brown bat is widely distributed over the eastern United States in a variety of habitats. They roost in buildings, caves, and mines and under loose tree bark. *Myotis lucifugus* has been captured less than 50 km from the Texas border in Beavers Bend State Park, McCurtain County, in southeastern Oklahoma (Caire et al. 1989). This species might be encountered in the Pineywoods of East Texas. It is a small bat with an average forearm size of 38 mm and no keel on the calcar. It has glossy brown fur with dark roots, and the skull lacks a sagittal crest (Fenton and Barclay 1980). In East Texas this species might be confused with the southeastern myotis (*Myotis austroriparius*). They are difficult to tell apart without careful examination. In *M. austroriparius*, the fur color is variable, but the overall appearance is wooly, and the skull has an abruptly sloping rostrum and a sagittal crest. The belly fur of *M. austroriparius* typically is dark at the base with white tips creating a contrast with the dorsal fur, whereas in *M. lucifugus* the belly hairs are tipped with a buffy color (Whitaker and Hamilton 1998).

Idionycteris phyllotis (G. M. Allen, 1916)
Allen's Big-Eared Bat

This species is similar to other big-eared bats but has distinctive flaps of skin, referred to as lappets, that project forward from the base of the long ears (> 34 mm) over the muzzle. The fur on the back is long, with blackish bases and tips a contrasting yellowish brown. The belly is cream colored. A tuft of white fur is present at the posterior base of the ears. The forearm averages 45 mm in length, and the calcar is keeled. Allen's big-eared bats occupy high-elevation pinyon-juniper woodlands near rocky cliffs and are known to roost in mine shafts (Czaplewski 1983). The closest-known specimens are found in New Mexico approximately 200 km from the nearest suitable habitat in the Guadalupe Mountains of Texas. This species also has been encountered in northern Chihuahua, Mexico. The low-frequency echolocation call is audible to humans and is similar to that of the spotted bat (*Euderma maculatum*).

Corynorhinus mexicanus (G. M. Allen, 1916)
Mexican Big-Eared Bat

This species is endemic to Mexico, and it has been encountered less than 50 km from the Texas border in the Sierra del Carmen in the state of Coahuila (Wilson et al. 1985). It is similar in size and appearance to the other 2 *Corynorhinus* species in Texas—*C. townsendii* and *C. rafinesquii*. The forearm length of *C. mexicanus* ranges from 39 to 47 mm, the ears average 33 mm, and the tragus is < 13 mm. The pelage is a sooty brown and darker than that of Townsend's big-eared bat (*Corynorhinus townsendii*), with less contrast between the bases and the tips of the dorsal hairs. In this respect, *C. mexicanus* is more like Rafinesque's big-eared bat (*Corynorhinus rafinesquii*) from eastern Texas. The Mexican big-eared bat would most likely be encountered in West Texas at high elevations (> 1400 m), such as the Chisos, Chinati, or Davis mountains, in pine-oak forests, but specimens also have been encountered in Mexico in transition zones below this elevation. They are known to roost in caves and mine shafts. The skull of *C. mexicanus* is generally smaller than that of *C. townsendii*; cranial measures that could be useful in distinguishing the 2 species in West Texas include skull length (< 15.9 mm in *C. mexicanus*), length of auditory bullae (< 4 mm), maxillary toothrow (< 5 mm), palatal length (< 5.2 mm); the first upper incisor has a prominent secondary cusp (Tumlison 1992).

Desmodus rotundus (Geoffroy, 1810)
Common Vampire Bat

The common vampire bat has been documented less than 200 km from the Texas border in Chihuahua, Nuevo León (south of Monterrey), and Tamaulipas, Mexico. However, it is not common at these locations, which represent the northernmost edge of their distribution. The common vampire is similar in appearance to the hairy-legged vampire bat (*Diphylla ecaudata*). Distinguishing characters are discussed in the account of *Diphylla*. *Desmodus* has a diet of mammal blood and is known to feed on livestock.

Literature and References

The following is a comprehensive bibliography listing all of the general references, textbooks, reports, articles, and technical papers that were used in the preparation of this book. All references that are cited in the text have been included, as well as many that were not directly cited but are nevertheless of importance to the study of Texas bats. The bibliography is divided into three sections: (1) general references; (2) mammalian species; and (3) technical papers.

General works include textbooks about bats, comprehensive guides to the mammalian fauna of geographical or political regions within and adjacent to Texas, general texts on mammalogy, and other bibliographies relevant to Texas bats. As this is not a comprehensive list of these types of publications, they and others are included in the technical papers section. The Mammalian Species series consists of comprehensive summaries of knowledge about individual species of mammals. These accounts are published by the American Society of Mammalogists and contain the most current and complete information and references available on the respective species. All Mammalian Species accounts that are available for Texas bats have been included in this section and are highly recommended as a good starting point for further study of specific Texas bats. The section on technical papers is the most extensive and includes all documented records and studies of Texas bats, as well as numerous references on bats from other parts of the United States and from Mexico that are useful in the study of Texas bats.

General References

Adams, R. A. 2003. Bats of the Rocky Mountain West: natural history, ecology, and conservation. University Press of Colorado, Boulder.

Altringham, J. D. 1996. Bats: biology and behaviour. Oxford University Press, New York.

Fenton, M. B. 1998. The bat: wings in the night sky. Firefly Books, Buffalo.

———. 2001. Bats. Checkmark Books, New York.

Hill, J. E., and J. D. Smith. 1984. Bats: a natural history. British Museum of Natural History, London.

Kunz, T. H., and M. B. Fenton. 2003. Bat ecology. University of Chicago Press, Chicago.

Kunz, T. H., and S. Parsons, eds. 2009. Ecological and behavioral methods for the study of bats, 2nd ed. Johns Hopkins University Press, Baltimore.

Kunz, T. H., and P. A. Racey, eds. 1998. Bat biology and conservation. Smithsonian Institution Press, Washington, D.C.

Lacki, M. J., J. P. Hayes, and A. Kurta, eds. 2007. Bats in forests: conservation and management. Johns Hopkins University Press, Baltimore.

Neuweiler, G. 2000. The biology of bats. Oxford University Press, New York.

Nowak, R. M. 1994. Walker's bats of the world. Johns Hopkins University Press, Baltimore.

Schmidly, D. J. 1991. Bats of Texas. Texas A&M University Press, College Station.

———. 2004. The mammals of Texas. University of Texas Press, Austin.

Tuttle, M. D. 1993. The bat house builder's handbook. Bat Conservation International, Austin.

———. 1998. America's neighborhood bats. University of Texas Press, Austin.

———. 2003. Texas bats. Bat Conservation International, Austin.

Williams, K., R. Mies, D. Stokes, and L. Stokes. 2002. Stokes beginner's guide to bats. Little, Brown, New York.

Wilson, D. E. 1997. Bats in question: the Smithsonian answer book. Smithsonian Institution Press, Washington, D.C.

———, and S. Ruff, eds. 1999. The Smithsonian book of North American mammals. Smithsonian Institution Press, Washington, D.C.

Zubaid, A., G. F. McCracken, and T. H. Kunz. 2006. Functional and evolutionary ecology of bats. Oxford University Press, New York.

Mammalian Species

1. Arroyo-Cabrales, J., R. R. Hollander, and J. K. Jones Jr. 1987. *Choeronycteris mexicana*. Mammalian Species 291:1–5.

2. Best, T. L., M. W. Kiser, and P. W. Freeman. 1996. *Eumops perotis*. Mammalian Species 534:1–8.

3. Caceres, M. C., and R. M. R. Barclay. 2000. *Myotis septentrionalis*. Mammalian Species 634:1–4.

4. Czaplewski, N. J. 1983. *Idionycteris phyllotis*. Mammalian Species 208:1–4.

5. Fenton, M. B., and R. M. R. Barclay. 1980. *Myotis lucifugus*. Mammalian Species 142:1–8.

6. Fitch, J. H., and K. A. Shump Jr. 1979. *Myotis keenii*. Mammalian Species 121:1–3.

7. Fitch, J. H., K. A. Shump Jr., and A. U. Shump. 1981. *Myotis velifer*. Mammalian Species 149:1–5.

8. Fujita, M. S., and T. H. Kunz. 1984. *Pipistrellus subflavus*. Mammalian Species 228:1–6.

9. Greenhall, A. M., G. Joermann, U. Schmidt, and M. R. Seidel. 1983. *Desmodus rotundus*. Mammalian Species 227:1–3.

10. Greenhall, A. M., U. Schmidt, and G. Joermann. 1984. *Diphylla ecaudata*. Mammalian Species 227:1–3.

11. Hensley, A. P., and K. T. Wilkins. 1988. *Leptonycteris nivalis*. Mammalian Species 307:1–4.

12. Hermanson, J. W., and T. J. O'Shea. 1983. *Antrozous pallidus*. Mammalian Species 213:1–8.

13. Holloway, G. L., and R. M. R. Barclay. 2001. *Myotis ciliolabrum*. Mammalian Species 670:1–5.

14. Jones, C. 1977. *Plecotus rafinesquii*. Mammalian Species 69:1–4.

15. Jones, C., and R. W. Manning. 1989. *Myotis austroriparius*. Mammalian Species 332:1–3.

16. Kumirai, A., and J. K. Jones Jr. 1990. *Nyctinomops femorosaccus*. Mammalian Species 349:1–5.

17. Kunz, T. H. 1982. *Lasionycteris noctivagans*. Mammalian Species 172:1–5.

18. Kunz, T. H., and R. A. Martin. 1982. *Plecotus townsendii*. Mammalian Species 175:1–6.

19. Kurta, A., and R. H. Baker. 1990. *Eptesicus fuscus*. Mammalian Species 356:1–10.

20. Kurta, A., and G. C. Lehr. 1995. *Lasiurus ega*. Mammalian Species 515:1–7.

21. Milner, J., C. Jones, and J. K. Jones Jr. 1990. *Nyctinomops macrotis*. Mammalian Species 351:1–4.

22. O'Farrell, M. J., and E. H. Studier. 1980. *Myotis thysanodes*. Mammalian Species 137:1–5.

23. Rezsutek, M., and G. N. Cameron. 1993. *Mormoops megalophylla*. Mammalian Species 448:1–5.

24. Shump, K. A., Jr., and A. U. Shump. 1982a. *Lasiurus cinereus*. Mammalian Species 185: 1–5.

25. Shump, K. A., Jr., and A. U. Shump. 1982b. *Lasiurus borealis*. Mammalian Species 183: 1–6.

26. Simpson, M. R. 1993. *Myotis californicus*. Mammalian Species 428:1–4.

27. Tumlison, R. 1992. *Plecotus mexicanus*. Mammalian Species 401:1–3.

28. Warner, R. M. 1982. *Myotis auriculus*. Mammalian Species 191:1–3.

29. Warner, R. M., and N. J. Czaplewski. 1984. *Myotis volans*. Mammalian Species 224:1–4.

30. Watkins, L. C. 1972. *Nycticeius humeralis*. Mammalian Species 23:1–4.

31. Watkins, L. C. 1977. *Euderma maculatum*. Mammalian Species 77:1–4.

32. Webster, W. D., J. K. Jones Jr., and R. J. Baker. 1980. *Lasiurus intermedius*. Mammalian Species 132:1–3.

33. Wilkins, K. T. 1987. *Lasiurus seminolus*. Mammalian Species 280:1–5.

34. Wilkins, K. T. 1989. *Tadarida brasiliensis*. Mammalian Species 331:1–10.

Technical Papers

35. Adam, M. D., M. J. Lacki, and T. G. Barnes. 1994. Foraging areas and habitat use of the Virginia big-eared bat in Kentucky. Journal of Wildlife Management 58:462–469.

36. Adams, D. B., and G. M. Bacr. 1966. Cesarian section and artificial feeding device for suckling bats. Journal of Mammalogy 47: 524–525.

37. Adams, R. A. 2003. Bats of the Rocky Mountain West: natural history, ecology, and conservation. University Press of Colorado, Boulder.

38. Adkins, R. M., and R. L. Honeycutt. 1991. Molecular phylogeny of the superorder Archonta. Proceedings of the National Academy of Sciences 88:10317–10321.

39. Agosta, S. J. 2002. Habitat use, diet, and roost selection by the big brown bat (*Eptesicus fuscus*) in North America: a case for conserving an abundant species. Mammal Review 32:179–198.

40. Alcorn, J. R. 1944. Notes on the winter occurrence of bats in Nevada. Journal of Mammalogy 25:308–310.

41. Aldridge, H. D. J. N., and R. M. Brigham. 1988. Load carrying and maneuverability in an insectivorous bat: a test of the 5% "rule" of radiotelemetry. Journal of Mammalogy 69:379–382.

42. Allan, P. F. 1947a. Notes on Mississippi kites in Hemphill County, Texas. Condor 49: 88–89.

43. Allan, P. F. 1947b. Blue jay attacks red bats. Journal of Mammalogy 28:180.

44. Allen, G. M. 1908. Notes on Chiroptera. Bulletin of the Harvard Museum of Comparative Zoology 52:25–62.

45. Allen, G. M. 1916. Bats of the genus *Corynorhinus*. Bulletin of the Harvard Museum of Comparative Zoology 60: 331–356.

46. Allen, G. M. 1922. Bats from New Mexico and Arizona. Journal of Mammalogy 3: 156–162.

47. Allen, J. A. 1891. On a collection of mammals from southern Texas and northeastern Mexico. Bulletin of the Harvard Museum of Comparative Zoology 3:219–228.

48. Allen, J. A. 1896a. On mammals collected in Bexar Country and vicinity, Texas, by

Mr. H.P. Atwater, with field notes by the collector. Bulletin of the Harvard Museum of Comparative Zoology 8:47–80.

49. Allen, J. A. 1896b. Descriptions of ten new North American mammals. Bulletin of the Harvard Museum of Comparative Zoology 8: 233–240.

50. Allen, L. C. 2009. Immunity, reproductive success and stress in cave- and bridge-roosting Brazilian free-tailed bats: implications for conservation. Ph.D. dissertation, Boston University, Boston.

51. Allen, L. C., A. S. Turmelle, M. T. Mendonça, K. J. Navara, T. H. Kunz, and G. F. McCracken. 2008. Roosting ecology and variation in adaptive and innate immune system function in the Brazilian free-tailed bat (*Tadarida brasiliensis*). Journal of Comparative Physiology B: Biochemical, Systemic, and Environmental Physiology 179:315–323.

52. Altenbach, J. S., and D. Dalton. 2009. Techniques for photographing bats. Pp. 78–90 *in* Ecological and behavioral methods for the study of bats (T. H. Kunz and S. Parsons, eds.). Johns Hopkins University Press, Baltimore.

53. Altringham, J. D. 1996. Bats: biology and behavior. Oxford University Press, New York.

54. Alvarez, T. 1963. The recent mammals of Tamaulipas, Mexico. University of Kansas Publications, Museum of Natural History 14: 363–473.

55. Amelon, S. K., D. Dalton, J. J. Millspaugh, and S. Wolf. 2009. Radiotelemetry: techniques and analysis. Pp. 55–77 *in* Ecological and behavioral methods for the study of bats (T. H. Kunz and S. Parsons, eds.). 2nd ed. Johns Hopkins University Press, Baltimore.

56. Ammerman, L. K. 2005. Noteworthy records of the eastern pipistrelle, *Perimyotis subflavus*, and the silver-haired bat, *Lasionycteris noctivagans* (Chiroptera: Vespertilionidae) from the Chisos Mountains, Texas. Texas Journal of Science 57:202–207.

57. Ammerman, L. K., and D. M. Hillis. 1992. A molecular test of bat relationships: monophyly or diphyly? Systematic Biology 41:222–232.

58. Ammerman, L. K., M. McDonough, N. I. Hristov, and T. H. Kunz. 2009. Census of the endangered Mexican long–nosed bat, *Leptonycteris nivalis*, in Texas, USA, using thermal imaging. Endangered Species Research 8:87–92.

59. Ammerman, L. K., R. M. Rodriguez, R. C. Dowler, and M. McDonough. 2008. Bat diversity and activity: a comparison among Texas Army National Guard sites. Occasional Papers, Museum of Texas Tech University 280:1–23.

60. Ammerman, L. K., R. M. Rodriguez, J. L. Higginbotham, and A. K. Matthews. 2002. Recent records of bats from the Lower Canyons of the Rio Grande river of west Texas. Texas Journal of Science 54: 369–374.

61. Ammerman, L. K., and R. Tabor. 2008. Monitoring the colony size and population fluctuations of the endangered Mexican long-nosed bat in Big Bend National Park using thermal imaging. Division of Science and Resource Management, Big Bend National Park. 8 pp.

62. Anderson, S. 1972. Mammals of Chihuahua: taxonomy and distribution. Bulletin of the American Museum of Natural History 148: 151–410.

63. Angelo, S. 2009. Determinant factors of summer roost selection for cave bats (*Myotis velifer*) in central Texas. M. S. Thesis, Antioch University New England, Keene, New Hampshire.

64. Anonymous. 1990. Animal heads diagnosed as positive (+) or negative (−) for rabies. Report of Texas Department of Health Laboratories, Austin. 18 pp.

65. Anonymous. 1991. Animal heads diagnosed as positive (+) or negative (−) for rabies. Report of Texas Department of Health Laboratories, Austin. 19 pp.

66. Anonymous. 1992. Animal heads diagnosed as positive (+) or negative (−) for rabies. Report of Texas Department of Health Laboratories, Austin. 17 pp.

67. Anonymous. 1993. Animal heads diagnosed as positive (+) or negative (−) for rabies. Report of Texas Department of Health Laboratories, Austin. 17 pp.

68. Anonymous. 1994. Animal heads diagnosed as positive (+) or negative (−) for rabies. Report of Texas Department of Health Laboratories, Austin. 16 pp.

69. Arita, H. T. 1991. Spatial segregation in long-nosed bats, *Leptonycteris nivalis* and *Leptonycteris curasoae*, in Mexico. Journal of Mammalogy 72:706–714.

70. Arita, H. T., and S. R. Humphrey. 1988. Revisión taxonómica de los murciélagos magueyeros del género *Leptonycteris* (Chiroptera: Phyllostomidae). Acta Zoologica Mexicana nueva serie 29:1–60.

71. Arita, H. T., and K. Santos-del-Prado. 1999. Conservation biology of nectar-feeding bats in Mexico. Journal of Mammalogy 80: 31–41.

72. Arlettaz, R., C. Ruchet, J. Aeschimann, E. Brun, M. Genoud, and P. Vogel. 2000. Physiological traits affecting the distribution and wintering strategy of the bat *Tadarida teniotis*. Ecology 81:1004–1014.

73. Armstrong, D. M., R. A. Adams, and K. E. Taylor. 2006. New record of the eastern pipistrelle (*Pipistrellus subflavus*) in Colorado. Western North American Naturalist 66: 268–269.

74. Arnett, E. B., et al. 2008. Patterns of bat fatalities at wind energy facilities in North America. Journal of Wildlife Management 72:61–78.

75. Avila-Flores, R., and R. A. Medellín. 2004. Ecological, taxonomic, and physiological correlates of cave use by Mexican bats. Journal of Mammalogy 85:675–687.

76. Axtell, R. W. 1961. An additional record for the bat *Tadarida molossa* from Trans-Pecos Texas. Southwestern Naturalist 6:52.

77. Baccus, J. T. 1971. The mammals of Baylor County, Texas. Texas Journal of Science 22: 177–185.

78. Baer, G. M., ed. 1991. The natural history of rabies, 2nd ed. CRC Press, Boca Raton.

79. Baer, G. M., and G. L. Bales. 1967. Experimental rabies infection in the Mexican free-tailed bat. Journal of Infectious Diseases 117:82–90.

80. Baer, G. M., and G. M. Holquin. 1971. Breeding Mexican freetail bats in captivity. American Midland Naturalist 85:515–517.

81. Baer, G. M., and R. G. McLean. 1972. A new method of bleeding small and infant bats. Journal of Mammalogy 53:231–232.

82. Baer, G. M., and J. S. Smith. 1991. Rabies in nonhematophagous bats. Pp. 341–366 *in* The natural history of rabies (G. M. Baer, ed.). 2nd ed. CRC Press, Boca Raton.

83. Baerwald, E. F., and R. M. R. Barclay. 2009. Geographic variation in activity and fatality

of migratory bats at wind energy facilities. Journal of Mammalogy 90:1341–1349.

84. Bagley, F., and J. Jacobs. 1985. Census technique for endangered big-eared bats proving successful. Endangered Species Technical Bulletin 10:5–7.

85. Bagley, F. M. 1984. A recovery plan for the Ozark big-eared bat and the Virginia big-eared bat. U.S. Fish and Wildlife Service, Bloomington, MN. 56 pp.

86. Bailey, V. 1905. Biological survey of Texas. North American Fauna 25:1–222.

87. Bain, J. R., and S. R. Humphrey. 1986. Social organization and biased primary sex ratio of the evening bat, *Nycticeius humeralis*. Florida Scientist 49:22–31.

88. Baker, J. K. 1962. The manner and efficiency of raptor predation on bats. Condor 64: 500–504.

89. Baker, M. D., and M. J. Lacki. 2006. Day-roosting habitat of female long-legged myotis in ponderosa pine forests. Journal of Wildlife Management 70:207–215.

90. Baker, M. D., M. J. Lacki, F. A. Falxa, P. L. Droppelman, R. A. Slack, and S. A. Slankard. 2008. Habitat use of pallid bats in coniferous forests of northern California. Northwest Science 82:269–275.

91. Baker, R. H. 1954. A new bat (genus *Pipistrellus*) from northeastern Mexico. University of Kansas Publications, Museum of Natural History 7:583–586.

92. Baker, R. H. 1956. Mammals of Coahuila, Mexico. University of Kansas Publications, Museum of Natural History 9:125–335.

93. Baker, R. H. 1964. *Myotis lucifugus lucifugus* (Le Conte) and *Pipistrellus hesperus maximus* Hatfield in Knox County, new to north-central Texas. Southwestern Naturalist 9:205.

94. Baker, R. H., and R. W. Dickerman. 1956. Daytime roost of the yellow bat in Veracruz. Journal of Mammalogy 37:443.

95. Baker, R. H., J. K. Jones Jr., and D. C. Carter, eds. 1976. Biology of bats of the New World family Phyllostomatidae. Part III. Special Publications, Museum of Texas Tech University.

96. Baker, R. J., and E. L. Cockrum. 1966. Geographic and ecological range of the long-nosed bats, *Leptonycteris*. Journal of Mammalogy 47:329–331.

97. Baker, R. J., R. L. Honeycutt, and R. A. Van Den Bussche. 1991a. Examination of monophyly of bats: restriction map of the ribosomal DNA cistron. Pp. 42–53 *in* Contributions to mammalogy in honor of Karl F. Koopman (T. A. Griffiths and D. Klingener, eds.). Bulletin of the American Museum of Natural History, Vol. 206.

98. Baker, R. J., and J. T. Mascarello. 1969. Chromosomes of some vespertilionid bats of the genera *Lasiurus* and *Plecotus*. Southwestern Naturalist 14:249–251.

99. Baker, R. J., T. Mollhagen, and G. Lopez. 1971. Notes on *Lasiurus ega*. Journal of Mammalogy 52:849–852.

100. Baker, R. J., M. J. Novacek, and N. B. Simmons. 1991b. On the monophyly of bats. Systematic Zoology 40:216–231.

101. Baker, R. J., J. C. Patton, H. H. Genoways, and J. W. Bickham. 1988. Genic studies of *Lasiurus* (Chiroptera: Vespertilionidae). Occasional Papers, Museum of Texas Tech University 117:1–15.

102. Baker, R. J., and J. L. Patton. 1967. Karyotypes and karyotypic variation of North American vespertilionid bats. Journal of Mammalogy 48:270–286.

103. Balcombe, J. P., and M. B. Fenton. 1988. Eavesdropping by bats: the influence of

echolocation call design and foraging strategy. Ethology 79:158–166.

104. Balin, L. 2009. Mexican long-tongued bat (*Choeronycteris mexicana*) in El Paso, Texas. Southwestern Naturalist 54:225–226.

105. Barber, J. R., and W. E. Conner. 2007. Acoustic mimicry in a predator-prey interaction. Proceedings of the National Academy of Sciences 104:9331–9334.

106. Barbour, R. W., and W. H. Davis. 1969. Bats of America. University Press of Kentucky, Lexington.

107. Barbour, R. W., and W. H. Davis. 1970. The status of *Myotis occultus*. Journal of Mammalogy 51:150–151.

108. Barclay, R. M. R. 1989. The effect of reproductive condition on the foraging behavior of female hoary bats, *Lasiurus cinereus*. Behavioral Ecology and Sociobiology 24:31–37.

109. Barclay, R. M. R. 1999. Bats are not birds—a cautionary note on using echolocation calls to identify bats: a comment. Journal of Mammalogy 80: 290–296.

110. Barclay, R. M. R., E. F. Baerwald, and J. C. Gruver. 2007. Variation in bat and bird fatalities at wind energy facilities: assessing the effects of rotor size and tower height. Canadian Journal of Zoology 85:381–387.

111. Barclay, R. M. R., and R. M. Brigham. 2001. Year-to-year reuse of tree-roosts by California bats (*Myotis californicus*) in southern British Columbia. American Midland Naturalist 146:80–85.

112. Barclay, R. M. R., P. A. Faure, and D. R. Farr. 1988. Roosting behavior and roost selection by migrating silver-haired bats (*Lasionycteris noctivagans*). Journal of Mammalogy 69:821–825.

113. Barclay, R. M. R., and L. D. Harder. 2003. Life histories of bats: life in the slow lane. Pp. 209–253 *in* Bat ecology (T. H. Kunz and M. B. Fenton, eds.). University of Chicago Press, Chicago.

114. Barclay, R. M. R., et al. 1996. Can external radiotransmitters be used to assess body temperature and torpor in bats? Journal of Mammalogy 77:1102–1106.

115. Barclay, R. M. R., and A. Kurta. 2007. Ecology and behavior of bats roosting in tree cavities and under bark. Pp. 17–59 *in* Bats in forests: conservation and management (M. J. Lacki, J. P. Hayes, and A. Kurta, eds.). Johns Hopkins University Press, Baltimore.

116. Barkalow, F. S., Jr. 1948. The status of the Seminole bat, *Lasiurus seminolus* (Rhoads). Journal of Mammalogy 29:415–416.

117. Bassett, J. E. 1984. Litter size and postnatal growth rate in the pallid bat, *Antrozous pallidus*. Journal of Mammalogy 65:317–319.

118. Bateman, G. C., and T. A. Vaughan. 1974. Nightly activities of Mormoopid bats. Journal of Mammalogy 55:45–65.

119. Baughman, J. L. 1951 The caves of Texas. Texas Game and Fish 9:2–7.

120. Beatty, L. D. 1955. The leafchin bat in Arizona. Journal of Mammalogy 36:290.

121. Beck, A. J., and R. L. Rudd. 1960. Nursery colonies in the pallid bat. Journal of Mammalogy 41:266–267.

122. Beer, J. R., and A. G. Richards. 1956. Hibernation of the big brown bat. Journal of Mammalogy 37:31–41.

123. Bell, G. P. 1982. Behavioral and ecological aspects of gleaning by a desert insectivorous bat, *Antrozous pallidus* (Chiroptera: Vespertilionidae). Behavioral Ecology and Sociobiology 10:217–223.

124. Bell, G. P. 1985. The sensory basis of

prey location by the California leaf-nosed bats *Macrotus californicus* (Chiroptera: Phyllostomidae). Behavioral Ecology and Sociobiology 16:343–347.

125. Bell, G. P., and M. B. Fenton. 1986. Visual acuity, sensitivity and binocularity in a gleaning insectivorous bat, *Macrotus californicus* (Chiroptera: Phyllostomidae). Animal Behaviour 34:409–414.

126. Bennett, F. M., S. C. Loeb, M. S. Bunch, and W. W. Bowerman. 2008. Use and selection of bridges as day roosts by Rafinesque's big-eared bats. American Midland Naturalist 160:386–389.

127. Benson, S. B. 1940. Notes on the pocketed free-tailed bat. Journal of Mammalogy 21: 26–29.

128. Berna, H. J. 1990. Seven bat species from the Kaibab Plateau, Arizona, with a new record of *Euderma maculatum*. Southwestern Naturalist 35:354–356.

129. Best, T. L. 1988. Morphological variation in the spotted bat *Euderma maculatum*. American Midland Naturalist 119:244–252.

130. Best, T. L., and K. N. Geluso. 2003. Summer foraging range of Mexican free-tailed bats (*Tadarida brasiliensis mexicana*) from Carlsbad Cavern, New Mexico. Southwestern Naturalist 48:590–596.

131. Betke, M., et al. 2008. Thermal imaging reveals significantly smaller Brazilian free-tailed bat colonies than previously estimated. Journal of Mammalogy 89:18–24

132. Betts, B. J. 1996. Roosting behaviour of silver-haired bats (*Lasionycteris noctivagans*) and big brown bats (*Eptesicus fuscus*) in northeast Oregon. Pp. 55–61 *in* Bats and forests symposium (R. M. R. Barclay and R. M. Brigham, eds.). British Columbia Ministry of Forestry, Victoria, British Columbia.

133. Betts, B. J. 1997. Microclimate in Hell's Canyon mines used by maternity colonies of *Myotis yumanensis*. Journal of Mammalogy 78:1240–1250.

134. Betts, B. J. 1998a. Variation in roost fidelity among reproductive female silver-haired bats in northeastern Oregon. Northwestern Naturalist 79:59–63.

135. Betts, B. J. 1998b. Roosts used by maternity colonies of silver-haired bats in northeastern Oregon. Journal of Mammalogy 79:643–650.

136. Bickham, J. W. 1979. Chromosomal variation and evolutionary relationships of vespertilionid bats. Journal of Mammalogy 60:350–363.

137. Bickham, J. W. 1987. Chromosomal variation among seven species of lasiurine bats (Chiroptera: Vespertilionidae). Journal of Mammalogy 68:837–842.

138. Bishop, S. C. 1947. Curious behavior of a hoary bat. Journal of Mammalogy 28: 293–294.

139. Black, H. L. 1972. Differential exploitation of moths by the bats *Eptesicus fuscus* and *Lasiurus cinereus*. Journal of Mammalogy 53: 598–601.

140. Black, H. L. 1974. A north temperate bat community: structure and prey populations. Journal of Mammalogy 55:138–157.

141. Black, H. L. 1976. American kestrel predation on the bats *Eptesicus fuscus*, *Euderma maculatum*, and *Tadarida brasiliensis*. Southwestern Naturalist 21: 250–251.

142. Blair, W. F. 1940. A contribution to the ecology and faunal relationships of the mammals of the Davis Mountain region, southwestern Texas. Miscellaneous Publications of the University of Michigan Museum of Zoology 46:1–39.

143. Blair, W. F. 1948. A color pattern aberration

and Valley of West Virginia. Northeastern Naturalist 10:83–88.

217. Carter, T. C., M. A. Menzel, and D. A. Saugey. 2003b. Population trends of solitary foliage roosting bats. Pp. 41–47 *in* Monitoring trends in bat populations of the United States and territories: problems and prospects (T. J. O'Shea and M. A. Bogan, eds.). U.S. Geological Survey, Biological Resources Discipline, Information and Technology Report, USGS/BRD/ITR-2003-0003.

218. Carter, T. D. 1950. On the migration of the red bat, *Lasiurus borealis borealis*. Journal of Mammalogy 31:349–350.

219. Carver, B. D., and N. Ashley. 2008. Roost tree use by sympatric Rafinesque's big-eared bats (*Corynorhinus rafinesquii*) and southeastern myotis (*Myotis austroriparius*). American Midland Naturalist 160:364–373.

220. Castle, K. T., and P. M. Cryan. 2010. White-nose syndrome in bats: a primer for resource managers. Park Science 27:20–25.

221. Castner, S. V., T. K. Snow, and D. C. Noel. 1994. Bat inventory and monitoring in Arizona 1992–1994. Arizona Game and Fish Department, Nongame and Endangered Wildlife Program. Technical Report 54:1–32.

222. Ceballos, G., and G. Oliva, eds. 2005. Los mamíferos silvestres de México FCE, CONABIO, Mexico.

223. Chapman, B. R., and S. G. Spencer. 1987. Distributional records for six Texas mammals. Texas Journal of Science 39: 379–380.

224. Chapman, K., K. McGuiness, and R. M. Brigham. 1994. Status of the pallid bat in British Columbia. Wildlife Branch, Ministry of Environment, Lands & Parks, Victoria, British Columbia. Wildlife Working Report No. WR-61. 32 pp.

225. Chapman, S. S. 1989. A survey of the distribution and ecological requirements of the bats of South Texas. M.S. Thesis, Corpus Christi State University, Corpus Christi.

226. Chapman, S. S., and B. R. Chapman. 1990. Bats from the coastal region of southern Texas. Texas Journal of Science 42:13–22.

227. Choate, J. R., and J. M. Anderson. 1997. Bats of Jewel Cave National Monument, South Dakota. Prairie Naturalist 29:39–47.

228. Choate, J. R., J. W. Dragoo, J. K. Jones Jr., and J. A. Howard. 1986. Sub-specific status of the big brown bat, *Eptesicus fuscus*, in Kansas. Prairie Naturalist 18:43–51.

229. Choate, J. R., and E. R. Hall. 1967. Two new species of bats, genus *Myotis*, from a Pleistocene deposit in Texas. American Midland Naturalist 78:531–534.

230. Choate, L. L., and F. C. Killebrew. 1991. Distributional records of the California myotis and the prairie vole in the Texas Panhandle. Texas Journal of Science 43: 214–215.

231. Choate, L. L., R. W. Manning, J. K. Jones Jr., C. Jones, and T. R. Mollhagen. 1991. Records of mammals from the Llano Estacado and adjacent areas of Texas and New Mexico. Occasional Papers, Museum of Texas Tech University 138:1–11.

232. Christian, J. J. 1956. The natural history of a summer aggregation of the big brown bat, *Eptesicus fuscus fuscus*. American Midland Naturalist 55:66–94.

233. Christiansen, B. W. 2003. In defense of bats. Journal of the American Veterinary Medical Association 222:1346–1347.

234. Chung-MacCoubrey, A. L. 1996. Bat species composition and roost use in pinyon-juniper woodlands of New Mexico, Pp. 118–123 in Bats and forests symposium (R. M. R. Barclay and R. M. Brigham, eds.).

British Columbia Ministry of Forestry, Victoria, British Columbia.

235. Clark, B. K., B. S. Clark, and D. M. Leslie Jr. 1990. Endangered Ozark big-eared bat eaten by eastern woodrat. Prairie Naturalist 22:273–274.

236. Clark, B. K., B. S. Clark, and D. M. Leslie Jr. 1997a. Seasonal variation in the use of caves by the endangered Ozark big-eared bat (*Corynorhinus townsendii ingens*) in Oklahoma. American Midland Naturalist 137:388–392.

237. Clark, B. S., B. K. Clark, and D. M. Leslie Jr. 2002. Seasonal variation in activity patterns of the endangered Ozark big-eared bat (*Corynorhinus townsendii ingens*). Journal of Mammalogy 83:590–598.

238. Clark, B. S., D. M. Leslie Jr., and T. S. Carter. 1993. Foraging activity of adult female Ozark big-eared bats (*Plecotus townsendii ingens*) in summer. Journal of Mammalogy 74:422–427.

239. Clark, B. S., W. L. Puckette, B. K. Clark, and D. M. Leslie, Jr. 1996a. Status of the Ozark big-eared bat (*Corynorhinus townsendii ingens*) in Oklahoma, 1957 to 1995. Southwestern Naturalist 42:20–24.

240. Clark, D. R., Jr. 1981. Bats and environmental contaminants: a review. U.S. Fish and Wildlife Service Resource Report 235:1–27.

241. Clark, D. R., Jr. 2001. DDT and the decline of free-tailed bats (*Tadarida brasiliensis*) at Carlsbad Cavern, New Mexico. Archives of Environmental Contamination and Toxicology 40:537–543.

242. Clark, D. R., Jr., and J. C. Kroll. 1977. Effects of DDE on experimentally poisoned free-tailed bats (*Tadarida brasiliensis*); lethal brain concentrations. Journal of Toxicology and Environmental Health 3:893–901.

243. Clark, D. R., Jr., A. Lollar, and D. F. Cowman. 1996b. Dead and dying Brazilian free-tailed bats (*Tadarida brasiliensis*) from Texas: rabies and pesticide exposure. Southwestern Naturalist 41:275–278.

244. Clark, D. R., Jr., C. O. Martin, and D. M. Swineford. 1975. Organochlorine insecticide residues in the free-tailed bat (*Tadarida brasiliensis*) at Bracken Cave, Texas. Journal of Mammalogy 56:429–443.

245. Clark, M. K. 1990. Roosting ecology of the eastern big-eared bat, *Plecotus rafinesquii* in North Carolina. M.S. Thesis, North Carolina State University, Raleigh.

246. Clark, M. K. 2003. Survey and monitoring of rare bats in bottomland hardwood forests. Pp. 79–90 *in* Monitoring trends in bat populations of the United States and territories: problems and prospects (T. J. O'Shea and M. A. Bogan, eds.). U.S. Geological Survey, Biological Resources Discipline, Information and Technology Report, USGS/BRD/ITR-2003–0003.

247. Clark, M. K., E. Hajnos, and A. Black. 1997b. Radio-tracking of *Corynorhinus rafinesquii* and *Myotis austroriparius* in South Carolina. Bat Research News 38:136–137.

248. Clem, P. D. 1993. Foraging patterns and the use of temporary roosts in female evening bats, *Nycticeius humeralis*, in an Indiana maternity colony. Proceedings of the Indiana Academy of Science 102: 201–206.

249. Cleveland, A. G., J. T. Baccus, and E. G. Zimmerman. 1984. Distributional records and notes for nine species of mammals in eastern Texas. Texas Journal of Science 35: 323–326.

250. Cleveland, C. J., et al. 2006. Economic value of the pest control service provided by Brazilian free-tailed bats in south

central Texas. Frontiers in Ecology and the Environment 5:238–243.

251. Cochran, S. M. 1999. Roosting and habitat use of Rafinesque's big-eared bat and other species in a bottomland hardwood forest ecosystem. M.S. Thesis, Arkansas State University, Jonesboro.

252. Cockerell, T. D. A. 1930. An apparently extinct Euglandina from Texas. Proceedings of the Colorado Museum of Natural History 9:52–53.

253. Cockrum, E. L. 1952. The big free-tailed bat in Oklahoma. Journal of Mammalogy 33:492.

254. Cockrum, E. L. 1956. Homing, movements, and longevity of bats. Journal of Mammalogy 37:48–57.

255. Cockrum, E. L. 1960. Distribution, habitat and habits of the mastiff bat, *Eumops perotis*, in North America. Journal of the Arizona-Nevada Academy of Science 1:79–84.

256. Cockrum, E. L. 1961. Southern yellow bat from Arizona. Journal of Mammalogy 42:97.

257. Cockrum, E. L. 1969. Migration in the guano bat, *Tadarida brasiliensis*. University of Kansas Publications, Museum of Natural History 51:303–336.

258. Cockrum, E. L. 1970. Insecticides and guano bats. Ecology 51:761–762.

259. Cockrum, E. L., and S. P. Cross. 1964. Time of bat activity over water holes. Journal of Mammalogy 45:635–636.

260. Cockrum, E. L., B. Musgrove, and Y. Petryszyn. 1996. Bats of Mohave County, Arizona: populations and movements. Occasional Papers, Museum of Texas Tech University 157:1–71.

261. Cockrum, E. L., and Y. Petryszyn. 1991. The long-nosed bat, *Leptonycteris*: an endangered species in the southwest? Occasional Papers, Museum of Texas Tech University 142:1–32.

262. Cokendolpher, J. C., D. L. Holub, and D. C. Parmley. 1979. Additional records of mammals from north-central Texas. Southwestern Naturalist 24:376–377.

263. Constantine, D. G. 1948. Great bat colonies attract predators. National Speleological Society Bulletin 10:100.

264. Constantine, D. G. 1957. Color variation and molt in *Tadarida brasiliensis* and *Myotis velifer*. Journal of Mammalogy 38:461–466.

265. Constantine, D. G. 1958a. Color variation and molt in *Mormoops megalophylla*. Journal of Mammalogy 39:344–47.

266. Constantine, D. G. 1958b. Bleaching of hair pigment in bats by the atmosphere in caves. Journal of Mammalogy 39:513–520.

267. Constantine, D. G. 1958c. An automatic bat-collecting device. Journal of Wildlife Management 22:17–22.

268. Constantine, D. G. 1958d. Ecological observations on lasiurine bats in Georgia. Journal of Mammalogy 39:64–70.

269. Constantine, D. G. 1958e. Remarks on external features of *Tadarida femorosacca*. Journal of Mammalogy 39:437.

270. Constantine, D. G. 1959. Ecological observations on Lasiurine bats in the North Bay area of California. Journal of Mammalogy 40:13–15.

271. Constantine, D. G. 1961a. Locality records and notes on western bats. Journal of Mammalogy 42:404–405.

272. Constantine, D. G. 1961b. Spotted bat and big free-tailed bat in northern New Mexico. Southwestern Naturalist 6:92–97.

273. Constantine, D. G. 1962. Rabies transmission by nonbite route. Public Health Reports 77:287–289.

274. Constantine, D. G. 1966a. Ecological observations on Lasiurine bats in Iowa. Journal of Mammalogy 47:34–41.

275. Constantine, D. G. 1966b. New bat locality records from Oaxaca, Arizona, and Colorado. Journal of Mammalogy 47:125–126.

276. Constantine, D. G. 1967a. Activity patterns of the Mexican free-tailed bat. University of New Mexico Publications in Biology 7:1–79.

277. Constantine, D. G. 1967b. Rabies transmission by air in bat caves. Public Health Services Publications 1617:1–51.

278. Constantine, D. G. 1970. Bats in relation to the health, welfare, and economy of man. Pp. 319–449 *in* Biology of bats (W. A. Wimsatt, ed.). Academic Press, New York.

279. Constantine, D. G. 1979. An updated list of rabies-infected bats in North America. Journal of Wildlife Diseases 15:347–349.

280. Constantine, D. G. 1987. Long-tongued bat and spotted bat at Las Vegas, Nevada. Southwestern Naturalist 32:392.

281. Constantine, D. G. 1988. Health precautions for bat researchers. Pp. 491–528 *in* Ecological and behavioral methods for the study of bats (T. H. Kunz, ed.). Smithsonian Institution Press, Washington, D.C.

282. Constantine, D. G., G. L. Humphrey, and T. S. Herbenick. 1979. Rabies in *Myotis thysanodes, Lasiurus ega, Euderma maculatum* and *Eumops perotis* in California. Journal of Wildlife Diseases 15:343–345.

283. Cook, J. A. 1986. The mammals of the Animas Mountains and adjacent areas, Hidalgo County, New Mexico. Occasional Papers, Museum of Southwestern Biology 4: 1–45.

284. Cope, J. B., W. Baker, and J. Confer. 1961. Breeding colonies of four species of bats of Indiana. Proceedings of the Indiana Academy of Science 70:262–266.

285. Cope, J. B., and S. R. Humphrey. 1972. Reproduction of the bats *Myotis keenii* and *Pipistrellus subflavus* in Indiana. Bat Research News 13:9–10.

286. Corbett, R. J. M., C. L. Chambers, and M. J. Herder. 2008. Roosts and activity areas of *Nyctinomops macrotis* in northern Arizona. Acta Chiropterologica 10:323–329.

287. Corcoran, A. J., J. R. Barber, and W. E. Conner. 2009. Tiger moths jam bat sonar. Science 325:325–327.

288. COSEWIC. 2004. COSEWIC assessment and update status report on the spotted bat *Euderma maculatum* in Canada. Committee on the Status of Endangered Wildlife in Canada, Ottawa. 26 pp.

289. Cox, T. J. 1965. Seasonal change in the behavior of the western pipistrelle because of lactation. Journal of Mammalogy 46:703.

290. Crampton, L. H., and R. M. R. Barclay. 1998. Selection of roosting and foraging habits by bats in different-aged aspen mixedwood stands. Conservation Biology 12: 1347–1358.

291. Crawley, C. 2009. Scientists tackle the mystery of white-nose syndrome in bats. Eurekalert. Retrieved from http://www.eurekalert.org/pub_releases/2009–06/nifm-stt060409.php.

292. Creel, G. C. 1963. Bat as a food item of *Rana pipiens*. Texas Journal of Science 15: 104–106.

293. Crichton, E. G. 2000. Sperm storage and fertilization. Pp. 295–320 *in* Reproductive biology of bats (E. G. Crichton and P. H. Krutzsch, eds.). Academic Press, London.

294. Crichton, E. G., F. Suzuki, P. H. Krutzsch, and R. H. Hammerstedt. 1993. Unique features of the cauda epididymidal epithelium of hibernating bats may promote sperm longevity. Anatomical Record 237:475–481.

295. Crnkovic, A. C. 2003. Discovery of northern long-eared myotis, *Myotis septentrionalis*

(Chiroptera: Vespertilionidae), in Louisiana. Southwestern Naturalist 48:715–717.

296. Cross, S. P. 1965. Roosting habits of *Pipistrellus hesperus*. Journal of Mammalogy 46:270–279.

297. Cryan, P. M. 2003. Seasonal distribution of migratory tree bats (*Lasiurus* and *Lasionycteris*) in North America. Journal of Mammalogy 84:579–93.

298. Cryan, P. M., and M. A. Bogan. 2003. Recurrence of Mexican long-tongued bats (*Choeronycteris mexicana*) at historical sites in Arizona and New Mexico. Western North American Naturalist 63:314–319.

299. Cryan, P. M., M. A. Bogan, R. O. Rye, G. P. Landis, and C. L. Kester. 2004. Stable hydrogen isotope analysis of bat hair as evidence for seasonal molt and long-distance migration. Journal of Mammalogy 85: 995–1001.

300. Cryan, P. M., M. A. Bogan, and G. M. Yanega. 2001. Roosting habits of four bat species in the Black Hills of South Dakota. Acta Chiropterologica 3:43–52.

301. Cryan, P. M., and B. O. Wolf. 2003. Sex differences in the thermoregulation and evaporative water loss of a heterothermic bat, *Lasiurus cinereus*, during its spring migration. Journal of Experimental Biology 206:3381–3390.

302. Cutter, W. L. 1959. The hoary bat in the panhandle of Texas. Journal of Mammalogy 40:442.

303. Czaplewski, N. J. 1993a. Late tertiary bats (Mammalia, Chiroptera) from the southwestern United States. Southeastern Naturalist 38:111–118.

304. Czaplewski, N. J. 1993b. *Myotis velifer* in the Quitaque local fauna, Motley County, Texas. Texas Journal of Science 45:97–100.

305. Czaplewski, N. J. 1993c. *Pizonyx wheeleri* Dalquest and Patrick (Mammalia: Chiroptera) from the Miocene of Texas referred to the genus *Antrozous* H. Allen. Journal of Vertebrate Paleontology 13:378–380.

306. Czaplewski, N. J., J. P. Farney, J. K. Jones Jr., and J. D. Druecker. 1979. Synopsis of bats of Nebraska. Occasional Papers, Museum of Texas Tech University 61:1–24.

307. da Silva, M. V., A. N. de Oliveira, M. C. R. Andrade, W. C. de Moura, and M. J. de Figueiredo. 1998. Feeding behavior of *Diphylla ecaudata* in captivity. Bat Research News 39:79.

308. Dalquest, W. W. 1947a. Notes on the natural history of the bat *Corynorhinus rafinesquii* in California. Journal of Mammalogy 28:17–30.

309. Dalquest, W. W. 1947b. Notes on the natural history of the bat, *Myotis yumanensis*, in California, with a description of a new race. American Midland Naturalist 38: 224–247.

310. Dalquest, W. W. 1954. Netting bats in tropical Mexico. Transactions of the Kansas Academy of Science 57:1–10.

311. Dalquest, W. W. 1955. Natural history of the vampire bats of eastern Mexico. American Midland Naturalist 53:79–87.

312. Dalquest, W. W. 1967. Mammals of the Pleistocene Slaton local fauna of Texas. Southwestern Naturalist 12:1–30.

313. Dalquest, W. W. 1968. Mammals of north-central Texas. Southwestern Naturalist 13: 13–22.

314. Dalquest, W. W. 1975. Vertebrate fossils from the Blanco fauna of Texas. Occasional Papers, Museum of Texas Tech University 30:1–52.

315. Dalquest, W. W. 1978. Early Blancan mammals of the Beck Ranch local fauna of Texas. Journal of Mammalogy 59:269–298.

316. Dalquest, W. W., and R. M. Carpenter. 1988. Early Pleistocene (Irvingtonian) mammals from the Seymour formation, Knox and Baylor counties, Texas, exclusive of Camelidae. Occasional Papers, Museum of Texas Tech University 124:1–28.

317. Dalquest, W. W., and E. R. Hall. 1947. Geographic range of the hairy-legged vampire in eastern Mexico. Transactions of the Kansas Academy of Science 50:315–317.

318. Dalquest, W. W., and N. V. Horner. 1984. Mammals of north-central Texas. Midwestern State University Press, Wichita Falls.

319. Dalquest, W. W., E. Roth, and F. Judd. 1969. The mammal fauna of Schulze Cave, Edwards County, Texas. Bulletin of the Florida Museum of Natural History 13: 205–276.

320. Dalquest, W. W., and E. L. Roth. 1970. The pallid bat (*Antrozous*) of the Edwards Plateau. Southwestern Naturalist 15:395–396.

321. Dalquest, W. W., and F. B. Stangl Jr. 1984a. The taxonomic status of *Myotis magnamolaris* Choate and Hall. Journal of Mammalogy 65: 485–486.

322. Dalquest, W. W., and F. B. Stangl Jr. 1984b. The Pleistocene mammals of Fowlkes Cave in southern Culberson County, Texas. Pp. 432–455 *in* Contributions in Quaternary vertebrate paleontology: a volume in memorial to John E. Guilday (H. H. Genoways and M. R. Dawson, eds.). Carnegie Museum of Natural History Special Publications, Vol. 8:1–538.

323. Dalquest, W. W., and F. B. Stangl Jr. 1986. Post-Pleistocene mammals of the Apache Mountains, Culberson County, Texas, with comments on zoogeography of the Trans-Pecos Front Range. Occasional Papers, Museum of Texas Tech University 104:1–35.

324. Dalquest, W. W., F. B. Stangl Jr., and J. K. Jones Jr. 1990. Mammalian zoogeography of a Rocky Mountain–Great Plains interface in New Mexico, Oklahoma, and Texas. Special Publications, Museum of Texas Tech University 34:1–78.

325. Davis, R. 1968. Wing defects in a population of pallid bats. American Midland Naturalist 79:388–395.

326. Davis, R. 1969. Growth and development of young pallid bats, *Antrozous pallidus.* Journal of Mammalogy 50:729–736.

327. Davis, R. 1970. Carrying of young by flying female North American bats. American Midland Naturalist 83:186–196.

328. Davis, R., and E. L. Cockrum. 1963a. Bridges utilized as day-roosts by bats. Journal of Mammalogy 44:428–430.

329. Davis, R., and E. L. Cockrum. 1963b. "Malfunction" of homing ability in bats. Journal of Mammalogy 44:131–132.

330. Davis, R., and S. E. Doster. 1972. Wing repair in pallid bats. Journal of Mammalogy 53:377–378.

331. Davis, R. B., C. F. Herreid, II, and H. L. Short. 1962. Mexican free-tailed bats in Texas. Ecological Monographs 32:311–346.

332. Davis, W. B. 1940. Mammals of the Guadalupe Mountains of western Texas. Occasional Papers of the Museum of Zoology, Louisiana State University 7:69–84.

333. Davis, W. B. 1944a. Notes on Mexican mammals. Journal of Mammalogy 25: 370–402.

334. Davis, W. B. 1944b. Status of *Myotis subulatus* in Texas. Journal of Mammalogy 25:201.

335. Davis, W. B. 1966a. Mammals of Texas.

Texas Parks and Wildlife Department, Bulletin No. 41, Austin.

336. Davis, W. B. 1974. Mammals of Texas. Texas Parks and Wildlife Department, Bulletin No. 41, Austin.

337. Davis, W. B., and D. C. Carter. 1962a. Review of the genus *Leptonycteris* (Mammalia: Chiroptera). Proceedings of the Biological Society of Washington 75: 193–198.

338. Davis, W. B., and D. C. Carter. 1962b. Notes on Central American bats with description of a new subspecies of *Mormoops*. Southwestern Naturalist 7:64–74.

339. Davis, W. B., and J. L. Robertson Jr. 1944. The mammals of Culberson County, Texas. Journal of Mammalogy 25:254–273.

340. Davis, W. H. 1959. Taxonomy of the eastern pipistrel. Journal of Mammalogy 40: 521–531.

341. Davis, W. H. 1966b. Population dynamics of the bat *Pipistrellus subflavus*. Journal of Mammalogy 47:383–396.

342. Davis, W. H., R. W. Barbour, and M. D. Hassell. 1968. Colonial behavior of *Eptesicus fuscus*. Journal of Mammalogy 49:44–50.

343. Davis, W. H., and W. Z. Lidicker Jr. 1956. Winter range of the red bat, *Lasiurus borealis*. Journal of Mammalogy 37:280–281.

344. de Fanis, E., and G. Jones. 1995. The role of odour in the discrimination of conspecifics by pipistrelle bats. Animal Behaviour 49: 835–839.

345. DeBaca, R. S., and C. Jones. 2002. The ghost-faced bat, *Mormoops megalophylla* (Chiroptera: Mormoopidae) from the Davis Mountains, Texas. Texas Journal of Science 54:89–91.

346. Debelica, A., A. K. Matthews, and L. K. Ammerman. 2006. Dietary study of big

free-tailed bats (*Nyctinomops macrotis*) in Big Bend National Park, Texas. Southwestern Naturalist 51:414–418.

347. Delpietro, V. H. A., and R. G. Russo. 2002. Observations of the common vampire bat (*Desmodus rotundus*) and the hairy-legged vampire bat (*Diphylla ecaudata*) in captivity. Mammalian Biology 67:65–78.

348. des Marias, D. J., J. M. Mitchell, W. G. Meinschein, and J. M. Hayes. 1980. The carbon isotope biogeochemistry of the individual hydrocarbons in bat guano and the ecology of the insectivorous bats in the region of Carlsbad, New Mexico. Geochimca et Cosmochimica Acta 44:2075–2086.

349. Di Salvo, A. F., H. N. Neuhauser, and R. E. Mancke. 1992. *Nyctinomops macrotis* in South Carolina. Bat Research News 33: 21–22.

350. Dixon, M. T., R. M. Rodriguez, and L. K. Ammerman. (in press). Comparison of two survey methods used for bats along the lower canyons of the Rio Grande in Big Bend National Park. Proceedings of the Sixth Symposium on the Natural Resources of the Chihuahuan Desert.

351. Dobkin, D. S., R. D. Gettinger, and M. G. Gerdes. 1995. Springtime movements, roost use, and foraging activity of Townsend's big-eared bat (*Plecotus townsendii*) in central Oregon. Great Basin Naturalist 55:315–321.

352. Dolman, R. W. 2009. Molecular systematics of *Nyctinomops* (Chiroptera: Molossidae). M.S. Thesis, Angelo State University, San Angelo.

353. Dominguez, S. R., T. J. O'Shea, L. M. Oko, and K. V. Holmes. 2007. Detection of Group 1 Coronaviruses in bats in North America. Emerging Infectious Diseases 13:1295–1300.

354. Dooley, T. J. 1974. Bats of El Paso County,

Texas, with notes on habitat, behavior, and ectoparasites. M.S. Thesis, University of Texas at El Paso, El Paso.

355. Dooley, T. J., J. R. Bristol, and A. G. Canaris. 1976. Ectoparasites from bats in extreme west Texas and south-central New Mexico. Journal of Mammalogy 57:189–191.

356. Dorsey, S. L. 1977. A reevaluation of two new species of fossil bats from Inner Space Caverns. Texas Journal of Science 28: 103–108.

357. Doutt, J. K., C. A. Heppenstall, and J. E. Guilday. 1966. Mammals of Pennsylvania. Pennsylvania Game Commission, Harrisburg.

358. Dowler, R. C., R. C. Dawkins, and T. C. Maxwell. 1999. Range extensions for the evening bat (*Nycticeius humeralis*) in west Texas. Texas Journal of Science 51:193–195.

359. Dowler, R. C., T. C. Maxwell, and D. S. Marsh. 1992. Noteworthy records of bats from Texas. Texas Journal of Science 44:121–123.

360. Drueker, J. D. 1972. Aspects of reproduction in *Myotis volans, Lasionycteris noctivagans*, and *Lasiurus cinereus*. M.S. Thesis, University of New Mexico, Albuquerque.

361. Duchamp, J. E., D. W. Sparks, and J. O. Whitaker Jr. 2004. Foraging-habitat selection by bats at an urban-rural interface: comparison between a successful and a less successful species. Canadian Journal of Zoology 82:1157–1164.

362. Duke, S. D., G. C. Bateman, and M. M. Bateman. 1979. Longevity record for *Myotis californicus*. Southwestern Naturalist 24:693.

363. Dunbar, M. B., and T. E. Tomasi. 2006. Arousal patterns, metabolic rate, and an energy budget of eastern red bats (*Lasiurus borealis*) in winter. Journal of Mammalogy 87:1069–1102.

364. Dunbar, M. B., J. O. Whitaker Jr., and L. W. Robbins. 2007. Winter feeding by bats in Missouri. Acta Chiropterologica 9: 305–310.

365. Dunnigan, P. B., and J. H. Fitch. 1967. Seasonal movements and population fluctuations of the cave bat (*Myotis velifer*) in south-central Kansas. Transactions of the Kansas Academy of Science 70:210–218.

366. Duszynski, D. W. 2002. Coccidia (Apicomplexa: Eimeriidae) of the mammalian order Chiroptera. Special Publication, The Museum of Southwestern Biology 5:1–45.

367. Dzal, Y., L. P. McGuire, N. Veselka, and M. B. Fenton. 2010. Going, going, gone: the impact of white-nose syndrome on the summer activity of the little brown bat (*Myotis lucifugus*). Biology Letters:doi: 10.1098/rsbl.2010.0859.

368. Eads, R. B., J. E. Grimes, and A. Conklin. 1957a. Additional Texas bat records. Journal of Mammalogy 38:514.

369. Eads, R. B., G. C. Menzies, and J. S. Wiseman. 1956. New locality records for Texas bats. Journal of Mammalogy 37:440.

370. Eads, R. B., J. S. Wiseman, J. E. Grimes, and G. C. Menzies. 1955a. Wildlife rabies in Texas. Public Health Reports 70:995–1000.

371. Eads, R. B., J. S. Wiseman, and G. C. Menzies. 1955b. Banding Mexican free-tailed bats. Journal of Mammalogy 36:120–121.

372. Eads, R. B., J. S. Wiseman, and G. C. Menzies. 1957b. Observations concerning the Mexican free-tailed bat, *Tadarida mexicana*, in Texas. Texas Journal of Science 9:227–242.

373. Easterla, D. A. 1965. The spotted bat in Utah. Journal of Mammalogy 46:665–668.

374. Easterla, D. A. 1967. Black rat snake preys upon gray myotis and winter observations of

red bats. American Midland Naturalist 77: 527–528.

375. Easterla, D. A. 1968. First records of the pocketed free-tailed bat for Texas. Journal of Mammalogy 49:515–516.

376. Easterla, D. A. 1970a. First records of the spotted bat in Texas and notes on its natural history. American Midland Naturalist 83: 306–308.

377. Easterla, D. A. 1970b. First record of the pocketed free-tailed bat for Coahuila, Mexico, and additional Texas records. Texas Journal of Science 22:92–93.

378. Easterla, D. A. 1971. Notes on young and adults of the spotted bat, *Euderma maculatum.* Journal of Mammalogy 52: 475–476.

379. Easterla, D. A. 1972a. A diurnal colony of big freetail bats, *Tadarida macrotis* (Gray), in Chihuahua, Mexico. American Midland Naturalist 88:468–470.

380. Easterla, D. A. 1972b. Status of *Leptonycteris nivalis* (Phyllostomatidae) in Big Bend National Park, Texas. Southwestern Naturalist 17:287–292.

381. Easterla, D. A. 1973. Ecology of the 18 species of Chiroptera at Big Bend National Park, Texas. Northwest Missouri State University Studies 34:1–165.

382. Easterla, D. A. 1975. The red bat in Big Bend National Park, Texas. Southwestern Naturalist 20:418–419.

383. Easterla, D. A. 1976. Notes on the second and third newborn of the spotted bat, *Euderma maculatum,* and comments on the species in Texas. American Midland Naturalist 96:499–501.

384. Easterla, D. A., and J. Baccus. 1973. A collection of bats from the Fronteriza Mountains, Coahuila, Mexico. Southwestern Naturalist 17:424–427.

385. Easterla, D. A., and P. Easterla. 1969. America's rarest mammal. National Wildlife 7:15–18.

386. Easterla, D. A., and J. O. Whitaker Jr. 1972. Food habits of some bats from Big Bend National Park, Texas. Journal of Mammalogy 53:887–890.

387. Easterla, P., and D. A. Easterla. 1974. Rare glimpses of newborn bats. Smithsonian 5: 104–107.

388. Edwards, C. W., et al. 2000. Records of mammals from northeast and south Texas. Occasional Papers, Museum of Texas Tech University 200:1–8.

389. Edwards, C. W., and S. A. Johnson. 2007. Report on a mammal survey at Camp Maxey, Lamar County, Texas (Texas Army National Guard facility). Occasional Papers, Museum of Texas Tech University 267:1–11.

390. Eger, J. L. 1977. Systematics of the genus *Eumops* (Chiroptera: Molossidae). Life Sciences Contributions, Royal Ontario Museum 110:1–69.

391. Eick, G. N., D. S. Jacobs, and C. A. Mathee. 2005. A nuclear DNA phylogenetic perspective on the evolution of echolocation and historical biogeography of extant bats (Chiroptera). Molecular Biology and Evolution 22:1869–1886.

392. Elder, W. H. 1945. Big-eared bat in Illinois. Journal of Mammalogy 26:433–434.

393. Elizalde-Arellano, C., J. C. Lopez-Vidal, J. Arroyo-Cabrales, R. A. Medellín, and J. W. Laundre. 2007. Food sharing behavior in the hairy-legged vampire bat *Diphylla ecaudata.* Acta Chiropterologica 9:314–319.

394. Elliot, W. R. 1993. Cave fauna conservation in Texas, Pp. 323–337 in Proceedings of the National Cave Management Symposium (D. L. Foster, ed.). American Cave Conservation Association, Bowling Green, KY.

395. Ellis, A. M., L. L. Patton, and S. B. Castleberry. 2002. Bat activity in upland and riparian habitats in the Georgia Piedmont. Proceedings of the Annual Conference of the Southeastern Association of Fish and Wildlife Agencies 56:210–218.

396. Ellison, L. E. 2008. Summary and analysis of the U.S. Government Bat Banding Program: U.S. Geological Survey Open-File Report 2008–1363. 117 pp.

397. Ellison, L. E., T. J. O'Shea, D. J. Neubaum, and R. A. Bowen. 2007a. Factors influencing movement probabilities of big brown bats (*Eptesicus fuscus*) in buildings. Ecological Applications 17:620–627.

398. Ellison, L. E., T. J. O'Shea, D. J. Neubaum, M. A. Neubaum, R. D. Pearce, and R. A. Bowen. 2007b. A comparison of conventional capture versus PIT reader techniques for estimating survival and capture probabilities of big brown bats (*Eptesicus fuscus*). Acta Chiropterologica 9:149–160.

399. Elmore, L. W., D. A. Miller, and F. J. Vilella. 2004. Selection of diurnal roosts by red bats (*Lasiurus borealis*) in an intensively managed pine forest in Mississippi. Forest Ecology and Management 199:11–20.

400. Elmore, L. W., D. A. Miller, and F. J. Vilella. 2005. Foraging area size and habitat use by red bats (*Lasiurus borealis*) in an intensively managed pine landscape in Mississippi. American Midland Naturalist 153:405–417.

401. England, D. R., D. A. Saugey, V. R. McDaniel, and S. M. Speight. 1990. Observations on the life history of Rafinesque's big-eared bat, *Plecotus rafinesquii,* in southern Arkansas. Bat Research News 30:62–63.

402. Engler, C. H. 1943. Carnivorous activities of big brown and pallid bats. Journal of Mammalogy 24:96–97.

403. Evelyn, M. J., D. A. Stiles, and R. A. Young. 2003. Conservation of bats in suburban landscapes: roost selection by *Myotis yumanensis* in a residential area in California. Biological Conservation 115:463–473.

404. Everette, A. L., T. J. O'Shea, L. E. Ellison, L. A. Stone, and J. L. McCance. 2001. Bat use of a high-plains urban wildlife refuge. Wildlife Society Bulletin 29:967–973.

405. Farney, J., and E. D. Fleharty. 1969. Aspect ratio, loading, wingspan, and membrane areas of bats. Journal of Mammalogy 50: 362–367.

406. Fassler, D. J. 1974. Red bat hibernating in a woodpecker hole. American Midland Naturalist 93:254.

407. Faure, P. A., J. H. Fullard, and J. W. Dawson. 1993. The gleaning attacks of the northern long-eared bat, *Myotis septentrionalis,* are relatively inaudible to moths. Journal of Experimental Biology 178: 173–189.

408. Feldham, R. 1984. Teichfledermaus: *Myotis dasycneme* (Boie, 1825). Pp. 107–111 in Die Säugetiere Westfalens. Münster: Westfälisches Museum für Naturkunde.

409. Feldhamer, G. A., T. C. Carter, A. T. Morzillo, and E. H. Nicholson. 2003. Use of bridges as day roosts by bats in southern Illinois. Transactions of the Illinois State Academy of Science 96:107–112.

410. Feldhamer, G. A., J. O. Whitaker Jr., J. K. Krejca, and S. J. Taylor. 1995. Food of the evening bat (*Nycticeius humeralis*) and red bat (*Lasiurus borealis*) from southern Illinois. Transactions of the Illinois State Academy of Science 88:139–143.

411. Fellers, G. M. 2000. Predation on

Corynorhinus townsendii by *Rattus rattus*. Southwestern Naturalist 45:524–527.

412. Fellers, G. M., and E. D. Pierson. 2002. Habitat use and foraging behavior of Townsend's big-eared bat (*Corynorhinus townsendii*) in coastal California. Journal of Mammalogy 82:167–177.

413. Fenton, M. B. 1969. The carrying of young by females of three species of bats. Canadian Journal of Zoology 47:158–159.

414. Fenton, M. B. 1997. Science and the conservation of bats. Journal of Mammalogy 78:1–14.

415. Fenton, M. B. 2003. Science and the conservation of bats: where to next? Wildlife Society Bulletin 31:6–15.

416. Fenton, M. B., D. C. Tennant, and J. Wyszecki. 1987. Using echolocation calls to measure the distribution of bats: the case of *Euderma maculatum*. Journal of Mammalogy 68:142–144.

417. Fernandez, M. K., S. A. Smith, and R. Escamilla. 2000. New county record for the Mexican long-tongued bat (*Choeronycteris mexicana*) from Texas. Texas Journal of Science 52:68–69.

418. Ferrara, F. J., and P. L. Leberg. 2005. Characteristics of positions selected by day-roosting bats under bridges in Louisiana. Journal of Mammalogy 86:729–735.

419. Findley, J. S. 1957. The hog-nosed bat in New Mexico. Journal of Mammalogy 27: 513–514.

420. Findley, J. S. 1969. Biogeography of southwestern boreal and desert mammals. University of Kansas Publications, Museum of Natural History 51:113–128.

421. Findley, J. S. 1972. Phenetic relationships among bats of the genus *Myotis*. Systematic Zoology 21:31–52.

422. Findley, J. S. 1987. The natural history of New Mexican mammals. University of New Mexico Press, Albuquerque.

423. Findley, J. S., A. H. Harris, D. E. Wilson, and C. Jones. 1975. Mammals of New Mexico. University of New Mexico Press, Albuquerque.

424. Findley, J. S., and C. Jones. 1964. Seasonal distribution of the hoary bat. Journal of Mammalogy 45:461–470.

425. Findley, J. S., and C. Jones. 1965. Comments on spotted bats. Journal of Mammalogy 46:679–680.

426. Findley, J. S., and C. Jones. 1967. Taxonomic relationships of bats of the species *Myotis fortidens*, *M. lucifugus*, and *M. occultus*. Journal of Mammalogy 48: 429–444.

427. Findley, J. S., E. H. Studier, and D. E. Wilson. 1972. Morphologic properties of bat wings. Journal of Mammalogy 53:429–444.

428. Findley, J. S., and G. L. Traut. 1970. Geographic variation in *Pipistrellus hesperus*. Journal of Mammalogy 51:741–765.

429. Flaquer, C., I. Torre, and A. Arrizabalaga. 2007. Comparison of sampling methods for inventory of bat communities. Journal of Mammalogy 88:526–533.

430. Fleming, T. H. 1995. The use of stable isotopes to study the diets of plant-visiting bats. Symposium of the Zoological Society of London 67:99–110.

431. Fleming, T. H., and P. Eby. 2003. Ecology of bat migration. Pp. 156–208 *in* Bat Ecology (T. H. Kunz and M. B. Fenton, eds.). University of Chicago Press, Chicago.

432. Fleming, T. H., T. Tibbitts, Y. Petryszyn, and V. Dalton. 2003. Current status of pollinating bats in southwestern North America. Pp. 63–98 *in* Monitoring trends in bat populations of

the United States and territories: problems and prospects (T. J. O'Shea and M. A. Bogan, eds.). U.S. Geological Survey, Biological Resources Discipline.

433. Ford, W. M., J. M. Menzel, M. A. Menzel, J. W. Edwards, and J. C. Kilgo. 2006a. Presence and absence of bats across habitat scales in the upper coastal plain of South Carolina. Journal of Wildlife Management 70:1200–1209.

434. Ford, W. M., M. A. Menzel, J. M. Menzel, and D. J. Welch. 2002. Influence of summer temperature on sex ratios in eastern red bats (*Lasiurus borealis*). American Midland Naturalist 147:179–84.

435. Ford, W. M., M. A. Menzel, J. L. Rodrigue, J. M. Menzel, and J. B. Johnson. 2005. Relating bat species presence to simple habitat measures in a central Appalachian forest. Biological Conservation 126:528–539.

436. Ford, W. M., S. F. Owen, J. W. Edwards, and J. L. Rodrigue. 2006b. *Robinia pseudacacia* (black locust) as day-roosts of male *Myotis septentrionalis* (northern bats) on the Fernow Experimental Forest, West Virginia. Northeastern Naturalist 13:15–24.

437. Foster, G. W., S. R. Humphrey, and P. P. Humphrey. 1978. Survival rate of young southeastern brown bats, *Myotis austroriparius*, in Florida. Journal of Mammalogy 59:299–304.

438. Foster, R. W., and A. Kurta. 1999. Roosting ecology of the northern bat (*Myotis septentrionalis*) and comparisons with the endangered Indiana bat (*Myotis sodalis*). Journal of Mammalogy 80:659–672.

439. Frank, J. D., T. H. Kunz, J. Horn, C. Cleveland, and S. Petronio. 2003. Advanced infrared detection and image processing for automated bat censusing. Proceedings of SPIE 5074:261–271.

440. Franka, R., et al. 2006. A new phylogenetic lineage of *Rabies virus* associated with western pipistrelle bats (*Pipistrellus hesperus*). Journal of General Virology 87:2309–2321.

441. Fraze, R. K., and K. T. Wilkins. 1990. Patterns of use of man-made roosts by *Tadarida brasiliensis mexicana* in Texas. Southwestern Naturalist 35:261–267.

442. Freeman, J., and L. Wunder. 1988. Observations at a colony of the Brazilian free-tailed bat (*Tadarida brasiliensis*) in southern Colorado. Southwestern Naturalist 33:102–104.

443. Freeman, P. W. 1981a. Correspondence of food habits and morphology in insectivorous bats. Journal of Mammalogy 62:166–173.

444. Freeman, P. W. 1981b. A multivariate study of the family Molossidae (Mammalia, Chiroptera): morphology, ecology, evolution. Fieldiana Zoology 7:1–173.

445. French, B., and C. Bunyard. 2002. Fall and winter records of *Nycticeius humeralis* in Texas and Oklahoma. Bat Research News 43:204.

446. French, B., and A. Lollar. 1998. Observations on the reproductive behavior of captive *Tadarida brasiliensis mexicana* (Chiroptera: Molossidae). Southwestern Naturalist 43:484–90.

447. Frick, W. F., P. A. Heady III, and J. P. Hayes. 2009. Facultative nectar-feeding behavior in a gleaning insectivorous bat (*Antrozous pallidus*). Journal of Mammalogy 90:1157–1164.

448. Frick, W. F., J. F. Pollock, A. C. Hicks, K. E. Langwig, D. S. Reynolds, G. G. Turner, et al. 2010. An emerging disease causes regional population collapse of a common North American bat species. Science 329:679–682.

449. Frick, W. F., W. E. Rainey, and E. D.

Pierson. 2007. Potential effects of environmental contamination on Yuma myotis demography and population growth. Ecological Applications 17:1213–1222.

450. Fries, J. N. 1981. *Pipistrellus hesperus* (Chiroptera) eating spiders. Southwestern Naturalist 26:215.

451. Frost, D. R., and R. M. Timm. 1992. Phylogeny of Plecotine bats (Chiroptera: "Vespertilionidae"): summary of the evidence and proposal of a logically consistent taxonomy. American Museum Novitates 3034:1–16.

452. Fullard, J. H., and J. W. Dawson. 1997. The echolocation calls of the spotted bat *Euderma maculatum* are relatively inaudible to moths. Journal of Experimental Biology 200:129–137.

453. Fullard, J. H., J. A. Simmons, and P. A. Saillant. 1994. Jamming bat echolocation: the dogbane tiger moth *Cynia tenera* times its clicks to the terminal attack call of the big brown bat *Eptesicus fuscus*. Journal of Experimental Biology 194:285–298.

454. Fuzessery, Z. M., P. Buttenhoff, B. Andrews, and J. M. Kennedy. 1993. Passive sound localization of prey by the pallid bat (*Antrozous p. pallidus*). Journal of Comparative Physiology B: Biochemical, Systemic, and Environmental Physiology 171:767–777.

455. Gannon, W. L., R. E. Sherwin, T. N. de Carvalho, and M. J. O'Farrell. 2001. Pinnae and echolocation call differences between *Myotis californicus* and *M. ciliolabrum* (Chiroptera: Vespertilionidae). Acta Chiropterologica 3:77–91.

456. Gannon, W. L., R. E. Sherwin, and S. Haymond. 2003. On the importance of articulating assumptions when conducting acoustic studies of habitat use by bats. Wildlife Society Bulletin 31:45–61.

457. Gardner, A. L. 2007. Mammals of South America, Vol. 1: Marsupials, Xenarthrans, Shrews, and Bats. University of Chicago Press, Chicago.

458. Gardner, A. L., and C. O. Handley, Jr. 2007. Genus *Lasiurus*. Pp. 457–468 *in* Mammals of South America, Vol. 1: Marsupials, Xenarthrans, Shrews, and Bats (A. L. Gardner, ed.). University of Chicago Press, Illinois.

459. Gardner, J. E., and V. R. McDaniel. 1978. Distribution of bats in the Delta Region of northeastern Arkansas. Proceedings of the Arkansas Academy of Science 32:46–48.

460. Gargas, A., M. T. Trest, M. Christensen, T. J. Volk, and D. S. Blehert. 2009. *Geomyces destructans* sp. nov. associated with bat white-nose syndrome. Mycotaxon 108: 147–154.

461. Garner, H. W., and J. W. Bluntzer. 1975. Mammals of the Kansas-Texas boundary in Texas: distributional records of mammals along the boundary. Texas Journal of Science 26:611–613.

462. Gates, W. H. 1941. A few notes on the evening bat, *Nycticeius humeralis* (Rafinesque). Journal of Mammalogy 22: 53–56.

463. Geggie, J. F., and M. B. Fenton. 1985. A comparison of foraging by *Eptesicus fuscus* (Chiroptera: Vespertilionidae) in urban and rural environments. Canadian Journal of Zoology 63:263–267.

464. Gellman, S. T., and W. J. Zielinski. 1996. Use by bats of old-growth redwood hollows on the north coast of California. Journal of Mammalogy 77:255–265.

465. Geluso, K. 2000. Distribution of the spotted bat (*Euderma maculatum*) in Nevada, including notes on reproduction. Southwestern Naturalist 45:347–352.

466. Geluso, K. 2007. Winter activity of bats over water and along flyways in New Mexico. Southwestern Naturalist 52:482–492.

467. Geluso, K., J. P. Damm, and E. W. Valdez. 2008. Late-seasonal activity and diet of the evening bat (*Nycticeius humeralis*) in Nebraska. Western North American Naturalist 68:21–24.

468. Geluso, K., and J. N. Mink. 2009. Use of bridges by bats (Mammalia: Chiroptera) in the Rio Grande Valley, New Mexico. Southwestern Naturalist 54:421–29.

469. Geluso, K., T. R. Mollhagen, J. M. Tigner, and M. A. Bogan. 2005. Westward expansion of the eastern pipistrelle (*Pipistrellus subflavus*) in the United States, including new records from New Mexico, South Dakota, and Texas. Western North American Naturalist 65:405–409.

470. Geluso, K. N. 1978. Urine concentrating ability and renal structure of insectivorous bats. Journal of Mammalogy 59:312–323.

471. Geluso, K. N., J. S. Altenbach, and D. E. Wilson. 1976. Bat mortality: pesticide poisoning and migratory stress. Science 194: 184–186.

472. Geluso, K. N., and K. Geluso. 2004. Mammals of Carlsbad Caverns National Park, New Mexico. Bulletin of the University of Nebraska State Museum 17:1–180.

473. Genoud, M. 1993. Temperature regulation in subtropical tree bats. Comparative Biochemistry and Physiology A 104: 321–331.

474. Genoways, H. H., and R. J. Baker. 1988. *Lasiurus blossevillii* (Chiroptera: Vespertilionidae) in Texas. Texas Journal of Science 40:111–113.

475. Genoways, H. H., R. J. Baker, and J. E. Cornely. 1979. Mammals of the Guadalupe Mountains National Park, Texas. Pp. 271–332 *in* Biological Investigations in the Guadalupe Mountains National Park, Texas (H. H. Genoways and R. J. Baker, eds.). Proceedings and Transactions of the National Park Service, Vol. 4:1–442.

476. Genoways, H. H., and J. K. Jones Jr. 1968. Notes on bats from the Mexican state of Zacatecas. Journal of Mammalogy 49: 743–745.

477. Genter, D. L. 1986. Wintering bats of the Upper Snake River Plain: occurrence in lava-tube caves. Great Basin Naturalist 46: 241–244.

478. George, J. E., and R. W. Strandtmann. 1960. New records of ectoparasites on bats in west Texas. Southwestern Naturalist 5: 228–229.

479. Gitzen, R. A., S. D. West, and J. A. Baumgardt. 2001. A record of the spotted bat (*Euderma maculatum*) from Crescent Bar, Washington. Northwestern Naturalist 82: 28–30.

480. Glass, B. P. 1958. Returns of Mexican freetail bats banded in Oklahoma. Journal of Mammalogy 39:435–437.

481. Glass, B. P. 1959. Additional returns from free-tailed bats banded in Oklahoma. Journal of Mammalogy 40:542–545.

482. Glass, B. P. 1966. Some notes on reproduction in the red bat, *Lasiurus borealis*. Proceedings of the Oklahoma Academy of Science 46:40–41.

483. Glass, B. P. 1982. Seasonal movements of Mexican freetail bats *Tadarida brasiliensis mexicana* banded in the Great Plains. Southwestern Naturalist 27:127–133.

484. Glass, B. P., and R. J. Baker. 1968. The status of the name of *Myotis subulatus* Say. Proceedings of the Biological Society of Washington 81:257–260.

485. Glass, B. P., and R. C. Morse. 1959. A new

pipistrel from Oklahoma and Texas. Journal of Mammalogy 40:531–534.

486. Goetze, J. R. 1998. The mammals of the Edwards Plateau, Texas. Special Publications, Museum of Texas Tech University 41:1–263.

487. Goetze, J. R., R. W. Manning, F. D. Yancey II, and C. Jones. 1996. The mammals of Kimble County, Texas. Occasional Papers, Museum of Texas Tech University 160:1–31.

488. Goetze, J. R., and A. Nelson. 2000. Distributional records and comments on mammals from six Texas counties. Occasional Papers, Museum of Texas Tech University 197:1–8.

489. Goetze, J. R., A. D. Nelson, and P. D. Sudman. 2003. Noteworthy records of bats from central and south Texas. Texas Journal of Science 55:365–367.

490. Goetze, J. R., F. D. Yancey II, C. Jones, and B. M. Gharaibeh. 1995. Noteworthy records of mammals from the Edwards Plateau of central Texas. Texas Journal of Science 47:3–8.

491. Goldman, E. A. 1926. Review of C. A. R. Campbell, *Bats, mosquitoes, and dollars.* Journal of Mammalogy 7:136–138.

492. Gooding, G., and J. R. Langford. 2004. Characteristics of tree roosts of Rafinesque's big-eared bat and southeastern bat in northeastern Louisiana. Southwestern Naturalist 49:61–67.

493. Goodpaster, W. W., and D. F. Hoffmeister. 1952. Notes on the mammals of western Tennessee. Journal of Mammalogy 33: 362–371.

494. Gore, J. A. 1992. Big brown bat (*Eptesicus fuscus fuscus*). Pp. 343–348 *in* Rare and endangered biota of Florida, Vol. 1: Mammals (S. R. Humphrey, ed.). University Press of Florida, Gainesville.

495. Gore, J. A., and J. A. Hovis. 1992. The southeastern bat: another cave-roosting species in peril. Bats 10:10–12.

496. Gould, P. J. 1961. Emergence time of *Tadarida* in relation to light intensity. Journal of Mammalogy 42:405–407.

497. Greenbaum, I. F., and C. J. Phillips. 1974. Comparative anatomy and general histology of tongues of long-nosed bats (*Leptonycteris sanborni* and *L. nivalis*) with reference to infestation of oral mites. Journal of Mammalogy 55:489–504.

498. Greenhall, A. M. 1982. House bat management. U.S. Fish and Wildlife Service Resource Report 143:1–33.

499. Greenhall, A. M. 1988. Feeding behavior. P. 246 *in* Natural history of vampire bats (A. M. Greenhall and U. Schmidt, eds.). CRC Press, Boca Raton.

500. Griffin, D. R. 1940. The migration of New England bats. Bulletin of the Museum of Comparative Zoology 86:217–246.

501. Griffin, D. R. 2004. The past and future history of bat detectors. Pp. 6–9 *in* Bat echolocation research: tools, techniques, and analysis (R. M. Brigham, E. K. V. Kalko, G. Jones, S. Parsons, and H. J. G. A. Limpens, eds.). Bat Conservation International, Austin.

502. Griffin, L. A., and J. E. Gates. 1985. Food habits of cave-dwelling bats in the central Appalachians. Journal of Mammalogy 66: 451–460.

503. Griffith, G. E., et al. 2004. Ecoregions of Texas. U.S. Environmental Protection Agency, Corvallis, Oregon.

504. Griffith, L. A., and J. E. Gates. 1985. Food habits of cave-dwelling bats in the central Appalachians. Journal of Mammalogy 66: 451–460.

505. Grindal, S. D., T. S. Collard, R. M.

Brigham, and R. M. R. Barclay. 1992. The influence of precipitation on reproduction by myotis bats in British Columbia. American Midland Naturalist 128:339–344.

506. Gruver, J. C. 2003. Assessment of bat faunal composition and roosting habitat preferences for the hoary bat (*Lasiurus cinereus*) near a wind power facility in southeastern Wyoming. Bat Research News 44:104–105.

507. Gunnell, G. F., and N. B. Simmons. 2005. Fossil evidence and the origin of bats. Journal of Mammalian Evolution 12: 209–246.

508. Gustin, M. K., and G. F. McCracken. 1987. Scent recognition between females and pups in the bat *Tadarida brasiliensis mexicana*. Animal Behaviour 35:13–19.

509. Guthrie, M. J. 1933. The reproductive cycles of some cave bats. Journal of Mammalogy 14:199–216.

510. Hafner, M. S., W. L. Gannon, J. Salazar-Bravo, and S. T. Alvarez-Castañeda. 1997. Mammal collections in the western hemisphere: a survey and directory of existing collections. American Society of Mammalogists, Lawrence, Kansas.

511. Hall, E. R. 1981. Mammals of North America. 2 vols. John Wiley and Sons, New York.

512. Hall, E. R., and W. W. Dalquest. 1950a. *Pipistrellus cinnamomeus* Miller 1902 referred to the genus *Myotis*. University of Kansas Publications, Museum of Natural History 1: 584–590.

513. Hall, E. R., and W. W. Dalquest. 1950b. A synopsis of the American bats of the genus *Pipistrellus*. University of Kansas Publications, Museum of Natural History 1:591–602.

514. Hall, E. R., and J. K. Jones Jr. 1961. North American yellow bats, "*Dasypterus*," and a list of the named kinds of the genus *Lasiurus* Gray. University of Kansas Publications, Museum of Natural History 14:73–98.

515. Hamilton, I. M., and R. M. R. Barclay. 1972. Patterns of daily torpor and day-roost selection by male and female big brown bats (*Eptesicus fuscus*). Canadian Journal of Zoology 72:744–749.

516. Hamilton, I. M., and R. M. R. Barclay. 1998. Diets of juvenile, yearling, and adult big brown bats (*Eptesicus fuscus*) in southeastern Alberta. Journal of Mammalogy 79:764–771.

517. Hamilton, W. J., Jr. 1933. The insect food of the big brown bat. Journal of Mammalogy 14:155–156.

518. Handley, C. O., Jr. 1959. A revision of American bats of the genera *Euderma* and *Plecotus*. Proceedings of the U.S. National Museum 110:95–246.

519. Handley, C. O., Jr. 1960. Descriptions of new bats from Panama. Proceedings of the U.S. National Museum 112:459–479.

520. Hargrave, L. L. 1944. A record of *Lasiurus borealis teliotis* from Arizona. Journal of Mammalogy 25:414.

521. Harris, A. H. 1974. *Myotis yumanensis* in interior southwestern North America with comments on *Myotis lucifugus*. Journal of Mammalogy 55:589–607.

522. Hart, J. A., G. L. Kirkland Jr., and S. C. Grossman. 1993. Relative abundance and habitat use by tree bats, *Lasiurus* spp., in southcentral Pennsylvania. Canadian Field-Naturalist 107:208–212.

523. Hartley, D. J., and R. A. Suthers. 1987. The sound emission pattern and the acoustical role of the noseleaf in the echolocating bat, *Carollia perspicillata*. Journal of the Acoustical Society of America 82:1892–1900.

524. Harvey, M. J. 1986. Arkansas bats: a valuable resource. Arkansas Game and Fish Commission, Federal Aid Project E-1.

525. Hatfield, D. M. 1936. A revision of the *Pipistrellus hesperus* group of bats. Journal of Mammalogy 17:257–262.

526. Hayes, J. P. 2003. Habitat ecology and conservation of bats in western coniferous forests. Pp. 81–119 *in* Mammal community dynamics: management and conservation in the coniferous forests of western North America (C. J. Zabel and R. G. Anthony, eds.). Cambridge University Press, New York.

527. Haynie, M. L., et al. 2005. Mammal records from Donley and Briscoe counties, Texas. Occasional Papers, Museum of Texas Tech University 247:1–4.

528. Hayward, B., and R. Davis. 1964. Flight speeds in western bats. Journal of Mammalogy 45:236–242.

529. Hayward, B. J. 1963. A maternity colony of *Myotis occultus*. Journal of Mammalogy 44:270.

530. Hayward, B. J. 1970. The natural history of the cave bat *Myotis velifer*. Western New Mexico University Research in Science 1: 1–74.

531. Hayward, B. J., and S. P. Cross. 1979. The natural history of *Pipistrellus hesperus* (Chiroptera: Vespertilionidae). Office of Research, Western New Mexico University 3:1–36.

532. Heath, D. R., G. A. Heidt, D. A. Saugey, and V. R. McDaniel. 1983. Arkansas range extensions of the Seminole bat (*Lasiurus seminolus*) and eastern big-eared bat (*Plecotus rafinesquii*) and additional county records for the hoary bat (*Lasiurus cinereus*), silver-haired bat (*Lasionycteris noctivagans*) and evening bat (*Nycticeius humeralis*). Proceedings of the Arkansas Academy of Science 37:90–91.

533. Hein, C. D., S. B. Castleberry, and K. V. Miller. 2005. Winter roost selection by Seminole bats in the Lower Coastal Plain of South Carolina. Southeastern Naturalist 4: 473–478.

534. Hein, C. D., S. B. Castleberry, and K. V. Miller. 2008a. Male Seminole bat winter roost-site selection in a managed forest. Journal of Wildlife Management 72: 1756–1764.

535. Hein, C. D., S. B. Castleberry, and K. V. Miller. 2008b. Sex-specific summer roost-site selection by Seminole bats in response to landscape-level forest management. Journal of Mammalogy 89:964–72.

536. Hein, C. D., K. V. Miller, and S. B. Castleberry. 2009. Evening bat summer roost-site selection on a managed pine landscape. Journal of Wildlife Management 73:511–517.

537. Hendricks, P., J. Johnson, S. Lenard, and C. Currier. 2005. Use of a bridge for day roosting by the hoary bat, *Lasiurus cinereus*. Canadian Field-Naturalist 119:132–133.

538. Henry, M., D. W. Thomas, R. Vaudry, and M. Carrier. 2002. Foraging distances and home range of pregnant and lactating little brown bats (*Myotis lucifugus*). Journal of Mammalogy 83:767–774.

539. Henshaw, R. D. 1959. Responses of free-tailed bats to increases in cave temperature. Journal of Mammalogy 41:396–398.

540. Herd, R. M. 1987. Electrophoretic divergence of *Myotis leibii* and *Myotis ciliolabrum* (Chiroptera: Vespertilionidae). Canadian Journal of Zoology 65:1857–1860.

541. Hermann, J. A. 1950. The mammals of the Stockton Plateau of northeastern Terrell County, Texas. Texas Journal of Science 3: 368–393.

542. Hermanson, J. W., and K. T. Wilkins. 1986.

Pre-weaning mortality in a Florida maternity roost of *Myotis austroriparius* and *Tadarida brasiliensis*. Journal of Mammalogy 67: 751–754.

543. Herreid, C. F., II. 1958a. Sexual dimorphism in teeth of the free-tailed bat. Journal of Mammalogy 40:538–541.

544. Herreid, C. F., II. 1958b. Four-thumbed free-tail bat. Journal of Mammalogy 39:587.

545. Herreid, C. F., II. 1959a. Notes on a baby free-tailed bat. Journal of Mammalogy 40: 609–610.

546. Herreid, C. F., II. 1959b. Roadrunner a predator of bats. Condor 62:67.

547. Herreid, C. F., II. 1959c. Sexual dimorphism of the free-tailed bat. Journal of Mammalogy 40:538–541.

548. Herreid, C. F., II. 1960. Comments on the odor of bats. Journal of Mammalogy 41:396.

549. Herreid, C. F., II. 1961a. Snakes as predators of bats. Herpetologica 17:271–272.

550. Herreid, C. F., II. 1961b. Notes on the pallid bat in Texas. Southwestern Naturalist 6:13–20.

551. Herreid, C. F., II. 1963a. Temperature regulation of Mexican free-tailed bats in cave habitats. Journal of Mammalogy 44:560–573.

552. Herreid, C. F., II. 1963b. Temperature regulation and metabolism in Mexican freetail bats. Science 142:1573–1574.

553. Herreid, C. F., II. 1963c. Metabolism of the Mexican free-tailed bat. Journal of Cellular Composition and Physiology 61:201–207.

554. Herreid, C. F., II. 1967a. Temperature regulation, temperature preference and tolerance, and metabolism of young and adult free-tailed bats. Physiological Zoology 40:1–22.

555. Herreid, C. F., II. 1967b. Mortality statistics of young bats. Ecology 48:310–312.

556. Herreid, C. F., II, and R. B. Davis. 1960.

Frequency and placement of white fur on free-tailed bats. Journal of Mammalogy 41: 117–119.

557. Herreid, C. F., II, and R. B. Davis. 1966. Flight patterns of bats. Journal of Mammalogy 47:78–86.

558. Herreid, C. F., II, R. B. Davis, and H. L. Short. 1960. Injuries due to bat banding. Journal of Mammalogy 41:398–400.

559. Herrera M., L. G., T. H. Fleming, and J. S. Findley. 1993. Geographic variation in carbon composition of the pallid bat, *Antrozous pallidus*, and its dietary implications. Journal of Mammalogy 74: 601–606.

560. Hickey, M. B. C. 1992. Effect of radiotransmitters on the attack success of hoary bats, *Lasiurus cinereus*. Journal of Mammalogy 73:344–346.

561. Hickey, M. B. C., L. Acharya, and S. Pennington. 1996. Resource partitioning by two species of vespertilionid bats (*Lasiurus cinereus* and *Lasiurus borealis*) feeding around street lights. Journal of Mammalogy 77:325–334.

562. Hickey, M. B. C., and M. B. Fenton. 1990. Foraging by red bats (*Lasiurus borealis*): do intraspecific chases mean territoriality? Canadian Journal of Zoology 68:2477–2482.

563. Hickey, M. B. C., and M. B. Fenton. 1996. Behavioural and thermoregulatory responses of female hoary bats, *Lasiurus cinereus* (Chiroptera: Vespertilionidae), to variations in prey availability. Ecoscience 3:414–422.

564. Higginbotham, J. L., and L. K. Ammerman. 2002. Chiropteran community structure and seasonal dynamics in Big Bend National Park. Special Publications, Museum of Texas Tech University 44:1–44.

565. Higginbotham, J. L., L. K. Ammerman, and M. T. Dixon. 1999. First record of *Lasiurus*

xanthinus (Chiroptera: Vespertilionidae) in Texas. Southwestern Naturalist 44:343–347.

566. Higginbotham, J. L., R. S. DeBaca, J. G. Brant, and C. Jones. 2002. Noteworthy records of bats from the Trans-Pecos region of Texas. Texas Journal of Science 54:277–282.

567. Higginbotham, J. L., M. T. Dixon, and L. K. Ammerman. 2000. Yucca provides roost for *Lasiurus xanthinus* (Chiroptera: Vespertilionidae) in Texas. Southwestern Naturalist 45:338–340.

568. Higginbotham, J. L., and C. Jones. 2001. The southeastern myotis, *Myotis austroriparius* (Chiroptera; Vespertilionidae), from Comanche County, Texas. Texas Journal of Science 53:193–195.

569. Hill, J. E., and J. D. Smith. 1984. Bats, a natural history. British Museum of Natural History, London.

570. Hirshfeld, J. R., Z. C. Nelson, and W. G. Bradley. 1977. Night roosting behavior in four species of desert bats. Southwestern Naturalist 22:427–433.

571. Hitchcock, H. B. 1965. Twenty-three years of bat banding in Ontario and Quebec. Canadian Field Naturalist 79:4–14.

572. Hoetker, G. M., and K. W. Gobalet. 1999. Predation on Mexican free-tailed bats by burrowing owls in California. Journal of Raptor Research 33:333–335.

573. Hoffmeister, D. F. 1957. Review of the long-nosed bats of the genus *Leptonycteris*. Journal of Mammalogy 38:454–461.

574. Hoffmeister, D. F. 1970. The seasonal distribution of bats in Arizona: a case for improving mammalian range maps. Southwestern Naturalist 15:11–22.

575. Hoffmeister, D. F. 1986. Mammals of Arizona. University of Arizona Press, Tucson.

576. Hofmann, J. E., J. E. Gardner, J. K. Krejca, and J. D. Garner. 1999. Summer records and a maternity roost of the southeastern myotis (*Myotis austroriparius*) in Illinois. Transactions of the Illinois State Academy of Science 92:95–107.

577. Holderied, M. W., and G. Jones. 2009. Flight dynamics of bats. Pp. 459–475 *in* Ecological and behavioral methods for the study of bats (T. H. Kunz and S. Parsons, eds.). 2nd ed. Johns Hopkins University Press, Baltimore.

578. Hollander, R. R., and J. K. Jones Jr. 1987. A record of the western small-footed myotis, *Myotis ciliolabrum* Merriam, from the Texas Panhandle. Texas Journal of Science 39:198.

579. Hollander, R. R., J. K. Jones Jr., R. W. Manning, and C. Jones. 1987. Noteworthy records of mammals from the Texas Panhandle. Texas Journal of Science 39: 97–102.

580. Holtcamp, W. 2010. Attack of the killer fungus. Texas Parks and Wildlife 68:36–40.

581. Hoofer, S. R., S. A. Reeder, E. W. Hansen, and R. A. Van Den Bussche. 2003. Molecular phylogenetics and taxonomic review of noctilionid and vespertilionid bats. Journal of Mammalogy 84:809–821.

582. Hoofer, S. R., and R. A. Van Den Bussche. 2001. Phylogenetic relationships of Plecotine bats and allies based on mitochondrial ribosomal sequences. Journal of Mammalogy 82:131–137.

583. Hoofer, S. R., and R. A. Van Den Bussche. 2003. Molecular phylogeny of the chiropteran family Vespertilionidae. Acta Chiropterologica 5 (Supplement):1–63.

584. Hoofer, S. R., R. A. Van Den Bussche, and I. Horacek. 2006. Generic status of the American pipistrelles (Vespertilionidae) with description of a new genus. Journal of Mammalogy 87:981–992.

585. Horn, J. W., and T. H. Kunz. 2008. Analyzing NEXRAD Doppler radar images to assess nightly dispersal patterns and population trends in Brazilian free-tailed bats (*Tadarida brasiliensis*). Integrative and Comparative Biology 48:24–39.

586. Horner, P., and R. Maxey. 1998. East Texas rare bat survey: 1997. Annual report. Texas Parks and Wildlife Department, Resource Protection Division, Austin. 14 pp.

587. Howell, A. B. 1920. Contribution to the life-history of the California mastiff bat. Journal of Mammalogy 1:111–117.

588. Howell, E. K., et al. 2009. Mammal records from Briscoe, Dickens, Hall, and Motley Counties, Texas. Occasional Papers, Museum of Texas Tech University 288: 1–10.

589. Hoying, K. M., and T. H. Kunz. 1998. Variation in size at birth and post-natal growth in the insectivorous bat *Pipistrellus subflavus* (Chiroptera: Vespertilionidae). Journal of Zoology (London) 245:15–27.

590. Hoyt, R. A., and J. S. Altenbach. 1981. Observations on *Diphylla ecaudata* in captivity. Journal of Mammalogy 62:215–216.

591. Hoyt, R. A., J. S. Altenbach, and D. J. Hafner. 1994. Observations on long-nosed bats (*Leptonycteris*) in New Mexico. Southwestern Naturalist 39:175–179.

592. Hristov, N. I., M. Betke, and T. H. Kunz. 2008. Applications of thermal infrared imaging for research in aeroecology. Integrative and Comparative Biology doi: 10.1093/icb/icn053.

593. Hristov, N. I., and W. E. Conner. 2005. Sound strategy: acoustic aposematism in the bat-tiger moth arms race. Naturwissenschaften 92:164–169.

594. Humphrey, G. C., G. E. Kemp, and E. G. Wood. 1960. A fatal case of rabies in a woman bitten by an insectivorous bat. Public Health Reports 75:317–325.

595. Humphrey, S. R., and J. B. Cope. 1968. Records of migration of the evening bat, *Nycticeius humeralis.* Journal of Mammalogy 49:329.

596. Humphrey, S. R., and J. B. Cope. 1970. Population samples of the evening bat, *Nycticeius humeralis.* Journal of Mammalogy 51:399–401.

597. Humphrey, S. R., and J. A. Gore. 1992. Southeastern brown bat (*Myotis austroriparius*). Pp. 335–342 *in* Rare and endangered biota of Florida, Vol. 1: Mammals (S. R. Humphrey, ed.). University Press of Florida, Gainesville.

598. Humphrey, S. R., and T. H. Kunz. 1976. Ecology of a Pleistocene relict, the western big-eared bat (*Plecotus townsendii*), in the southern Great Plains. Journal of Mammalogy 57:470–494.

599. Hurst, T. E., and M. J. Lacki. 1997. Food habits of Rafinesque's big-eared bat in southeastern Kentucky. Journal of Mammalogy 78:525–528.

600. Hurst, T. E., and M. J. Lacki. 1999. Roost selection, population size and habitat use by a colony of Rafinesque's big-eared bats (*Corynorhinus rafinesquii*). American Midland Naturalist 142:363–371.

601. Hutcheon, J. M., and J. A. W. Kirsch. 2006. A moveable face: deconstructing the Microchiroptera and a new classification for extant bats. Acta Chiropterologica 8:1–10.

602. Hutchinson, J. T. 2001. Observations on use of coastal scrub habitat by evening bats. Bat Research News 42:44–46.

603. Hutchinson, J. T. 2006. Bats of Archbold Biological Station and notes on some roost sites. Florida Field Naturalist 34:48–51.

604. Hutchinson, J. T., and M. J. Lacki. 1999.

Foraging behavior and habitat use of red bats in mixed mesophytic forests of the Cumberland Plateau, Kentucky. Pp. 171–177 *in* Proceedings, 12th Central Hardwood Forest Conference (J. W. Stringer and D. L. Loftis, eds.). U.S. Forest Service, Southern Experimental Station, Asheville, North Carolina.

605. Hutchinson, J. T., and M. J. Lacki. 2000a. Roosting behavior and foraging activity of a female red bat with nonvolant young. Bat Research News 41:36–38.

606. Hutchinson, J. T., and M. J. Lacki. 2000b. Selection of day roosts by red bats in mixed mesophyte forests. Journal of Wildlife Management 64:87–94.

607. Hutchinson, J. T., and M. J. Lacki. 2001. Possible microclimate benefits of roost site selection in the red bat, *Lasiurus borealis,* in mixed mesophytic forests of Kentucky. Canadian Field-Naturalist 115:205–209.

608. Hutchinson, J. T., and M. Meisenburg. 2004. Two winter roost sites of lasiurines in north-central Florida. Bat Research News 45:90–91.

609. IPCC. 2007. Climate change 2007: synthesis report. United Nations Intergovernmental Panel on Climate Change. 73 pp.

610. Irons, J. V., R. B. Eads, J. E. Grimes, and A. Conklin. 1957. The public health importance of bats. Texas Reports on Biology and Medicine 15:292–298.

611. Irons, J. V., R. B. Eads, T. Sullivan, and J. E. Grimes. 1954. The current status of rabies in Texas. Texas Reports on Biology and Medicine 12:489–499.

612. Izor, R. J. 1979. Winter range of the silver-haired bat. Journal of Mammalogy 60: 641–643.

613. Jackson, H. H. T. 1961. Mammals of Wisconsin. University of Wisconsin Press, Madison.

614. Jackson, J. A., B. J. Schardien, C. D. Cooley, and B. E. Rowe. 1982. Cave myotis roosting in barn swallow nests. Southwestern Naturalist 47:463–464.

615. Jagnow, D. H. 1998. Bat usage and cave management of Torgac Cave, New Mexico. Journal of Cave and Karst Studies 66:33–38.

616. James, P., and A. Hayse. 1962. Sparrow hawk preys on Mexican free-tailed bat at Falcon Reservoir. Journal of Mammalogy 44:574.

617. Jameson, D. K. 1959. A survey of parasites of five species of bats. Southwestern Naturalist 4:61–65.

618. Jennings, W. L. 1958. The ecological distribution of bats in Florida. Ph.D. dissertation, University of Florida, Gainesville.

619. Jiménez Guzmán, M. A. A. 1982. Rabia en murciélagos de la Cueva del Guano, Santa Catarina, Nuevo Leon, México. Texas Memorial Museum Bulletin 28:245–248.

620. Johnson, G. D., W. P. Erickson, M. D. Strickland, M. F. Shepherd, and D. A. Shepherd. 2003. Mortality of bats at a large-scale wind power development at Buffalo Ridge, Minnesota. American Midland Naturalist 150:332–342.

621. Johnson, J. B., J. W. Edwards, W. M. Ford, and J. E. Gates. 2009. Roost tree selection by northern myotis (*Myotis septentrionalis*) maternity colonies following prescribed fire in a Central Appalachian Mountains hardwood forest. Forest Ecology and Management 258:233–242.

622. Johnson, J. S., M. J. Lacki, and M. D. Baker. 2007. Foraging ecology of long-legged myotis (*Myotis volans*) in north-central Idaho. Journal of Mammalogy 88:1261–1270.

623. Johnson, S. A., A. P. Bradstreet, and C. W. Edwards. 2005. Noteworthy records of mammals housed in the Stephen F. Austin State University Vertebrate Natural History Collection. Texas Journal of Science 57: 289–294.

624. Johnston, D. S., and M. B. Fenton. 2001. Individual and population-level variability in diets of pallid bats (*Antrozous pallidus*). Journal of Mammalogy 82:362–373.

625. Jones, C. 1965. Ecological distribution and activity periods of bats of the Mogollon Mountains area of New Mexico and adjacent Arizona. Tulane Studies in Zoology 12: 93–100.

626. Jones, C. 1966. Changes in populations of some western bats. American Midland Naturalist 76:522–528.

627. Jones, C., and R. D. Bradley. 1999. Notes on red bats, *Lasiurus* (Chiroptera: Vespertilionidae), of the Davis Mountains, Texas. Texas Journal of Science 51:267–269.

628. Jones, C., L. Hedges, and K. Bryan. 1999. The western yellow bat, *Lasiurus xanthinus* (Chiroptera: Vespertilionidae), from the Davis Mountains, Texas. Texas Journal of Science 51:267–269.

629. Jones, C., and M. W. Lockwood. 2008. Additions to the mammalian fauna of Big Bend Ranch State Park, Texas. Occasional Papers, Museum of Texas Tech University 282:1–3.

630. Jones, C., and J. Pagels. 1968. Notes on a population of *Pipistrellus subflavus* in southern Louisiana. Journal of Mammalogy 49:134–139.

631. Jones, C., and R. D. Suttkus. 1973. Colony structure and organization of *Pipistrellus subflavus* in southern Louisiana. Journal of Mammalogy 54:962–968.

632. Jones, C., and R. D. Suttkus. 1975. Notes on the natural history of *Plecotus rafinesquii*. Occasional Papers of the Museum of Zoology, Louisiana State University 47:1–14.

633. Jones, C., R. D. Suttkus, and M. A. Bogan. 1987a. Notes on some mammals of North-Central Texas. Occasional Papers, Museum of Texas Tech University 115:1–21.

634. Jones, G., and J. Rydell. 2003. Attack and defense: interactions between echolocating bats and their insect prey. Pp. 301–345 *in* Bat ecology (T. H. Kunz and M. B. Fenton, eds.). University of Chicago Press, Chicago.

635. Jones, J. K., Jr., E. D. Fleharty, and P. B. Dunnigan. 1967. The distributional status of bats in Kansas. University of Kansas Publications, Museum of Natural History 46:1–33.

636. Jones, J. K., Jr., and H. H. Genoways. 1969. Holotypes of Recent mammals in the Museum of Natural History, The University of Kansas. University of Kansas Publications, Museum of Natural History 51:129–146.

637. Jones, J. K., Jr., R. R. Hollander, and R. W. Manning. 1987b. The fringed myotis, *Myotis thysanodes*, in west-central Texas. Southwestern Naturalist 32:149.

638. Jones, J. K., Jr., and M. R. Lee. 1962. Three species of mammals from western Texas. Southwestern Naturalist 7:77–78.

639. Jones, J. K., Jr., and R. W. Manning. 1990. Additional comments on big brown bats (*Eptesicus fuscus*) from northwestern Texas. Southeastern Naturalist 35:342–343.

640. Jones, J. K., Jr., R. W. Manning, R. R. Hollander, and C. Jones. 1987c. Annotated checklist of Recent mammals of Northwestern Texas. Occasional Papers, Museum of Texas Tech University 111:1–14.

641. Jones, J. K., Jr., R. W. Manning, C. Jones, and R. R. Hollander. 1988. Mammals of the northern Texas Panhandle. Occasional

Papers, Museum of Texas Tech University
126:1–54.

642. Jones, J. K., Jr., J. D. Smith, and H. H.
Genoways. 1973. Annotated checklist of
mammals of the Yucatan Peninsula, Mexico.
I. Chiroptera. Occasional Papers, Museum of
Texas Tech University 13:1–31.

643. Jones, K. E., O. R. P. Bininda-Emonds,
and J. L. Gittleman. 2005. Bats, clocks, and
rocks: diversification patterns in Chiroptera.
Evolution 59:2243–2255.

644. Jones, K. E., and E. C. Teeling. 2009.
Phylogenetic tools for examining character
and clade evolution in bats. Pp. 715–738 in
Ecological and behavioral methods for the
study of bats (T. H. Kunz and S. Parsons,
eds.). Johns Hopkins University Press,
Baltimore.

645. Jones, L. L. C., and W. B. Gillespie. 2009.
Sceloporus magister (desert spiny lizard) prey.
Herpetological Review 40:349–350.

646. Jones, R. S., and W. F. Hettler. 1959. Bat
feeding by green sunfish. Texas Journal of
Science 11:48.

647. Judd, F. W. 1967. Notes on some mammals
from Big Bend National Park. Southwestern
Naturalist 12:192–94.

648. Judd, F. W., and D. J. Schmidly. 1969.
Distributional notes for some mammals
from western Texas and eastern New
Mexico. Texas Journal of Science 20:381–83.

649. Julian, S., and J. D. Altringham. 1994. Bat
predation by a tawny owl. Naturalist 119:
49–56.

650. Jung, T. S., I. D. Thompson, R. D. Titman,
and A. P. Applejohn. 1999. Habitat selection
by forest bats in relation to mixed-wood
stand types and structure in central Ontario.
Journal of Wildlife Management 63:
1306–1319.

651. Kalcounis-Rueppell, M. C., V. H. Payne,

S. R. Huff, and A. L. Boyko. 2007. Effects of
wastewater treatment plant effluent on bat
foraging ecology in an urban stream system.
Biological Conservation 138:120–130.

652. Kalcounis, M. C., and R. M. Brigham.
1998. Secondary use of aspen cavities by
tree-roosting big brown bats. Journal of
Wildlife Management 62:603–611.

653. Kalko, E. K. V. 1995. Insect pursuit, prey
capture, and echolocation in pipistrelle bats
(Microchiroptera). Animal Behaviour 50:
861–880.

654. Kalko, E. K. V., D. Friemel, C. O. Handley
Jr., and H.-U. Schnitzler. 1999. Roosting
and foraging behavior of two neotropical
gleaning bats, Tonatia silvicola and Trachops
cirrhosus (Phyllostomidae). Biotropica 31:
344–353.

655. Kalko, E. K. V., and H.-U. Schnitzler. 1989.
The echolocation and hunting behavior
of Daubenton's bat, Myotis daubentoni.
Behavioral Ecology and Sociobiology 24:
225–238.

656. Kannan, K., S. H. Yun, R. J. Rudd, and
M. Behr. 2010. High concentrations of
persistent organic pollutants including
PCBs, DDT, PBDEs, and PFOs in little brown
bats with white-nose syndrome in New York,
USA. Chemosphere 80:613–618.

657. Kastberger, G., and R. Stachl. 2003.
Infrared imaging technology and biological
applications. Behavior Research Methods,
Instruments, & Computers 35:429–439.

658. Keeley, A. T. H., and B. W. Keeley. 2004.
The mating system of Tadarida brasiliensis
(Chiroptera: Molossidae) in a large highway
bridge colony. Journal of Mammalogy 85:
113–119.

659. Keeley, B. W., M. B. Fenton, and E. Arnett.
2003. A North American partnership
for advancing research, education, and

management for the conservation of bats and their habitats. Wildlife Society Bulletin 31:80–86.

660. Keeley, B. W., and M. D. Tuttle. 1999. Bats in American bridges. Bat Conservation International, Austin.

661. Kennedy, M. L., P. K. Kennedy, and G. D. Baumgardner. 1984. First record of the Seminole bats (*Lasiurus seminolus*) in Tennessee. Journal of the Tennessee Academy of Science 59:89–90.

662. Kern, W. H., Jr. 1992. Northern yellow bat. Pp. 349–356 *in* Rare and endangered biota of Florida, Vol. I: Mammals (S. R. Humphrey, ed.). University of Florida Press, Gainesville.

663. Kirkwood, J. J., and A. Cartwright. 1991. Behavioral observations in thermal imaging of the big brown bat, *Eptesicus fuscus*. Pp. 369–371 *in* Thermosense XIII (G. Baird, ed.). SPIE 1467, Bellingham, Washington.

664. Kirkwood, J. J., and A. Cartwright. 1993. Comparison of two systems for viewing bat behavior in the dark. Proceedings of the Indiana Academy of Science 102:133–137.

665. Knowles, B. 1992. Bat hibernacula on Lake Superior's north shore, Minnesota. Canadian Field-Naturalist 106:252–254.

666. Koehler, C. E., and R. M. R. Barclay. 2000. Post-natal growth and breeding biology of the hoary bat (*Lasiurus cinereus*). Journal of Mammalogy 81:234–44.

667. Koford, C. B., and M. R. Koford. 1948. Breeding colonies of bats, *Pipistrellus hesperus* and *Myotis subulatus melanorhinus*. Journal of Mammalogy 29:417–418.

668. Kohls, G. M., and W. L. Jellison. 1948. Ectoparasites and other arthropods occurring in Texas bat caves. National Speleological Society Bulletin 10:116–117.

669. Koontz, T., and W. H. Davis. 1991. Winter roosting of the red bat, *Lasiurus borealis*. Bat Research News 32:3–4.

670. Krebs, J. W., E. J. Mandel, D. L. Swerdlow, and C. E. Rupprecht. 2004. Rabies surveillance in the United States during 2003. Journal of the American Veterinary Medical Association 225:1837–1849.

671. Krebs, J. W., E. J. Mandel, D. L. Swerdlow, and C. E. Rupprecht. 2005. Rabies surveillance in the United States during 2004. Journal of the American Veterinary Medical Association 227:1912–1925.

672. Krebs, J. W., A. M. Mondul, C. E. Rupprecht, and J. E. Childs. 2001. Rabies surveillance in the United States during 2000. Journal of the American Veterinary Medical Association 219:1687–1699.

673. Krebs, J. W., H. R. Noll, C. E. Rupprecht, and J. E. Childs. 2002. Rabies surveillance in the United States during 2001. Journal of the American Veterinary Medical Association 221:1690–1701.

674. Krebs, J. W., C. E. Rupprecht, and J. E. Childs. 2000. Rabies surveillance in the United States during 1999. Journal of the American Veterinary Medical Association 217:1799–1811.

675. Krebs, J. W., J. T. Wheeling, and J. E. Childs. 2003. Rabies surveillance in the United States during 2002. Journal of the American Veterinary Medical Association 223:1736–1748.

676. Krishon, D. M., M. A. Menzel, T. C. Carter, and J. Laerm. 1997. Notes on the home range of four species of vespertilionid bats (*Chiroptera*) on Sapelo Island, Georgia. Georgia Journal of Science 55:215–223.

677. Krutzsch, P. H. 1944. Notes on the little known pocketed bat. Journal of Mammalogy 25:196.

678. Krutzsch, P. H. 1945. Observations on a

colony of molossids. Journal of Mammalogy 26:196.

679. Krutzsch, P. H. 1954. Notes on the habits of the bat, *Myotis californicus*. Journal of Mammalogy 35:539–545.

680. Krutzsch, P. H. 1955. Observations on the California mastiff bat. Journal of Mammalogy 36:407–414.

681. Krutzsch, P. H. 1975. Reproduction of the canyon bat, *Pipistrellus hesperus*, in the southwestern United States. American Journal of Anatomy 143:163–200.

682. Krutzsch, P. H. 2000. Anatomy, physiology, and cyclicity of the male reproductive tract. Pp. 92–155 *in* Reproductive biology of bats (E. G. Crichton and P. H. Krutzsch, eds.). Academic Press, London.

683. Krutzsch, P. H., T. H. Fleming, and E. G. Crichton. 2002. Reproductive biology of male Mexican free-tailed bats (*Tadarida brasiliensis mexicana*). Journal of Mammalogy 83:489–500.

684. Krutzsch, P. H., and A. H. Hughes. 1959. Hematological changes with torpor in the bat. Journal of Mammalogy 40:547–554.

685. Krutzsch, P. H., and S. E. Sulkin. 1958. The laboratory care of the Mexican free-tailed bat. Journal of Mammalogy 39:62–65.

686. Kuenzi, A. J., G. T. Downard, and M. I. Morrison. 1999. Bat distribution and hibernacula use in west central Nevada. Great Basin Naturalist 59:213–220.

687. Kunz, T. H. 1971. Reproduction of some vespertilionid bats in central Iowa. American Midland Naturalist 86:477–486.

688. Kunz, T. H. 1973a. Resource utilization: temporal and spatial components of bat activity in central Iowa. Journal of Mammalogy 54:14–32.

689. Kunz, T. H. 1973b. Population studies of the cave bat (*Myotis velifer*): reproduction, growth, and development. University of Kansas Publications, Museum of Natural History 15:1–43.

690. Kunz, T. H. 1974a. Feeding ecology of a temperate insectivorous bat (*Myotis velifer*). Ecology 55:693–711.

691. Kunz, T. H. 1974b. Reproduction, growth, and mortality of the vespertilionid bat, *Eptesicus fuscus*, in Kansas. Journal of Mammalogy 55:1–13.

692. Kunz, T. H. 1988. Ecological and behavioral methods for the study of bats. Smithsonian Institution Press, Washington, D.C.

693. Kunz, T. H. 2003. Censusing bats: challenges, solutions, and sampling biases. Pp. 9–19 *in* Monitoring trends in bat populations of the United States and territories: problems and prospects (T. J. O'Shea and M. A. Bogan, eds.). U.S. Geological Survey, Biological Resources Discipline, Information and Technology Report, USGS/BRD/ITR-2003-0003.

694. Kunz, T. H. 2004. Foraging habits of North American insectivorous bats. Pp. 13–25 *in* Bat echolocation research: tools, techniques, and analysis (R. M. Brigham, E. K. V. Kalko, G. Jones, S. Parsons, and H. J. G. A. Limpens, eds.). Bat Conservation International, Austin.

695. Kunz, T. H., and E. L. P. Anthony. 1977. On the efficiency of the Tuttle bat trap. Journal of Mammalogy 58:309–315.

696. Kunz, T. H., et al. 2007. Ecological impacts of wind energy development on bats: questions, research needs, and hypotheses. Frontiers in Ecology and the Environment 5: 315–324.

697. Kunz, T. H., and E. R. Arnett. 2009. Bats: going . . . going . . . gone with the wind? International Congress of Speleology Proceedings 15:760.

698. Kunz, T. H., M. Betke, N. I. Histrov, and M. J. Vonhof. 2009a. Methods for assessing colony size, population size, and relative abundance of bats. Pp. 133–157 *in* Ecological and behavioral methods for the study of bats (T. H. Kunz and S. Parsons, eds.). Johns Hopkins University Press, Baltimore.

699. Kunz, T. H., D. S. Blehert, P. M. Cryan, J. T. H. Coleman, A. Hicks, and M. D. Tuttle. 2009b. White-nose syndrome in hibernating bats: are these affected bats the next "canary in the mine"? International Congress of Speleology Proceedings 15:1291.

700. Kunz, T. H., and M. B. Fenton. 2003. Bat Ecology. University of Chicago Press, Chicago.

701. Kunz, T. H., et al. 2008. Aeroecology: probing and modeling the aerosphere. Integrative and Comparative Biology doi: 10.1093/icb/icn037.

702. Kunz, T. H., and L. F. Lumsden. 2003. Ecology of cavity and foliage roosting bats. Pp. 3–89 *in* Bat Ecology (T. H. Kunz and M. B. Fenton, eds.). University of Chicago Press, Chicago.

703. Kunz, T. H., and S. Parsons, eds. 2009. Ecological and behavioral methods for the study of bats, 2nd ed. Johns Hopkins University Press, Baltimore.

704. Kunz, T. H., J. O. Whitaker Jr., and M. D. Wadanoli. 1995. Dietary energetics of the insectivorous Mexican free-tailed bat (*Tadarida brasiliensis*) during pregnancy and lactation. Oecologia 101:407–415.

705. Kurta, A. 1994. Bark roost of a male big brown bat, *Eptesicus fuscus*. Bat Research News 35:63.

706. Kurta, A., and T. H. Kunz. 1987. Size of bats at birth and maternal investment during pregnancy. Symposium of the Zoological Society of London 57:79–106.

707. Kurta, A., T. H. Kunz, and K. A. Nagy. 1990. Energetics and water flux of free-ranging big brown bats (*Eptesicus fuscus*) during pregnancy and lactation. Journal of Mammalogy 71:59–65.

708. Kurta, A., and S. W. Murray. 2002. Philopatry and migration of banded Indiana bats (*Myotis sodalis*) and effects of radio transmitters. Journal of Mammalogy 83: 585–589.

709. Kurta, A., and J. A. Teramino. 1994. A novel hibernaculum and noteworthy records of the Indiana bat and eastern pipistrelle (Chiroptera: Vespertilionidae). American Midland Naturalist 132:410–413.

710. Kurta, A., L. Winhold, J. O. Whitaker Jr., and R. Foster. 2007. Range expansion and changing abundance of the eastern pipistrelle (Chiroptera: Vespertilionidae) in the central Great Lakes region. American Midland Naturalist 157:404–411.

711. Kurtén, B., and E. Anderson. 1980. Pleistocene mammals of North America. Columbia University Press, New York.

712. Lacki, M. J. 2000. Effect of trail users at a maternity roost of Rafinesque's big-eared bats. Journal of Cave and Karst Studies 62: 163–168.

713. Lacki, M. J., M. D. Adam, and L. G. Shoemaker. 1994. Observations on seasonal cycle, population patterns and roost selection in summer colonies of *Plecotus townsendii virginianus* in Kentucky. American Midland Naturalist 131:34–42.

714. Lacki, M. J., S. K. Amelon, and M. D. Baker. 1996. Foraging ecology of bats in forests. Pp. 83–127 *in* Bats and forests symposium (R. M. R. Barclay and R. M. Brigham, eds.). British Columbia Ministry of Forestry, Victoria, British Columbia.

715. Lacki, M. J., and M. D. Baker. 2007. Day

roosts of female fringed myotis (*Myotis thysanodes*) in xeric forests of the Pacific Northwest. Journal of Mammalogy 88: 967–973.

716. Lacki, M. J., D. R. Cox, and M. B. Dickinson. 2009a. Meta-analysis of summer roosting characteristics of two species of Myotis bats. American Midland Naturalist 162:318–326.

717. Lacki, M. J., D. R. Cox, L. E. Dodd, and M. B. Dickinson. 2009b. Response of northern bats (*Myotis septentrionalis*) to prescribed fires in eastern Kentucky forests. Journal of Mammalogy 90:1165–1175.

718. Lacki, M. J., J. S. Johnson, L. E. Dodd, and M. D. Baker. 2007. Prey consumption of insectivorous bats in coniferous forests of north-central Idaho. Northwest Science 81: 199–205.

719. Lacki, M. J., and J. H. Schwierjohann. 2001. Day-roost characteristics of northern bats in mixed mesophytic forest. Journal of Wildlife Management 65:482–488.

720. LaDuc, T. J. 1993. Accidental death by web entanglement in the western pipistrelle, *Pipistrellus hesperus*. Bat Research News 34:58.

721. Lance, R. F., B. T. Hardcastle, A. Talley, and P. L. Leberg. 2001. Day-roost selection by Rafinesque's big-eared bats (*Corynorhinus rafinesquii*) in Louisiana forests. Journal of Mammalogy 82:166–172.

722. Land, T. A. 2001. Population size and contaminant exposure of bats using caves on Fort Hood Military Base. M. S. Thesis, Texas A&M University, College Station.

723. Lausen, C. L., and R. M. R. Barclay. 2002. Roosting behaviour and roost selection of female big brown bats (*Eptesicus fuscus*) roosting in rock crevices in southeastern Alberta. Canadian Journal of Zoology 80: 1069–1076.

724. Lausen, C. L., and R. M. R. Barclay. 2003. Thermoregulation and roost selection by reproductive female big brown bats (*Eptesicus fuscus*) roosting in rock crevices. Journal of Zoology, London 260:235–244.

725. Lausen, C. L., and R. M. R. Barclay. 2006. Benefits of living in a building: Big brown bats (*Eptesicus fuscus*) in rocks versus buildings. Journal of Mammalogy 87: 362–370.

726. LaVal, R. K. 1970. Infraspecific relationships of bats of the species *Myotis austroriparius*. Journal of Mammalogy 51: 542–552.

727. LaVal, R. K. 1973. Occurrence, ecological distribution, and relative abundance of bats in McKittrick Canyon, Culberson County, Texas. Southwestern Naturalist 17:357–364.

728. LaVal, R. K., and M. L. LaVal. 1979. Notes on reproduction, behavior, and abundance of the red bat, *Lasiurus borealis*. Journal of Mammalogy 60:209–212.

729. LaVal, R. K., and M. L. LaVal. 1980. Ecological studies and management of Missouri bats, with emphasis on cave-dwelling species. Missouri Department of Conservation, Terrestrial Series 8:1–53.

730. LaVal, R. K., and W. A. Shifflet. 1971. *Choeronycteris mexicana* from Texas. Bat Research News 12:40.

731. Lawrence, B. D., and J. A. Simmons. 1982. Echolocation in bats: the external ear and perception of the vertical positions of targets. Science 218:481–483.

732. Layne, J. N. 1955. Seminole bat, *Lasiurus seminolus*, in central New York. Journal of Mammalogy 36:453.

733. LeConte, J. L. 1855. Observations on the American species of bats. Proceedings of the Academy of Natural Science of Philadelphia 7:437.

734. Lee, T. E., Jr. 1987. Distributional record of *Lasiurus seminolus* (Chiroptera: Vespertilionidae). Texas Journal of Science 39:193.

735. Lee, Y. F., and Y. M. Kuo. 2001. Predation on Mexican free-tailed bats by peregrine falcons and red-tailed hawks. Journal of Raptor Research 35:115–123.

736. Lee, Y. F., and G. F. McCracken. 2001. Timing and variation in the emergence and return of Mexican free-tailed bats, *Tadarida brasiliensis mexicana*. Zoological Studies 40: 309–316.

737. Lee, Y. F., and G. F. McCracken. 2005. Dietary variation of Brazilian free-tailed bats links to migratory populations of pest insects. Journal of Mammalogy 86:67–76.

738. Leggett, K. 2010. Spread of white-nose syndrome forces closure of caves. Refuge Update 7:7.

739. Lenhart, P. A., V. Mata-Silva, and J. D. Johnson. 2010. Foods of the pallid bat, *Antrozous pallidus* (Chiroptera: Vespertilionidae), in the Chihuahuan Desert of western Texas. Southwestern Naturalist 55:110–115.

740. Leonard, M. L., and M. B. Fenton. 1983. Habitat use by spotted bats (*Euderma maculatum*, Chiroptera: Vespertilionidae): roosting and foraging behaviour. Canadian Journal of Zoology 61:1487–1491.

741. Leonard, M. L., and M. B. Fenton. 1984. Echolocation calls of *Euderma maculatum* (Vespertilionidae): use in orientation and communication. Journal of Mammalogy 65: 122–126.

742. Leput, D. W. 2004. Eastern red bat (*Lasiurus borealis*) and eastern pipistrelle (*Pipistrellus subflavus*) maternal roost selection: implications for forest management. M.S. Thesis, Clemson University, South Carolina.

743. Lewis-Oritt, N., C. A. Porter, and R. J. Baker. 2001. Molecular systematics of the family Mormoopidae (Chiroptera) based on cytochrome b and recombination activating gene 2 sequences. Molecular Phylogenetics and Evolution 20:426–436.

744. Lewis, S. E. 1993. Effect of climatic variation on reproduction by pallid bats (*Antrozous pallidus*). Canadian Journal of Zoology 71:1429–1433.

745. Lewis, S. E. 1994. Night roosting ecology of pallid bats (*Antrozous pallidus*) in Oregon. American Midland Naturalist 132:219–226.

746. Lewis, S. E. 1995. Roost fidelity of bats: a review. Journal of Mammalogy 76:481–496.

747. Lewis, S. E. 1996. Low roost-site fidelity in pallid bats: associated factors and effect on group stability. Behavioral Ecology and Sociobiology 39:335–344.

748. Limpens, H. J. G. A., and G. F. McCracken. 2004. Choosing a bat detector: theoretical and practical aspects. Pp. 28–37 *in* Bat echolocation research: tools techniques and analysis (R. M. Brigham, E. K. V. Kalko, G. Jones, S. Parsons, and H. J. G. A. Limpens, eds.). Bat Conservation International, Austin.

749. Limpert, D. L., D. L. Birch, M. S. Scott, M. Andre, and E. Gillam. 2007. Tree selection and landscape analysis of eastern red bat day roosts. Journal of Wildlife Management 71:478–486.

750. Lindner, D. L., A. Gargas, J. M. Lorch, M. T. Banik, J. Glaeser, T. H. Kunz, et al. 2010. DNA-based detection of the fungal pathogen *Geomyces destructans* in soils from bat hibernacula. Mycologia doi:10.3852/ 10–262.

751. Little, L. 1920. Some notes concerning the mastiff bat. Journal of Mammalogy 1:182.

752. Loeb, S. C., and T. A. Waldrop. 2008. Bat activity in relation to fire and fire surrogate treatments in southern pine stands. Forest Ecology and Management 255:3185–3192.

753. Logan, L. E., and C. C. Black. 1979. The Quaternary vertebrate fauna of Upper Sloth Cave, Guadalupe Mountains National Park, Texas. Pp. 141–158 in Biological investigations in the Guadalupe Mountains National Park, Texas (H. H. Genoways and R. J. Baker, eds.). Transactions of the National Park Service, Vol. 4.

754. López-González, C., and T. L. Best. 2006. Current status of wintering sites of Mexican free-tailed bats Tadarida brasiliensis mexicana (Chiroptera: Molossidae) from Carlsbad Cavern, New Mexico. Vertebrata Mexicana 18:13–22.

755. López-González, C., and L. Torres-Morales. 2004. Use of abandoned mines by long-eared bats, genus Corynorhinus (Chiroptera: Vespertilionidae) in Durango, Mexico. Journal of Mammalogy 85:989–994.

756. López-Wilchis, R., G. López-Ortega, and R. D. Owen. 1994. Noteworthy record of the western small-footed myotis (Mammalia: Chiroptera: Myotis ciliolabrum). Southwestern Naturalist 39:211–212.

757. Loughry, W. J., and G. F. McCracken. 1991. Factors influencing female-pup scent recognition in Mexican free-tailed bats. Journal of Mammalogy 72:624–626.

758. Loukes, L. M. S., and W. Caire. 2007. Sex ratio variation of Myotis velifer (Chiroptera: Vespertilionidae) in Oklahoma. Southwestern Naturalist 52:67–74.

759. Love, J. P. 2009. Notes on mortality of eastern red bats (Lasiurus borealis), including a copulating pair, in Great Smoky Mountains National Park, Tennessee. Bat Research News 50:19–21.

760. Lowery, G. H., Jr. 1974. The mammals of Louisiana and its adjacent waters. Louisiana State University Press, Baton Rouge.

761. Lundelius, E. L., Jr. 1967. Late-Pleistocene and Holocene faunal history of central Texas. Pp. 287–319 in Pleistocene extinctions, the search for a cause (P. S. Martin and H. E. Wright, eds.). Yale University Press, New Haven.

762. Mackiewicz, J., and R. H. Backus. 1956. Oceanic records of Lasionycteris noctivagans and Lasiurus borealis. Journal of Mammalogy 37:442–443.

763. Mager, K. J., and T. A. Nelson. 2001. Roost-site selection by eastern red bats (Lasiurus borealis). American Midland Naturalist 145:120–126.

764. Mann, S. L., R. J. Steidl, and V. M. Dalton. 2002. Effects of cave tours on breeding Myotis velifer. Journal of Wildlife Management 66:618–624.

765. Manning, R. W., C. Jones, R. R. Hollander, and J. K. Jones Jr. 1987a. An unusual number of fetuses in the pallid bat. Prairie Naturalist 19:261.

766. Manning, R. W., C. Jones, J. K. Jones Jr., and R. R. Hollander. 1988. Subspecific status of the pallid bat, Antrozous pallidus, in the Texas Panhandle and adjacent areas. Occasional Papers, Museum of Texas Tech University 118:1–5.

767. Manning, R. W., C. Jones, and F. D. Yancey II. 2008. Annotated checklist of Recent land mammals of Texas, 2008. Occasional Papers, Museum of Texas Tech University 278:1–18.

768. Manning, R. W., J. K. Jones Jr., R. R. Hollander, and C. Jones. 1987b. Notes on

distribution and natural history of some bats on the Edwards Plateau and in adjacent areas of Texas. Texas Journal of Science 39: 279–285.

769. Manning, R. W., J. K. Jones Jr., and C. Jones. 1989. Comments on distribution and variation in the big brown bat, *Eptesicus fuscus*, in Texas. Texas Journal of Science 41: 95–101.

770. Marks, C. S., and G. E. Marks. 2006. Bats of Florida. University Press of Florida, Gainesville.

771. Martin, C. O. 1974. Systematics, ecology, and life history of *Antrozous* (Chiroptera: Vespertilionidae). M.S. Thesis, Texas A&M University, College Station.

772. Martin, C. O. 1977. A noteworthy record of the silver-haired bat in southeast Texas. Texas Journal of Science 28:356–357.

773. Martin, C. O., and D. J. Schmidly. 1982. Taxonomic review of the pallid bat, *Antrozous pallidus* (Le Conte). Special Publications, Museum of Texas Tech University 18:1–48.

774. Martin, R. A. 1972. Synopsis of late Pliocene and Pleistocene bats of North America and the Antilles. American Midland Naturalist 87:326–335.

775. Martin, R. L. 1959. A history of chiropteran rabies with special reference to occurrence and importance in the United States. Wildlife Diseases 3:1–75.

776. Matthews, A. K., and L. K. Ammerman. 2003. Recapture of a banded pocketed free-tailed bat (*Nyctinomops femorosaccus*) in Big Bend National Park, Texas. Bat Research News 44:4.

777. Matthews, A. K., S. A. Neiswenter, and L. K. Ammerman. 2010. Trophic ecology of the free-tailed bats *Nyctinomops femorosaccus* and *Tadarida brasiliensis* (Chiroptera:

Molossidae) in Big Bend National Park, Texas. Southwestern Naturalist 55:340–346.

778. Mattson, T. A., S. W. Buskirk, and N. L. Stanton. 1996. Roost sites of the silver-haired bat (*Lasionycteris noctivagans*) in the Black Hills, South Dakota. Great Basin Naturalist 56:247–253.

779. McAllister, C. T., C. R. Bursey, and R. C. Dowler. 2007. *Acanthatrium alicatai* (Tremotoda: Lecithodendriidae) from two species of bats (Chiroptera: Vespertilionidae) in southwestern Texas. Southwestern Naturalist 52:597–600.

780. McAllister, C. T., Z. D. Ramsey, and N. E. Solley. 2004. Noteworthy records of the Seminole bat, *Lasiurus seminolus* (Chiroptera: Vespertilionidae), from southwestern Arkansas and northeastern Texas. Journal of the Arkansas Academy of Science 58:137–138.

781. McCarley, W. H. 1959. The mammals of eastern Texas. Texas Journal of Science 11: 385–425.

782. McCarley, W. H., and W. N. Bradshaw. 1953. New locality records for some mammals of eastern Texas. Journal of Mammalogy 34:515–516.

783. McClure, H. E. 1942. Summer activities of bats (Genus *Lasiurus*) in Iowa. Journal of Mammalogy 23:430–434.

784. McCracken, G. F. 1984. Communal nursing in Mexican free-tailed bat maternity colonies. Science 223:1090–1091.

785. McCracken, G. F. 2003. Estimates of population sizes in summer colonies of Brazilian free-tailed bats (*Tadarida brasiliensis*). Pp. 21–30 *in* Monitoring trends in bat populations of the United States and territories: problems and prospects (T. J. O'Shea and M. A. Bogan, eds.). U.S. Geological Survey, Biological Resources

Discipline, Information and Technology Report, USGS/BRD/ITR-2003–0003.

786. McCracken, G. F., and M. F. Gassel. 1997. Genetic structure in migratory and nonmigratory populations of Brazilian free-tailed bats. Journal of Mammalogy 78: 348–357.

787. McCracken, G. F., E. H. Gillam, J. K. Westbrook, Y. F. Lee, M. L. Jensen, and B. B. Balsley. 2008. Brazilian free-tailed bats (*Tadarida brasiliensis*: Molossidae, Chiroptera) at high altitude: links to migratory insect populations. Integrative and Comparative Biology 48:107–118.

788. McCracken, G. F., Y. F. Lee, J. K. Westbrook, B. B. Balsley, and M. L. Jensen. 1997. High altitude foraging by Mexican free-tailed bats: vertical profiling using kites and hot air balloons. Bat Research News 38:117.

789. McCracken, G. F., M. K. McCracken, and A. T. Vawter. 1994. Genetic structure in migratory populations of the bat *Tadarida brasiliensis mexicana*. Journal of Mammalogy 75:500–514.

790. McGee, B. K., and R. W. Manning. 2000. Mammals of Lost Maples State Natural Area, Texas. Occasional Papers, Museum of Texas Tech University 198:1–24.

791. McWilliams, L. A. 2005. Variation in the diet of the Mexican free-tailed bat (*Tadarida brasiliensis mexicana*). Journal of Mammalogy 86:599–605.

792. Mead, J. I., and D. G. Mikesic. 2001. First fossil record of *Euderma maculatum* (Chiroptera: Vespertilionidae), eastern Grand Canyon, Arizona. Southwestern Naturalist 46:380–383.

793. Mearns, E. A. 1900. On the occurrence of a bat of the genus *Mormoops* in the United States. Proceedings of the Biological Society of Washington 13:166.

794. Medellín, R. A. 2003. Diversity and conservation of bats in Mexico: research priorities, strategies, and actions. Wildlife Society Bulletin 31:87–97.

795. Medellín, R. A., and W. López-Forment. 1986. Las cuevas, un recurso copartido. Anales del Instituto de Biología, Universidad Nacional Autónoma de México, Serie Zoológica 56:1027–1034.

796. Mendonça, M. T., and W. A. Hopkins. 1997. Effects of arousal from hibernation and plasma androgen levels on mating behavior in the male big brown bat, *Eptesicus fuscus*. Physiological Zoology 70:556–562.

797. Menu, H. 1984. Revision du statut de *Pipistrellus subflavus* (F. Cuvier, 1832), Proposition d'un taxon generique nouveau: *Perimyotis* nov. gen. Mammalia 48:409–416.

798. Menzel, J. M., M. A. Menzel, J. C. Kilgo, W. M. Ford, J. W. Edwards, and G. F. McCracken. 2005. Effect of habitat and foraging height on bat activity in the coastal plain of South Carolina. Journal of Wildlife Management 69:235–245.

799. Menzel, M. A., T. C. Carter, B. R. Chapman, and J. Laerm. 1998. Quantitative comparison of tree roost use by red bats (*Lasiurus borealis*) and Seminole bats (*L. seminolus*). Canadian Journal of Zoology 76:630–634.

800. Menzel, M. A., T. C. Carter, W. M. Ford, and B. R. Chapman. 2001a. Tree-roost characteristics of subadult and female adult evening bats (*Nycticeius humeralis*) in the upper coastal plain of South Carolina. American Midland Naturalist 145:112–119.

801. Menzel, M. A., T. C. Carter, W. M. Ford, B. R. Chapman, and J. Ozier. 2000. Summer roost tree selection by eastern red, Seminole, and evening bats in the upper coast plain of South Carolina. Proceedings of the Annual

Conference of the Southeastern Association of Fish and Wildlife Agencies 54:304–313.

802. Menzel, M. A., T. C. Carter, L. R. Jablonowski, B. L. Mitchell, J. M. Menzel, and B. R. Chapman. 2001b. Home range size and habitat use of big brown bats (*Eptesicus fuscus*) in a maternity colony located on a rural-urban interface in the southeast. Journal of the Elisha Mitchell Scientific Society 117:36–45.

803. Menzel, M. A., D. M. Krishon, T. C. Carter, and J. Laerm. 1999. Notes on the tree roost characteristics of the northern yellow bat (*Lasiurus intermedius*), the Seminole bat (*L. seminolus*), the evening bat (*Nycticeius humeralis*), and the eastern pipistrelle (*Pipistrellus subflavus*). Florida Scientist 62: 185–193.

804. Menzel, M. A., L. T. Lepardo, and J. Laerm. 1996. Possible use of a basal cavity as a maternity roost by the eastern pipistrelle, *Pipistrellus subflavus*. Bat Research News 37: 115–116.

805. Menzel, M. A., J. M. Menzel, T. C. Carter, J. O. Whitaker Jr., and W. M. Ford. 2002a. Notes on the late summer diet of male and female eastern pipistrelles (*Pipistrellus subflavus*) at Fort Mountain State Park, Georgia. Georgia Journal of Science 60: 170–179.

806. Menzel, M. A., et al. 2001c. Home range and habitat use of male Rafinesque's big-eared bats (*Corynorhinus rafinesquii*). American Midland Naturalist 145:401–408.

807. Menzel, M. A., et al. 2002b. Roost tree selection by northern long-eared bat (*Myotis septentrionalis*) maternity colonies in an industrial forest of the central Appalachian mountains. Forest Ecology and Management 155:107–114.

808. Merriam, C. H. 1897. A new bat of the genus *Antrozous* from California. Proceedings of the Biological Society of Washington 11:179–180.

809. Messenger, S. L., C. E. Rupprecht, and J. S. Smith. 2003. Bats, emerging virus infections, and the rabies paradigm. Pp. 622–679 *in* Bat ecology (T. H. Kunz and M. B. Fenton, eds.). University of Chicago Press, Chicago.

810. Meteyer, C. U., E. L. Buckles, D. S. Blehert, A. C. Hicks, D. E. Green, V. Shearn-Bochsler, et al. 2009. Histopathologic criteria to confirm white-nose syndrome in bats. Journal of Veterinary Diagnostic Investigations 21:411–414.

811. Metheny, J. D., M. C. Kalcounis-Rueppell, C. K. R. Willis, K. A. Kolar, and R. M. Brigham. 2008. Genetic relationships between roost-mates in a fission-fusion society of tree-roosting big brown bats (*Eptesicus fuscus*). Behavioral Ecology and Sociobiology 62:1043–1051.

812. Michael, E. D., and J. B. Birch. 1967. First Texas record of *Plecotus rafinesquii*. Journal of Mammalogy 48:672.

813. Michael, E. D., R. L. Wisennand, and G. Anderson. 1970. A Recent record of *Myotis austroriparius* from Texas. Journal of Mammalogy 51:620.

814. Mickey, A. B. 1961. Record of the spotted bat from Wyoming. Journal of Mammalogy 42:401–402.

815. Milam-Dunbar, M. B. 2005. Ecophysiology of hibernating eastern red bats (*Lasiurus borealis*). M.S. Thesis, Missouri State University, Springfield.

816. Miles, A. C., S. B. Castleberry, D. A. Miller, and L. M. Conner. 2006. Multi-scale roost-site selection by evening bats on pine-dominated landscapes in southwest Georgia. Journal of Wildlife Management 70:1191–1199.

817. Miller, D. A. 2003. Species diversity, reproduction, and sex ratios of bats in managed pine forest landscapes of Mississippi. Southeastern Naturalist 2: 59–72.

818. Miller, D. A., E. B. Arnett, and M. J. Lacki. 2003. Habitat management for forest-roosting bats of North America: a critical review of habitat studies. Wildlife Society Bulletin 31:30–44.

819. Miller, F. W. 1948. The Mexican free-tailed bat in Tarrant County, Texas. Journal of Mammalogy 29:418–419.

820. Miller, G. S., Jr. 1902. Note on the *Vespertilio incautus* of J. A. Allen. Proceedings of the Biological Society of Washington 15:155.

821. Miller, G. S., Jr. 1903. A second specimen of *Euderma maculatum*. Proceedings of the Biological Society of Washington 16: 165–166.

822. Mills, R. S., G. W. Barrett, and M. P. Farrell. 1975. Population dynamics of the big brown bat (*Eptesicus fuscus*) in southwestern Ohio. Journal of Mammalogy 56:591–604.

823. Milstead, W. W., and D. W. Tinkle. 1959. Seasonal occurrence and abundance of bats (Chiroptera) in northwestern Texas. Southwestern Naturalist 4:134–142.

824. Miner, K. L., and D. C. Stokes. 2005. Bats in the South Coast ecoregion: status, conservation issues, and research needs. USDA Forest Service General Technical Report PSW-GTR-195:211–227.

825. Mirowsky, K. M. 1997. Bats in palms: precarious habitat. Bats 15:3–6.

826. Mirowsky, K. M., P. A. Horner, R. W. Maxey, and S. A. Smith. 2004. Distributional records and roosts of southeastern myotis and Rafinesque's big-eared bat in eastern Texas. Southwestern Naturalist 49:294–298.

827. Mistry, S. 1990. Characteristics of the visually guided escape response of the Mexican free-tailed bat, *Tadarida brasiliensis mexicana*. Animal Behaviour 39:314–320.

828. Mistry, S., and G. F. McCracken. 1990. Behavioural response of the Mexican free-tailed bat, *Tadarida brasiliensis mexicana*, to visible and infra-red light. Animal Behaviour 39:598–599.

829. Mizutani, H., D. A. McFarlane, and Y. Kabaya. 1992. Nitrogen and carbon isotope study of bat guano core from Eagle Creek Cave, Arizona, U.S.A. Mass Spectroscopy 40:57–65.

830. Mohr, C. E. 1948a. Texas cave bats. National Speleological Society Bulletin 10: 103–105.

831. Mohr, C. E. 1948b. Texas bat caves served in three wars. National Speleological Society Bulletin 10:89–96.

832. Mohr, C. E. 1972. The status of threatened species of cave dwelling bats. National Speleological Society Bulletin 34:33–47.

833. Mollhagen, T. 1970. A key to the bats of Texas and adjacent regions, with an annotated list. Texas Speleological Survey 3: 1–26.

834. Mollhagen, T. 1973. Distributional and taxonomic notes on some west Texas bats. Southwestern Naturalist 17:427–430.

835. Mollhagen, T., and R. H. Baker. 1972. *Myotis volans interior* in Knox County, Texas. Southwestern Naturalist 17:97.

836. Mondul, A. M., J. W. Krebs, and J. E. Childs. 2003. Trends in national surveillance for rabies among bats in the United States (1993–2000). Journal of the American Veterinary Medical Association 222: 633–639.

837. Moor, K., F. Stay, and W. G. Bradley. 1965. Mexican free-tailed bat and western

pipistrelle found roosting in sedges. Journal of Mammalogy 46:507.

838. Moore, W. 1949. Bat caves and bat bombs. Turtox News 26:262–265.

839. Moorman, C. E., K. R. Russell, M. A. Menzel, S. M. Lohr, J. E. Ellenberger, and D. H. V. Lear. 1999. Bats roosting in deciduous leaf litter. Bat Research News 40: 74–75.

840. Morales, J. C., S. W. Ballinger, J. W. Bickham, I. F. Greenbaum, and D. A. Schlitter. 1991. Genetic relationships among eight species of *Eptesicus* and *Pipistrellus* (Chiroptera: Vespertilionidae). Journal of Mammalogy 72:286–291.

841. Morales, J. C., and J. W. Bickham. 1995. Molecular systematics of the genus *Lasiurus* (Chiroptera: Vespertilionidae) based on restriction-site maps of the mitochondrial ribosomal genes. Journal of Mammalogy 76: 730–749.

842. Moreno-Valdez, A. 1996. Noteworthy records of two species of *Myotis* (Chiroptera: Vespertilionidae) from northeastern Mexico. Texas Journal of Science 48:329–330.

843. Moreno-Valdez, A., R. L. Honeycutt, and W. E. Grant. 2004. Colony dynamics of *Leptonycteris nivalis* (Mexican long-nosed bat) to flowering Agave in northern Mexico. Journal of Mammalogy 85:453–459.

844. Morimoto, K., et al. 1996. Characterization of a unique variant of bat rabies virus responsible for newly emerging human cases in North America. Proceedings of the National Academy of Science 93:5653–5658.

845. Mormann, B. M., and L. W. Robbins. 2007. Winter roosting ecology of eastern red bats in southwest Missouri. Journal of Wildlife Management 71:213–217.

846. Morrell, T. E., M. J. Rabe, J. C. deVos Jr., H. Green, and C. R. Miller. 1999. Bats captured in two ponderosa pine habitats in north-central Arizona Southwestern Naturalist 44:501–506.

847. Morse, R. C., and B. P. Glass. 1960. The taxonomic status of *Antrozous bunkeri*. Journal of Mammalogy 41:10–15.

848. Muliak, S. 1943. Notes on some bats of the southwest. Journal of Mammalogy 24:269.

849. Mumford, R. E. 1973. Natural history of the red bat (*Lasiurus borealis*) in Indiana. Periodicals in Biology 75:155–158.

850. Mumford, R. E., L. L. Oakley, and D. A. Zimmerman. 1964. June bat records from Guadalupe Canyon, New Mexico. Southwestern Naturalist 9:43–45.

851. Mumford, R. E., and J. O. Whitaker Jr. 1982. Mammals of Indiana. Indiana University Press, Bloomington.

852. Mumford, R. E., and D. A. Zimmerman. 1963. The southern yellow bat in New Mexico. Journal of Mammalogy 44:417–418.

853. Muñiz-Martínez, R., C. López-González, J. Arroyo-Cabrales, and M. Ortiz Gómez. 2003. Noteworthy records of free-tailed bats (Chiroptera: Molossidae) from Durango, Mexico. Southwestern Naturalist 48:138–144.

854. Murphy, W. J., et al. 2001. Resolution of the early placental mammal radiation using Bayesian phylogenetics. Science 294: 2348–2351.

855. Myers, R. F. 1960. *Lasiurus* from Missouri caves. Journal of Mammalogy 41:114–117.

856. Nader, I. A., and D. F. Hoffmeister. 1983. Bacula of big-eared bats *Plecotus*, *Corynorhinus*, and *Idionycteris*. Journal of Mammalogy 64:528–529.

857. Nagorsen, D. W., A. A. Bryant, D. Kerridge, G. Roberts, A. Roberts, and M. J. Sarell. 1993. Winter bat records for British Columbia. Northwestern Naturalist 74: 61–66.

858. Navo, K. W., J. A. Gore, and G. T. Skiba. 1992. Observations on the spotted bat, *Euderma maculatum*, in northwestern Colorado. Journal of Mammalogy 73:547–551.

859. Nedbal, M. A., D. J. Schmidly, and R. D. Bradley. 1994. Records of three bat species in southeast Texas. Texas Journal of Science 46: 195–196.

860. Neill, W. T. 1952. Hoary bat in a squirrel's nest. Journal of Mammalogy 33:113.

861. Neubaum, D. J., M. A. Neubaum, L. E. Ellison, and T. J. O'Shea. 2005. Survival and condition of big brown bats (*Eptesicus fuscus*) after radiotagging. Journal of Mammalogy 86:95–98.

862. Neubaum, D. J., K. R. Wilson, and T. J. O'Shea. 2007. Urban maternity-roost selection by big brown bats in Colorado. Journal of Wildlife Management 71:728–736.

863. Neuweiler, G. 2000. The biology of bats. Oxford University Press, New York.

864. Nielsen, L. T., D. K. Eaton, D. W. Wright, and B. Schmidt-French. 2006. Characteristic odors of *Tadarida brasiliensis mexicana* Chiroptera: Molossidae. Journal of Cave and Karst Studies 68:27–31.

865. Noah, D. L., et al. 1998. Epidemiology of human rabies in the United States, 1980 to 1996. Annals of Internal Medicine 128: 922–930.

866. Norberg, U. M. 1987. Wing form and flight mode in bats. Pp. 43–56 *in* Recent advances in the study of bats (M. B. Fenton, P. A. Racey, and J. M. V. Rayner, eds.). Cambridge University Press, New York.

867. Norberg, U. M. 1994. Wing design, flight performance, and habitat use in bats. Pp. 205–239 *in* Ecological morphology: integrative organismal biology (P. C. Wainwright and S. M. Reilly, eds.). University of Chicago Press, Chicago.

868. Norberg, U. M. 1998. Morphological adaptations for flight in bats. Pp. 93–108 *in* Bat conservation and biology (T. H. Kunz and P. A. Racey, eds.). Smithsonian Institution Press, Washington, D.C.

869. Nowak, R. M. 1999. Walker's mammals of the world, 6th ed. 2 vols. Johns Hopkins University Press, Baltimore.

870. Noyes, A., and G. W. Pierce. 1937. Apparatus for acoustic research in the supersonic frequency range. Journal of the Acoustical Society of America 9:205–211.

871. O'Farrell, M. J. 1999. *Myotis thysanodes*. Pp. 98–100 *in* The Smithsonian book of North American mammals (D. E. Wilson and S. Ruff, eds.). Smithsonian Institution Press, Washington, D. C.

872. O'Farrell, M. J., and W. G. Bradley. 1970. Activity patterns of bats over a desert spring. Journal of Mammalogy 51:18–26.

873. O'Farrell, M. J., and W. G. Bradley. 1977. Comparative thermal relationships of flight for some bats in the southwestern United States. Comparative Biochemistry and Physiology A 58:223–227.

874. O'Farrell, M. J., C. Corben, and W. L. Gannon. 2000. Geographic variation in the echolocation calls of the hoary bat (*Lasiurus cinereus*). Acta Chiropterologica 2:185–195.

875. O'Farrell, M. J., C. Corben, W. L. Gannon, and B. W. Miller. 1999a. Confronting the dogma: a reply. Journal of Mammalogy 80: 297–302.

876. O'Farrell, M. J., and W. L. Gannon. 1999. A comparison of acoustic versus capture techniques for the inventory of bats. Journal of Mammalogy 80:24–30.

877. O'Farrell, M. J., B. W. Miller, and W. L. Gannon. 1999b. Qualitative identification of free-flying bats using the Anabat detector. Journal of Mammalogy 80:11–23.

878. O'Farrell, M. J., and E. H. Studier. 1973. Reproduction, growth, and development in *Myotis thysanodes* and *M. lucifugus* (Chiroptera: Vespertilionidae). Ecology 54: 18–30.

879. O'Farrell, M. J., and E. H. Studier. 1975. Population structure and emergence activity patterns in *Myotis thysanodes* and *M. lucifugus* (Chiroptera: Vespertilionidae) in northeastern New Mexico. American Midland Naturalist 93:368–376.

880. O'Farrell, M. J., J. A. Williams, and B. Lund. 2004. Western yellow bat (*Lasiurus xanthinus*) in southern Nevada. Southwestern Naturalist 49:514–518.

881. O'Keefe, J. M., S. C. Loeb, J. D. Lanham, and H. S. Hill Jr. 2009. Macrohabitat factors affect day roost selection by eastern red bats and eastern pipistrelles in the southern Appalachian Mountains, USA. Forest Ecology and Management 257:1757–1763.

882. O'Shea, T. J., M. A. Bogan, and L. E. Ellison. 2003. Monitoring trends in bat populations of the United States and Territories: status of the science and recommendations for the future. Wildlife Society Bulletin 31:16–29.

883. O'Shea, T. J., and T. A. Vaughan. 1977. Nocturnal and seasonal activities of the pallid bat, *Antrozous pallidus*. Journal of Mammalogy 58:269–284.

884. O'Shea, T. J., and T. A. Vaughan. 1999. Population changes in bats from central Arizona: 1972 and 1997. Southwestern Naturalist 44:495–500.

885. Obrist, M. K. 1995. Flexible bat echolocation: the influence of individual, habitat and conspecifics on sonar signal design. Behavioral Ecology and Sociobiology 36:207–219.

886. Ohlendorf, H. M. 1972. Observations on a colony of *Eumops perotis* (Molossidae). Southwestern Naturalist 17:297–300.

887. Ormsbee, P. C. 1996. Characteristics, use, and distribution of day roosts selected by female *Myotis volans* (long-legged myotis) in forested habitat of the central Oregon Cascades. Pp. 124–131 *in* Bats and forests symposium (R. M. R. Barclay and R. M. Brigham, eds.). British Columbia Ministry of Forestry, Victoria, British Columbia.

888. Ormsbee, P. C., and W. C. McComb. 1998. Selection of day roosts by female long-legged myotis in the central Oregon Cascade range. Journal of Wildlife Management 62:596–603.

889. Orr, R. T. 1950. Unusual behavior and occurrence of a hoary bat. Journal of Mammalogy 31:456–457.

890. Orr, R. T. 1954. Natural history of the pallid bat, *Antrozous pallidus* (Le Conte). Proceedings of the California Academy of Sciences 28:165–246.

891. Orr, R. T., and G. S. Taboada. 1960. A new species of bat of the genus *Antrozous* from Cuba. Proceedings of the Biological Society of Washington 73:83–86.

892. Owen, R. D., R. K. Chesser, and D. C. Carter. 1990. The systematic status of *Tadarida brasiliensis cynocephala* and Antillean members of the *Tadarida brasiliensis* group, with comments on the generic name *Rhizomops* Legendre. Occasional Papers, Museum of Texas Tech University 133:1–18.

893. Owen, S. F., et al. 2003. Home-range size and habitat used by the northern myotis (*Myotis septentrionalis*). American Midland Naturalist 150:352–359.

894. Owen, S. F., et al. 2002. Roost tree selection by maternal colonies of the northern long-eared myotis in an intensively managed forest. General Technical Report

NE-292. Newtown Square, Pennsylvania, U.S. Department of Agriculture, Northeastern Research Station. 6 pp.

895. Packard, R. L. 1966. *Myotis austroriparius* in Texas. Journal of Mammalogy 47:128.

896. Packard, R. L., and H. W. Garner. 1964. Records of some mammals from the Texas High Plains. Texas Journal of Science 16: 387–390.

897. Packard, R. L., and F. W. Judd. 1968. Comments on some mammals from western Texas. Journal of Mammalogy 49:535–38.

898. Padgett, T. M., and R. K. Rose. 1991. Bats (Chiroptera: Vespertilionidae) of the Great Dismal Swamp of Virginia and North Carolina. Brimleyana 17:17–25.

899. Pagels, J. F., and C. Jones. 1974. Growth and development of the free-tailed bat, *Tadarida brasiliensis cynocephala* (Le Conte). Southwestern Naturalist 19:267–276.

900. Painter, M. L., C. L. Chambers, M. Siders, R. R. Doucett, J. O. Whitaker Jr., and D. L. Phillips. 2009. Diet of spotted bats (*Euderma maculatum*) in Arizona as indicated by fecal analysis and stable isotopes. Canadian Journal of Zoology 87:865–875.

901. Palmeirim, J. M., and L. Rodrigues. 1993. The 2-minute harp trap for bats. Bat Research News 34:60–64.

902. Paradiso, J. L., and A. M. Greenhall. 1967. Longevity records for American bats. American Midland Naturalist 78:251–252.

903. Parlos, J. A. 2008. Population genetic structure of a cave-dwelling bat, *Myotis velifer*. M.S. Thesis, Texas State University, San Marcos.

904. Parsons, H. J., D. A. Smith, and R. F. Whittam. 1986. Maternity colonies of silver-haired bats, *Lasionycteris noctivagans*, in Ontario and Saskatchewan. Journal of Mammalogy 67:598–600.

905. Parsons, S., and J. M. Szewczak. 2009. Detecting, recording, and analyzing the vocalizations of bats. Pp. 91–111 *in* Ecological and behavioral methods for the study of bats (T. H. Kunz and S. Parsons, eds.). Johns Hopkins University Press, Baltimore.

906. Patton, T. H. 1963. Fossil vertebrates from Miller's Cave, Llano County, Texas. Texas Memorial Museum Bulletin 7:1–41.

907. Pavey, C. R., and C. J. Burwell. 1998. Bat predation on eared moths: a test of the allotonic frequency hypothesis. Oikos 81: 143–151.

908. Pearson, O. P., M. R. Koford, and A. K. Pearson. 1952. Reproduction of the lump-nosed bat (*Corynorhinus rafinesquii*) in California. Journal of Mammalogy 33: 273–320.

909. Pekins, C. E. 2006. Cave myotis bat population and cave microclimate monitoring at a maternity site on Fort Hood, Texas during 2005–2006. *In* Endangered species monitoring and management at Fort Hood, Texas: 2006 annual report. The Nature Conservancy, Fort Hood, Texas.

910. Pekins, C. E. 2007. Cave myotis bat roost switch following prolonged torrential rainfall on Fort Hood, Texas. *In* Endangered species monitoring and management at Fort Hood, Texas: 2007 annual report. The Nature Conservancy, Fort Hood, Texas.

911. Pekins, C. E. 2008. Cave myotis (*Myotis velifer incautus*) monitoring at Harrell's Cave, San Saba County, Texas during 2008. Unpublished report for Conservation 1 Wildlife and Management Services. 17 pp.

912. Pekins, C. E. 2009a. Polydactyly in the cave myotis (*Myotis velifer*) in north-central Texas. Southwestern Naturalist 54:222–225.

913. Pekins, C. E. 2009b. Cave myotis (*Myotis velifer incautus*) population and cave

microclimate monitoring at a maternity site on Fort Hood, Texas during 2009. *in* Endangered species monitoring and management at Fort Hood, Texas: 2009 annual report. The Nature Conservancy, Fort Hood Project, Fort Hood, Texas.

914. Perkins, J. M., J. M. Barss, and J. Peterson. 1990. Winter records of bats in Oregon and Washington. Northwestern Naturalist 71: 59–62.

915. Perkins, J. M., and S. P. Cross. 1988. Differential use of some coniferous forest habitats by hoary and silver-haired bats in Oregon. Murrelet 69:21–24.

916. Perry, R. W., and R. E. Thill. 2007a. Roost characteristics of hoary bats in Arkansas. American Midland Naturalist 158:132–138.

917. Perry, R. W., and R. E. Thill. 2007b. Roost selection by male and female northern long-eared bats in a pine dominated landscape. Forest Ecology and Management 247: 220–226.

918. Perry, R. W., and R. E. Thill. 2007c. Tree roosting by male and female eastern pipistrelles in a forested landscape. Journal of Mammalogy 88:974–981.

919. Perry, R. W., and R. E. Thill. 2007d. Summer roosting by adult male Seminole bats in the Ouachita Mountains, Arkansas. American Midland Naturalist 158:361–368.

920. Perry, R. W., and R. E. Thill. 2008a. Diurnal roosts of male evening bats (*Nycticeius humeralis*) in diversely managed pine-hardwood forests. American Midland Naturalist 160:374–385.

921. Perry, R. W., and R. E. Thill. 2008b. Roost selection by big brown bats in forests of Arkansas: importance of pine snags and open forest habitats to males. Southeastern Naturalist 7:607–618.

922. Perry, R. W., R. E. Thill, and S. A. Carter.

2007a. Sex-specific roost selection by adult red bats in a diverse forested landscape. Forest Ecology and Management 253:48–55.

923. Perry, R. W., R. E. Thill, and D. M. Leslie Jr. 2007b. Selection of roosting habitat by forest bats in a diverse forested landscape. Forest Ecology and Management 238: 156–166.

924. Perry, R. W., R. E. Thill, and D. M. Leslie Jr. 2008. Scale-dependent effects of landscape structure and composition on diurnal roost selection by forest bats. Journal of Wildlife Management 72:913–925.

925. Perry, T. W., P. M. Cryan, S. R. Davenport, and M. A. Bogan. 1997. New locality for *Euderma maculatum* (Chiroptera: Vespertilionidae) in New Mexico. Southwestern Naturalist 42:99–101.

926. Peterson, R. L. 1946. Recent and Pleistocene mammalian fauna of Brazos County, Texas. Journal of Mammalogy 27: 162–169.

927. Petit, M. G. 1978. Imperiled bats of Eagle Creek Cave. Natural History 87:50–55.

928. Pettigrew, J. D. 1991a. A fruitful, wrong hypothesis? Response to Baker, Novacek, and Simmons. Systematic Zoology 40: 231–239.

929. Pettigrew, J. D. 1991b. Wings or brain? Convergent evolution in the origins of bats. Systematic Zoology 40:199–216.

930. Pettigrew, J. D., B. G. M. Jamieson, S. K. Robson, L. S. Hall, K. I. McNally, and H. M. Cooper. 1989. Phylogenetic relations between microbats, megabats, and primates (Mammalia: Chiroptera and Primates). Philosophical Transactions of the Royal Society of London B 325 (1229):489–559.

931. Phelps, K. L., C. J. Schmidt, and J. R. Choate. 2008. Presence of the evening bat (*Nycticeius humeralis*) in westernmost

Kansas. Transactions of the Kansas Academy of Science 111:159–160.

932. Phillips, C. J. 1971. The dentition of glossophagine bats: development, morphological characteristics, variation, pathology, and evolution. University of Kansas Publications, Museum of Natural History 54:1–138.

933. Phillips, C. J., J. K. Jones Jr., and F. J. Radovsky. 1969. Macronyssid mites in oral mucosa of long-nosed bats: occurrence and associated pathology. Science 165:1368–1369.

934. Phillips, G. L. 1966. Ecology of the big brown bat (Chiroptera: Vespertilionidae) in northeastern Kansas. American Midland Naturalist 75:168–198.

935. Piaggio, A. J., and S. L. Perkins. 2005. Molecular phylogeny of North American long-eared bats (Vespertilionidae: Corynorhinus); inter- and intraspecific relationships inferred from mitochondrial and nuclear DNA sequences. Molecular Phylogenetics and Evolution 37:762–775.

936. Piaggio, A. J., E. W. Valdez, M. A. Bogan, and G. S. Spicer. 2002. Systematics of Myotis occultus (Chiroptera: Vespertilionidae) inferred from sequences of two mitochondrial genes. Journal of Mammalogy 83:386–395.

937. Pierce, G. W., and D. R. Griffin. 1938. Experimental determination of supersonic notes emitted by bats. Journal of Mammalogy 19:454–455.

938. Pierson, E. D. 1998. Tall trees, deep holes, and scarred landscapes: conservation biology of North American bats. Pp. 309–325 in Bat biology and conservation (T. H. Kunz and P. A. Racey, eds.). Smithsonian Institution Press, Washington, D.C.

939. Pierson, E. D., and W. E. Rainey. 1998. Distribution of the spotted bat, Euderma maculatum, in California. Journal of Mammalogy 79:1296–1305.

940. Pierson, E. D., W. E. Rainey, and C. Corben. 2006. Distribution and status of western red bats (Lasiurus blossevillii) in California. California Department of Fish and Game, Habitat Conservation Planning Branch, Species Conservation and Recovery Program Report 2006–04, Sacramento. 45 pp.

941. Pine, R. H., D. C. Carter, and R. K. LaVal. 1971. Status of Bauerus Van Gelder and its relationship to other nyctophiline bats. Journal of Mammalogy 52:663–669.

942. Pinto, C. M. 2006. An accidental record of the northern yellow bat, Lasiurus intermedius, in Illinois. Bat Research News 47:37–38.

943. Piorkowski, M. D. 2006. Breeding bird habitat use and turbine collisions of birds and bats located at a wind farm in Oklahoma mixed-grass prairie. M.S. Thesis, Oklahoma State University, Stillwater.

944. Pitts, R. M., and J. J. Scharninghausen. 1986. Use of cliff swallow and barn swallow nests by the cave bat, Myotis velifer, and the free-tailed bat, Tadarida brasiliensis. Texas Journal of Science 38:265–266.

945. Pitts, R. M., G. T. Schimmelphenig, and J. R. Choate. 1996. The big free-tailed bat, Nyctinomops macrotis, in Missouri. Southwestern Naturalist 41:86–87.

946. Poche, R. M. 1975. New record of Euderma maculatum from Arizona. Journal of Mammalogy 56:931–933.

947. Poche, R. M. 1981. Ecology of the spotted bat (Euderma maculatum) in southwest Utah. Report No. 81–1. Utah Division of Wildlife Resources, Salt Lake City. 63 pp.

948. Poche, R. M., and G. L. Bailie. 1974. Notes on the spotted bat (Euderma maculatum)

from southwest Utah. Great Basin Naturalist 34:254–256.

949. Poche, R. M., and G. A. Ruffner. 1975. Roosting behavior of male *Euderma maculatum* from Utah. Great Basin Naturalist 35:121–122.

950. Polaco, O. J., J. Arroyo-Cabrales, and J. K. Jones Jr. 1992. Noteworthy records of some bats from Mexico. Texas Journal of Science 44:331–338.

951. Poole, E. L. 1932. *Lasiurus seminolus* in Pennsylvania. Journal of Mammalogy 13:162.

952. Poole, E. L. 1949. A second Pennsylvania specimen of *Lasiurus seminolus* (Rhoads). Journal of Mammalogy 30:80.

953. Ports, M. A., and P. V. Bradley. 1996. Habitat affinities of bats from northeastern Nevada. Great Basin Naturalist 56:48–53.

954. Priday, J., and B. Luce. 1999. New distributional records for spotted bat (*Euderma maculatum*) in Wyoming. Great Basin Naturalist 59:97–101.

955. Puechmaille, S. J., G. Mathy, and E. J. Petit. 2007. Good DNA from bat droppings. Acta Chiropterologica 9:269–276.

956. Quay, W. B., and J. S. Miller. 1955. Occurrence of the red bat, *Lasiurus borealis,* in caves. Journal of Mammalogy 36:454–455.

957. Rabe, M. J., T. E. Morrell, H. Green, J. C. DeVos Jr., and C. R. Miller. 1998a. Characteristics of ponderosa pine snag roosts used by reproductive bats in northern Arizona. Journal of Wildlife Management 62:612–621.

958. Rabe, M. J., M. S. Siders, C. R. Miller, and T. K. Snow. 1998b. Long foraging distance for spotted bat (*Euderma maculatum*) in northern Arizona. Southwestern Naturalist 43:266–269.

959. Rabinowitz, A. 1981. Thermal preference of the eastern pipistrelle bat (*Pipistrellus subflavus*) during hibernation. Journal of the Tennessee Academy of Science 56:113–114.

960. Racey, P. A. 1979. The prolonged storage and survival of spermatozoa in Chiroptera. Journal of Reproduction and Fertility 56:391–402.

961. Racey, P. A., and A. C. Entwistle. 2003. Conservation of bats. Pp. 680–743 *in* Bat Ecology (T. H. Kunz and M. B. Fenton, eds.). University of Chicago Press, Chicago.

962. Raesly, R. L., and J. E. Gates. 1987. Winter habitat selection by north temperate cave bats. American Journal of Anatomy 118:15–31.

963. Rainey, W. E., and E. D. Pierson. 1992. Bats in hollow redwoods: seasonal use and role in nutrient transfer into old growth communities. Bat Research News 35:111.

964. Rancourt, S. J., M. I. Rule, and M. A. O'Connell. 2007. Maternity roost site selection of big brown bats in ponderosa pine forests of the Channeled Scablands of northeastern Washington State, USA. Forest Ecology and Management 248:183–192.

965. Randolph, N. M., and R. B. Eads. 1946. An ectoparasitic survey of mammals from Lavaca County, Texas. Annals of the Entomological Society America 39:597–601.

966. Rasweiler, J. J., IV, and N. K. Badwaik. 2000. Anatomy and physiology of the female reproductive tract. Pp. 157–219 *in* Reproductive biology of bats (E. G. Crichton and P. H. Krutzsch, eds.). Academic Press, London.

967. Ratcliffe, J. M., and J. W. Dawson. 2003. Behavioural flexibility: the little brown bat, *Myotis lucifugus,* and the northern long-eared bat, *M. septentrionalis,* both glean and hawk prey. Animal Behaviour 66:847–856.

968. Raun, G. G. 1960. A mass die-off of the

Mexican brown bat, *Myotis velifer,* in Texas. Southwestern Naturalist 5:104–105.

969. Raun, G. G. 1961. The big free-tailed bat in southern Texas. Journal of Mammalogy 42:253.

970. Raun, G. G. 1966. A heretofore unnoted collection of Texas mammals. Texas Journal of Science 18:225–227.

971. Raun, G. G., and J. K. Baker. 1958. Some observations of Texas cave bats. Southwestern Naturalist 3:102–106.

972. Ray, C. E., O. J. Linares, and G. S. Morgan. 1988. Paleontology. Pp. 19–30 *in* Natural history of vampire bats (A. M. Greenhall and S. Schmidt, eds.). CRC Press, Inc., Boca Raton.

973. Ray, C. E., and D. E. Wilson. 1979. Evidence for *Macrotus californicus* from Terlingua, Texas. Occasional Papers, Museum of Texas Tech University 57:1–10.

974. Razak, K. A., W. Shen, T. Zumsteg, and Z. M. Fuzessery. 2007. Parallel thalamocortical pathways for echolocation and passive sound localization in a gleaning bat, *Antrozous pallidus.* Journal of Comparative Neurology 500:322–338.

975. Reddell, J. R., ed. 1967a. The caves of Medina County, Texas, Vol. 3. Texas Speleological Survey.

976. Reddell, J. R. 1967b. A checklist of the cave fauna of Texas. III: Vertebrata. Texas Journal of Science 19:184–226.

977. Reddell, J. R. 1968. The hairy-legged vampire, *Diphylla ecaudata,* in Texas. Journal of Mammalogy 49:769.

978. Reddell, J. R. 1994. The cave fauna of Texas with special reference to the western Edwards Plateau. Pp. 31–50 *in* The caves and karst of Texas (W. R. Elliot and G. Veni, eds.). National Speleological Society, Huntsville, Alabama.

979. Rehn, J. A. 1902. A revision of the genus *Mormoops.* Proceedings of the Academy of Natural Science of Philadelphia 54:160–172.

980. Reichard, J. D., and T. H. Kunz. 2009. White-nose syndrome inflicts lasting injuries to the wings of little brown myotis (*Myotis lucifugus*). Acta Chiropterologica 11: 457–464.

981. Reynolds, L. A., and W. A. Mitchell. 1998. Species profile: southeastern myotis (*Myotis austroriparius*) on military installations in the southeastern United States, Technical Report SERDP-98–8. U.S. Army Engineer Waterways Experiment Station, Vicksburg, MS. 26 pp.

982. Reynolds, R. P. 1981. Elevational record for *Euderma maculatum* (Chiroptera: Vespertilionidae). Southwestern Naturalist 26:91–92.

983. Rice, D. W. 1957. Life history and ecology of *Myotis austroriparius* in Florida. Journal of Mammalogy 38:15–32.

984. Ridlehuber, K. T., and N. J. Silvy. 1981. Texas rat snake feeds on Mexican freetail bat and wood duck eggs. Southwestern Naturalist 26:70–71.

985. Rieger, J. F., and E. M. Jakob. 1988. The use of olfaction in food location by frugivorous bats. Biotropica 20:161–164.

986. Rippy, C. L., and M. J. Harvey. 1965. Notes on *Plecotus townsendii virginianus* in Kentucky. Journal of Mammalogy 46:499.

987. Riskin, D. K., J. W. Bahlman, T. Y. Hubel, J. M. Ratcliffe, T. H. Kunz, and S. M. Swartz. 2009. Bats go head-under-heels: the biomechanics of landing on a ceiling. Journal of Experimental Biology 212:945–953.

988. Ritzi, C. M., L. K. Ammerman, M. T. Dixon, and J. V. Richerson. 2001. Bat ectoparasites from the Trans-Pecos region of Texas, including notes from Big

Bend National Park. Journal of Medical Entomology 38:400–404.

989. Ritzi, C. M., C. W. Walker, and R. L. Honeycutt. 1998. Utilization of cave swallow nests by the cave myotis, *Myotis velifer,* in central Texas. Texas Journal of Science 50: 175–176.

990. Roberts, K. J., F. D. Yancey II, and C. Jones. 1997a. Distributional records of small mammals from the Texas panhandle. Texas Journal of Science 49:57–64.

991. Roberts, K. J., F. D. Yancey II, and C. Jones. 1997b. Predation by great horned owls on Brazilian free-tailed bats in north Texas. Texas Journal of Science 49:215–218.

992. Rodeck, H. G. 1961. Another spotted bat from New Mexico. Journal of Mammalogy 42:401.

993. Rodhouse, T. J., M. F. McCaffrey, and R. G. Wright. 2005. Distribution, foraging behavior, and capture results of the spotted bat (*Euderma maculatum*) in central Oregon. Western North American Naturalist 65: 215–222.

994. Rodriguez, R. M., and L. K. Ammerman. 2004. Mitochondrial DNA divergence does not reflect morphological difference between *Myotis californicus* and *Myotis ciliolabrum.* Journal of Mammalogy 85:842–851.

995. Roehrs, Z. P., et al. 2008. New records of mammals from western Oklahoma. Occasional Papers, Museum of Texas Tech University 273:1–15.

996. Rohde, R. E., B. C. Mayes, J. S. Smith, and S. U. Neill. 2004. Bat rabies, Texas, 1996–2000. Emerging Infectious Diseases 10: 948–952.

997. Rolseth, S. L., C. E. Koehler, and R. M. R. Barclay. 1994. Differences in diets of juvenile and adult hoary bats, *Lasiurus cinereus.* Journal of Mammalogy 75:394–398.

998. Ross, A. J. 1961. Notes on food habits of bats. Journal of Mammalogy 42:66–71.

999. Ross, A. J. 1967. Ecological aspects of the food habits of insectivorous bats. Proceedings of the Western Foundation of Vertebrate Zoology 1:204–263.

1000. Roth, E. L. 1970. Silver-haired bat at Wichita Falls, Texas. Southwestern Naturalist 14:449–450.

1001. Roth, E. L. 1972. Late Pleistocene mammals from Klein Cave, Kerr County, Texas. Texas Journal of Science 24:75–84.

1002. Ruffner, G. A., R. M. Poche, M. Meierkord, and J. A. Neal. 1979. Winter bat activity over a desert wash in southwestern Utah. Southwestern Naturalist 24:447–453.

1003. Ruhl, M. W., and F. B. Stangl Jr. 1997. Noteworthy records of mammals from Stonewall County, Texas. Texas Journal of Science 49:345–348.

1004. Rupprecht, C. E. 1979. Bats (Chiroptera) as constituents of the food of barn owls *Tyto alba* in Poland. Ibis 121:489–494.

1005. Russell, A. L., and G. F. McCracken. 2006. Population genetic structure of very large populations: the Brazilian free-tailed bat. Pp. 227–247 *in* Functional and evolutionary ecology of bats (A. Zubaid, G. F. McCracken, and T. H. Kunz, eds.). Oxford University Press, New York.

1006. Russell, A. L., R. A. Medellín, and G. F. McCracken. 2005. Genetic variation and migration in the Mexican free-tailed bat (*Tadarida brasiliensis mexicana*). Molecular Ecology 14:2207–2222.

1007. Sabol, B. M., and M. K. Hudson. 1995. Technique using thermal infrared-imaging for estimating populations of gray bats. Journal of Mammalogy 76:1242–1248.

1008. Safford, M. 1989. Bat population study, Fort Stanton Cave. NSS Bulletin 51:42–46.

1009. Sample, B. E., and R. C. Whitmore. 1993. Food habits of the endangered Virginia big-eared bat in West Virginia. Journal of Mammalogy 74:428–435.

1010. Sanborn, C. C. 1932. The bats of the genus *Eumops*. Journal of Mammalogy 13:347–357.

1011. Sánchez, R., and R. A. Medellín. 2007. Food habits of the threatened bat *Leptonycteris nivalis* (Chiroptera: Phyllostomidae) in a mating roost in Mexico. Journal of Natural History 41:1753–1764.

1012. Sandel, J. K., G. R. Benatar, K. M. Burke, C. W. Walker, T. E. Lacher Jr., and R. L. Honeycutt. 2001. Use and selection of winter hibernacula by the eastern pipistrelle (*Pipistrellus subflavus*) in Texas. Journal of Mammalogy 82:173–178.

1013. Santos, N., V. Fagundes, Y. Yonenaga-Yassuda, and M. J. de Souza. 2001. Comparative karyology of Brazilian bats *Desmodus rotundus* and *Diphylla ecaudata* (Phyllostomidae, Chiroptera) banding patterns, base-specific flourochromes, and FISH of ribosomal genes. Hereditas 134: 189–194.

1014. Sarkozi, D. L., and D. M. Brooks. 2003. Eastern red bat (*Lasiurus borealis*) impaled by a loggerhead shrike (*Lanius ludovicianus*). Southwestern Naturalist 48:301–303.

1015. Sasse, D. B., and P. J. Pekins. 1996. Summer roosting ecology of northern long-eared bats (*Myotis septentrionalis*) in the White Mountain National Forest. Pp. 91–101 *in* Bats and forests symposium (R. M. R. Barclay and R. M. Brigham, eds.). Research Branch, Ministry of Forests, Victoria, British Columbia.

1016. Saugey, D. A., B. G. Crump, R. L. Vaughn, and G. A. Heidt. 1998. Notes on the natural history of *Lasiurus borealis* in Arkansas. Journal of the Arkansas Academy of Science 52:92–98.

1017. Saugey, D. A., D. R. Heath, and G. A. Heidt. 1989. The bats of the Ouachita Mountains. Proceedings of the Arkansas Academy of Science 42:81–83.

1018. Saugey, D. A., V. R. McDaniel, D. R. England, M. C. Rowe, L. R. Chandler-Mozisek, and B. G. Cochran. 1993. Arkansas range extensions of the eastern small-footed bat (*Myotis leibii*) and northern long-eared bat (*Myotis septentrionalis*) and additional county records for the silver-haired bat (*Lasionycteris noctivagans*), hoary bat (*Lasiurus cinereus*), southeastern bat (*Myotis austroriparius*), and Rafinesque's big-eared bat (*Plecotus rafinesquii*). Proceedings of the Arkansas Academy of Science 47:102–106.

1019. Saugey, D. A., D. G. Saugey, G. A. Heidt, and D. R. Heath. 1988. The bats of Hot Springs National Park, Arkansas. Proceedings of the Arkansas Academy of Science 42:81–83.

1020. Scales, J. A., and K. T. Wilkins. 2007. Seasonality and fidelity in roost use of the Mexican free-tailed bat, *Tadarida brasiliensis*, in an urban setting. Western North American Naturalist 67:402–408.

1021. Scarbrough, D. L. 1989. Big free-tailed bat, *Tadarida macrotis* (Gray, 1839) from Brazos County, Texas. Texas Journal of Science 41:109.

1022. Scatterday, J. E., and M. G. Galton. 1954. Bat rabies in Florida. Veterinary Medicine 49:133–135.

1023. Scheel, D., T. L. S. Vincent, and G. N. Cameron. 1996. Global warming and the species richness of bats in Texas. Conservation Biology 10:452–464.

1024. Schmidly, D. J. 1977. The mammals of Trans-Pecos Texas. Texas A&M University Press, College Station.

1025. Schmidly, D. J. 1983. Texas mammals

east of the Balcones Fault zone. Texas A&M University Press, College Station.

1026. Schmidly, D. J. 1991. The bats of Texas. Texas A&M University Press, College Station.

1027. Schmidly, D. J. 2002. Texas natural history: a century of change. Texas Tech University Press, Lubbock.

1028. Schmidly, D. J. 2004. The mammals of Texas, 6th ed. University of Texas Press, Austin.

1029. Schmidly, D. J., and F. S. Hendricks. 1984. Mammals of the San Carlos Mountains of Tamaulipas, Mexico. Pp. 15–69 *in* Contributions in mammalogy in honor of Robert L. Packard (R. E. Martin and B. R. Chapman, eds.). Special Publications, Museum Texas Tech University, Vol. 22.

1030. Schmidly, D. J., K. T. Wilkins, R. L. Honeycutt, and B. C. Weynand. 1977. The bats of east Texas. Texas Journal of Science 28:127–143.

1031. Schmidt, C. A. 2003. Conservation assessment for the small-footed myotis in the Black Hills National Forest, South Dakota and Wyoming. United State Department of Agriculture, Forest Service, Custer, South Dakota. 15 pp.

1032. Schnitzler, H.-U., and E. K. V. Kalko. 2001. Echolocation by insect-eating bats. BioScience 51:557–568.

1033. Schnitzler, H.-U., E. K. V. Kalko, I. Kaipf, and A. D. Grinnell. 1994. Fishing and echolocation behavior of the greater bulldog bat, *Noctilio leporinus*, in the field. Behavioral Ecology and Sociobiology 35:327–345.

1034. Schowalter, D. B., L. D. Harder, and B. H. Treichel. 1978. Age composition of some vespertilionid bats as determined by dental annuli. Canadian Journal of Zoology 56:355–358.

1035. Schultz, J. G., C. D. Fisher, and S. Hightower. 1975. Recent records of the eastern big-eared bat (*Plecotus rafinesquii*) in eastern Texas. Southwestern Naturalist 20:144–145.

1036. Schutt, W. A., Jr., and J. S. Altenbach. 1997. A sixth digit in *Diphylla ecaudata*, the hairy-legged vampire bat (Chiroptera: Phyllostomidae). Mammalia 61:280–285.

1037. Schutt, W. A., Jr., W. Uieda, and A. Bredt. 1998. Roosting and feeding behavior of *Diphylla ecaudata*, the hairy-legged vampire bat: field observations using infrared videography. Bat Research News 39:186.

1038. Schwartz, A. 1955. The status of the species of the *brasiliensis* group of the genus *Tadarida*. Journal of Mammalogy 36:106–109.

1039. Schwartz, C., J. Tressler, H. Keller, M. Vanzant, S. Ezell, and M. Smotherman. 2007. The tiny difference between foraging and communication buzzes uttered by the Mexican free-tailed bat, *Tadarida brasiliensis*. Journal of Comparative Physiology A: Neuroethology, Sensory, Neural, and Behavioral Physiology 193:853–863.

1040. Schwartz, C. W., and E. R. Schwartz. 1981. The wild mammals of Missouri. University of Missouri Press, Columbia.

1041. Scudday, J. F. 1976a. Vertebrate fauna of the Fresno Canyon area. Pp. 97–110 *in* Fresno Canyon a natural survey area no. 10 Lyndon B. Johnson School of Public Affairs, University of Texas at Austin.

1042. Scudday, J. F. 1976b. Vertebrate fauna of the Colorado Canyon area, Presidio County, Texas. Pp. 99–114 *in* Colorado Canyon a natural survey area no. 11 Lyndon B. Johnson School of Public Affairs, University of Texas at Austin.

1043. Seager, R., et al. 2007. Model projections

of an imminent transition to a more arid climate in southwestern North America. Science Express doi: 10.1126/science.1139601.

1044. Sealander, J. A. 1979. A guide to Arkansas mammals. River Road Press, Conway, Arkansas.

1045. Selander, R. K., and J. K. Baker. 1957. The cave swallow in Texas. Condor 59:345–363.

1046. Semken, H. A., Jr. 1961. Fossil vertebrates from Longhorn Cavern, Burnett County, Texas. Texas Journal of Science 13:290–310.

1047. Serrano, H., J.-J. Pérez-Rivero, A. Aguilar-Setién, O. de-Paz, and A. Villa-Godoy. 2007. Vampire bat reproduction control by a naturally occurring phytoestrogen. Reproduction, Fertility, and Development 19:470–472.

1048. Sgro, M. P., and K. T. Wilkins. 2003. Roosting behavior of the Mexican free-tailed bat (Tadarida brasiliensis) in a highway overpass. Western North American Naturalist 63:366–373.

1049. Shamel, H. H. 1931. Notes on the American bats of the genus Tadarida. Proceedings of the U.S. National Museum 78:1–27.

1050. Sheeler-Gordon, L. L., and J. S. Smith. 2001. Survey of bat populations from Mexico and Paraguay for rabies. Journal of Wildlife Diseases 37:582–593.

1051. Sherman, H. B. 1930. Birth of the young of Myotis austroriparius. Journal of Mammalogy 11:495–503.

1052. Sherman, H. B. 1937. Breeding habits of the free-tailed bat. Journal of Mammalogy 18:176–187.

1053. Sherman, H. B. 1939. Notes on the food of some Florida bats. Journal of Mammalogy 20:103–140.

1054. Sherman, H. B. 1944. The Florida yellow bat, Dasypterus floridanus. Proceedings of the Florida Academy of Science 7:193–197.

1055. Sherwin, R. E., J. S. Altenbach, and D. L. Waldien. 2009. Managing abandoned mines for bats. Bat Conservation International, Austin.

1056. Sherwin, R. E., and W. L. Gannon. 2005. Documentation of an urban winter roost of the spotted bat (Euderma maculatum). Southwestern Naturalist 50:402–407.

1057. Sherwin, R. E., W. L. Gannon, and J. S. Altenbach. 2003. Managing complex systems simply: understanding inherent variation in the use of roosts by Townsend's big-eared bat. Wildlife Society Bulletin 31:62–72.

1058. Sherwin, R. E., W. L. Gannon, J. S. Altenbach, and D. Stricklan. 2000a. Roost fidelity of Townsend's big-eared bat in Utah and Nevada. Transactions of the Western Section of the Wildlife Society 36:15–20.

1059. Sherwin, R. E., D. Stricklan, and D. S. Rogers. 2000b. Roosting affinities of Townsend's big-eared bat (Corynorhinus townsendii) in northern Utah. Journal of Mammalogy 81:939–947.

1060. Short, H. L. 1961a. Growth and development of Mexican free-tailed bats. Southwestern Naturalist 6:156–163.

1061. Short, H. L. 1961b. Age at sexual maturity of Mexican free-tailed bats. Journal of Mammalogy 42:533–536.

1062. Short, H. L., R. B. Davis, and I. C. F. Herreid. 1960. Movements of the Mexican free-tailed bat in Texas. Southwestern Naturalist 5:208–216.

1063. Shryer, J. S., and D. S. Flath. 1980. First record of the pallid bat (Antrozous pallidus) from Montana. Great Basin Naturalist 40:115.

1064. Shull, A. J. 1988. Endangered and threatened wildlife and plants; determination of endangered status for two long-nosed bats. Federal Register 53:3846–3860.

1065. Simmons, N. B. 1994. The case of chiropteran monophyly. American Museum Novitates 3103:1–46.

1066. Simmons, N. B. 2005. Order Chiroptera. Pp. 312–529 in Mammal species of the world: a taxonomic and geographic reference (D. E. Wilson and D. M. Reeder, eds.). 3rd ed. Johns Hopkins University Press, Baltimore.

1067. Simmons, N. B., and T. M. Conway. 2003. Evolution of ecological diversity in bats. Pp. 493–535 in Bat Ecology (T. H. Kunz and M. B. Fenton, eds.). University of Chicago Press, Chicago.

1068. Simmons, N. B., and J. H. Geisler. 1998. Phylogenetic relationships of Icaronycteris, Archaeonycteris, Hassianycteris, and Palaeochiropteryx to extant bat lineages, with comments on the evolution of echolocation and foraging strategies in Microchiroptera. Bulletin of the American Museum of Natural History 25:1–182.

1069. Simmons, N. B., M. J. Novacek, and R. J. Baker. 1991. Approaches, methods, and future of the chiropteran monophyly controversy: a reply to J. D. Pettigrew. Systematic Zoology 40:239–243.

1070. Simmons, N. B., K. L. Seymour, J. Habersetzer, and G. F. Gunnell. 2008. Primitive Early Eocene bat from Wyoming and the evolution of flight and echolocation. Science 451:818–822.

1071. Simpson, L. A., and T. C. Maxwell. 1989. The mammal fauna of Coke County, Texas. Texas Journal of Science 41:177–192.

1072. Slaughter, B. H., and W. L. McClure. 1965. The Sims Bayou local fauna: Pleistocene of Houston, Texas. Texas Journal of Science 17:404–417.

1073. Smith, D. D. 1975. Record of the red bat in Brewster County, Texas. Texas Journal of Science 26:601–602.

1074. Smith, J. D. 1972. Systematics of the Chiropteran family Mormoopidae. University of Kansas Publications, Museum of Natural History 56:1–132.

1075. Smith, S. J., D. M. Leslie Jr., M. J. Hamilton, J. B. Lack, and R. A. Van Den Bussche. 2008. Subspecific affinities and conservation genetics of western big-eared bats (Corynorhinus townsendii pallescens) at the edge of their distributional range. Journal of Mammalogy 89:799–814.

1076. Smith, T., and R. Tumlison. 2004. An evaluation of geographic variation within an isolated population of big-eared bats (Corynorhinus townsendii) in Oklahoma, Kansas and Texas. Proceeding of the Oklahoma Academy of Sciences 84:1–7.

1077. Smotherman, M., and A. Guillén-Servent. 2008. Doppler-shift compensation behavior by Wagner's mustached bat, Pteronotus personatus. Journal of the Acoustical Society of America 123:4331–4339.

1078. Snow, T. K. 1996. Western red bat. Wildlife Views 39:5.

1079. Sparks, D. W., M. T. Simmons, C. L. Gummer, and J. E. Duchamp. 2003. Disturbance of roosting bats by woodpeckers and raccoons. Northeastern Naturalist 10: 105–108.

1080. Sparks, D. W., and E. W. Valdez. 2003. Food habits of Nyctinomops macrotis at a maternity roost in New Mexico, as indicated by analysis of guano. Southwestern Naturalist 48:132–135.

1081. Speakman, J. R. 1991. The impact of predation by birds on bat populations in the British Isles. Mammal Review 21:123–142.

1082. Spencer, S. G., P. C. Choucair, and B. R.

Chapman. 1988. Northward expansion of the southern yellow bat, *Lasiurus ega*, in Texas. Southwestern Naturalist 33:493.

1083. Spenrath, C. A., and R. K. LaVal. 1974. An ecological study of a resident population of *Tadarida brasiliensis* in eastern Texas. Occasional Papers, Museum of Texas Tech University 21:1–14.

1084. Spradling, K. D., F. B. Stangl Jr., and W. B. Cook. 2003. Evidence for a case of multiple paternity in the red bat (*L. borealis*) as indicated by DNA fingerprinting. Occasional Papers, Museum of Texas Tech University 224:1–10.

1085. Sprunt, A., Jr. 1950. Hawk predation at the bat caves of Texas. Texas Journal of Science 2:463–470.

1086. Stager, K. E. 1941. A group of bat-eating duck hawks. Condor 43:137–139.

1087. Stager, K. E. 1942. A new free-tailed bat from Texas. Bulletin of the Southern California Academy of Sciences 41:49–50.

1088. Stager, K. E. 1948. Falcons prey on Ney Cave bats. National Speleological Society Bulletin 10:49–50.

1089. Stains, H. J. 1957. A new bat (genus *Leptonycteris*) from Coahuila. University of Kansas Publications, Museum of Natural History 9:353–356.

1090. Stallcup, W. B. 1956. Notes on mammals of Dallas County, Texas. Field and Laboratory 24:96–101.

1091. Stangl, F. B., Jr., P. G. Christiansen, and E. J. Galbraith. 1993. Abbreviated guide to pronunciation and etymology of scientific names for North American land mammals north of Mexico. Occasional Papers, Museum of Texas Tech University 154:1–28.

1092. Stangl, F. B., Jr., W. W. Dalquest, and J. V. Grimes. 1996. Observations on the early life history, growth, and development of the red bat, *L. borealis* (Chiroptera: Vespertilionidae) in North Texas. Pp. 139–148 *in* Contributions in mammalogy: a memorial volume honoring Dr. J. Knox Jones Jr. (H. H. Genoways and R. J. Baker, eds.). Museum of Texas Tech University.

1093. Stangl, F. B., Jr., W. W. Dalquest, and R. R. Hollander. 1994. Evolution of a desert fauna: a 10,000-year history of mammals from Culberson and Jeff Davis counties, Trans-Pecos, Texas. Midwestern State University Press, Wichita Falls, Texas.

1094. Stangl, F. B., Jr., S. Kasper, and T. S. Schafer. 1989. Noteworthy range extensions and marginal distributional records for five species of Texas mammals. Texas Journal of Science 41:436–437.

1095. Steece, R., and J. S. Altenbach. 1989. Prevalence of rabies specific antibodies in the Mexican free-tailed bat (*Tadarida brasiliensis mexicana*) at Lava Cave, New Mexico. Journal of Wildlife Diseases 25:490–496.

1096. Stevenson, D. E., and M. D. Tuttle. 1981. Survivorship in the endangered gray bat (*Myotis grisescens*). Journal of Mammalogy 62:244–257.

1097. Stokstad, E. 2010. Europe's bats resist fungal scourge of North America. Science 327:132.

1098. Storer, T. I. 1926. Bats, bat towers and mosquitoes. Journal of Mammalogy 7:85–91.

1099. Storz, J. F. 1995. Local distribution and foraging behavior of the spotted bat (*Euderma maculatum*) in northwestern Colorado and adjacent Utah. Great Basin National Park 55:78–83.

1100. Straney, D. O., M. H. Smith, R. J. Baker, and I. F. Greenbaum. 1976. Biochemical variation and genic similarity of *Myotis velifer* and *Macrotus californicus*. Comparative Biochemistry and Physiology B 54:43–48.

1101. Strecker, J. K. 1910. Notes on the fauna of northwestern Texas. Baylor University Bulletin 18:1–31.

1102. Strecker, J. K. 1924. The mammals of McLennon County, Texas. Baylor University Bulletin 27:1–20.

1103. Studier, E. M., and A. A. Fresquez. 1969. Carbon dioxide retention: a mechanism of ammonia tolerance in mammals. Ecology 50:492–494.

1104. Stuewer, F. W. 1948. A record of red bats mating. Journal of Mammalogy 29:180–181.

1105. Suchecki, J. R., et al. 2003. *Lasiurus ega* and other small mammals records from Dimmit and La Salle counties, Texas. Occasional Papers, Museum of Texas Tech University 225:1–3.

1106. Sulkin, S. E., and M. J. Greve. 1954. Human rabies caused by bat bite. Texas State Journal of Medicine 50:620–621.

1107. Sullivan, T. D., J. E. Grimes, R. B. Eads, G. C. Menzies, and J. V. Irons. 1954. Recovery of rabies virus from colonial, insectivorous bats in Texas. Public Health Reports 69:766–768.

1108. Svoboda, R. L., J. R. Choate, and R. K. Chesser. 1985. Genetic relationships among southwestern populations of the Brazilian free-tailed bat. Journal of Mammalogy 66: 444–450.

1109. Szewczak, J. M., S. M. Szewczak, M. L. Morrison, and L. S. Hall. 1998. Bats of the White and Inyo Mountains of California-Nevada. Great Basin Naturalist 58:66–75.

1110. Tamsitt, J. R. 1954. The mammals of two areas in the Big Bend region of Trans-Pecos Texas. Texas Journal of Science 6:33–61.

1111. Tatarian, G. 1999. Use of buildings and tolerance of disturbance by pallid bats, *Antrozous pallidus*. Bat Research News 40: 11–12.

1112. Taylor, W. P., and W. B. Davis. 1947. The mammals of Texas. Texas Game, Fish, and Oyster Commission, Bulletin No. 27, Austin.

1113. Teeling, E. C., O. Madsen, R. A. Van Den Bussche, W. Jong, M. Stanhope, and M. Springer. 2002. Microbat monophyly and the convergent evolution of a key innovation in old world rhinolophoid microbats. Proceedings of the National Academy of Sciences 99:1431–1436.

1114. Teeling, E. C., M. S. Springer, O. Madsen, P. Bates, S. J. O'Brien, and W. J. Murphy. 2005. A molecular phylogeny for bats illuminates biogeography and the fossil record. Science 307:580–584.

1115. Téllez, J. G. 2001. Migración de los murciélagos-hocicudos (*Leptonycteris*) en el trópico mexicano. Ph.D. dissertation, Universidad Nacional Autónoma de México, Mexico City.

1116. Thies, M. 1994. A new record for *Plecotus rafinesquii* (Chiroptera: Vespertilionidae) from east Texas. Texas Journal of Science 225:1–3.

1117. Thies, M. L., and K. M. Thies. 1997. Organochlorine residues in bats from Eckert James River Cave, Texas. Bulletin of Environmental Contamination and Toxicology 58:673–680.

1118. Thomas, D. W. 1993. Lack of evidence for a biological alarm clock in bats (*Myotis* spp.) hibernating under natural conditions. Canadian Journal of Zoology 71:1–3.

1119. Thomas, D. W. 1995. Hibernating bats are sensitive to nontactile human disturbance. Journal of Mammalogy 76:940–946.

1120. Tibbets, T. 1956. Homing instincts of two bats, *Eptesicus fuscus* and *Tadarida mexicana* (Mammalia: Chiroptera). Southwestern Naturalist 1:194.

1121. Timpone, J. C., J. C. Boyles, and L. W.

Robbins. 2006. Potential for niche overlap in roosting sites between *Nycticeius humeralis* (evening bats) and *Eptesicus fuscus* (big brown bats). Northeastern Naturalist 13: 597–602.

1122. Tinkle, D. W., and W. W. Milstead. 1960. Sex ratios and population density in hibernating *Myotis*. American Midland Naturalist 63:327–334.

1123. Tinkle, D. W., and I. G. Patterson. 1965. A study of hibernating populations of *Myotis velifer* in northwestern Texas. Journal of Mammalogy 46:612–633.

1124. Tipps, T. M., B. Mayes, and L. K. Ammerman. (in press). New county records for six species of bats (Vespertilionidae and Molossidae) in Texas. Texas Journal of Science.

1125. Trousdale, A. W., and D. C. Beckett. 2004. Seasonal use of bridges by Rafinesque's big-eared bat, *Corynorhinus rafinesquii*, in southern Mississippi. Southeastern Naturalist 3:103–112.

1126. Trousdale, A. W., and D. C. Beckett. 2005. Characteristics of tree roosts of Rafinesque's big-eared bat (*Corynorhinus rafinesquii*) in southeastern Mississippi. American Midland Naturalist 154:442–449.

1127. Trune, D. R., and C. N. Slobodchikoff. 1976. Social effects of roosting on the metabolism of the pallid bat (*Antrozous pallidus*). Journal of Mammalogy 57:656–663.

1128. Trune, D. R., and C. N. Slobodchikoff. 1978. Position of immatures in pallid bat clusters: a case of reciprocal altruism? Journal of Mammalogy 59:193–195.

1129. Tumlison, R., and M. E. Douglas. 1992. Parsimony analysis and the phylogeny of the Plecotine bats (Chiroptera: Vespertilionidae). Journal of Mammalogy 73: 276–285.

1130. Turner, G. G., and D. M. Reeder. 2009. Update of white-nose syndrome in bats, September 2009. Bat Research News 50: 47–53.

1131. Tuttle, M. D. 1974. An improved trap for bats. Journal of Mammalogy 55:475–477.

1132. Tuttle, M. D. 1988. America's neighborhood bats. University of Texas Press, Austin.

1133. Tuttle, M. D. 2003. Texas bats. Bat Conservation International, Austin.

1134. Tuttle, M. D., and L. R. Heaney. 1974. Maternity habits of *Myotis leibii* in South Dakota. Bulletin of Southern California Academy of Science 73:80–83.

1135. Tuttle, M. D., and D. Stevenson. 1982. Growth and survival of bats. Pp. 105–150 *in* Ecology of bats (T. H. Kunz, ed.). Plenum, New York.

1136. Twente, J. W., Jr. 1954. Habitat selection of cavern-dwelling bats as illustrated by four vespertilionids. Ph.D. dissertation, University of Michigan, Ann Arbor.

1137. Twente, J. W., Jr. 1955a. Some aspects of habitat selection and other behavior of cavern-dwelling bats. Ecology 36:706–732.

1138. Twente, J. W., Jr. 1955b. Aspects of a population study of cavern-dwelling bats. Journal of Mammalogy 36:379–390.

1139. U.S. Fish and Wildlife Service. 1994. Mexican long-nosed bat (*Leptonycteris nivalis*) recovery plan. U.S. Fish and Wildlife Service, Albuquerque, New Mexico. 91 pp.

1140. U.S. Fish and Wildlife Service. 2010. A national plan for assisting states, federal agencies, and tribes in managing white-nose syndrome in bats. 16 pp.

1141. Ubelaker, J. E., R. D. Specian, and D. W. Duszynski. 1977. Endoparasites. Pp. 7–56 *in* Biology of bats of the New World family Phyllostomatidae, part II (R. J. Baker, J. K.

Jones Jr., and D. C. Carter, eds.). Special Publications, Museum of Texas Tech University.

1142. Valdez, E. W. 2006. Geographic variation in morphology, diet, and ectoparasites of the bat *Myotis occultus* in New Mexico and southern Colorado. Ph.D. dissertation, University of New Mexico, Albuquerque.

1143. Valdez, E. W., J. R. Choate, M. A. Bogan, and T. L. Yates. 1999a. Taxonomic status of *Myotis occultus*. Journal of Mammalogy 80: 545–552.

1144. Valdez, E. W., and P. M. Cryan. 2009. Food habits of the hoary bat (*Lasiurus cinereus*) during spring migration through New Mexico. Southwestern Naturalist 54: 195–200.

1145. Valdez, E. W., K. Geluso, J. Foote, G. Allison-K., and D. M. Roemer. 2009a. Spring and winter records of the eastern pipistrelle (*Perimyotis subflavus*) in southeastern New Mexico. Western North American Naturalist 69:396–398.

1146. Valdez, E. W., C. M. Ritzi, and J. O. Whitaker Jr. 2009b. Ectoparasites of the occult bat, *Myotis occultus* (Chiroptera: Vespertilionidae). Western North American Naturalist 69:364–370.

1147. Valdez, E. W., J. N. Stuart, and M. A. Bogan. 1999b. Additional records of bats from the middle Rio Grande Valley New Mexico. Southwestern Naturalist 44:398–400.

1148. Valiente-Banuet, A., M. del Coro Arizmendi, A. Rojas-Martínez, and L. Domínguez-Canseco. 1996. Ecological relationships between columnar cacti and nectar-feeding bats in Mexico. Journal of Tropical Ecology 12:103–119.

1149. Van Deusen, H. M. 1961. Yellow bat collected over south Atlantic. Journal of Mammalogy 42:530–531.

1150. Van Devender, T. R., G. L. Bradley, and A. H. Harris. 1987. Late Quaternary mammals from the Hueco Mountains, El Paso and Hudspeth counties, Texas. Southwestern Naturalist 32:179–195.

1151. van Zyll de Jong, C. G. 1979. Distribution and systematic relationships of long-eared *Myotis* in western Canada. Canadian Journal of Zoology 57:987–994.

1152. van Zyll de Jong, C. G. 1984. Taxonomic relationships of Nearctic small-footed bats of the *Myotis leibii* group (Chiroptera: Vespertilionidae). Canadian Journal of Zoology 62:2519–2526.

1153. van Zyll de Jong, C. G. 1985. Handbook of Canadian mammals, Vol. 2. National Museum of Canada, Ottawa.

1154. Vaughan, T. A. 1959. Functional morphology of three bats: *Eumops, Myotis, Macrotus*. University of Kansas Publications, Museum of Natural History 12:1–153.

1155. Vaughan, T. A. 1966. Morphology and flight characteristics of molossid bats. Journal of Mammalogy 47:249–260.

1156. Vaughan, T. A. 1980. Opportunistic feeding by two species of *Myotis*. Journal of Mammalogy 61:118–119.

1157. Vaughan, T. A., and G. C. Bateman. 1970. Functional morphology of the forelimb of Mormoopid bats. Journal of Mammalogy 51: 217–235.

1158. Vaughan, T. A., and T. J. O'Shea. 1976. Roosting ecology of the pallid bat, *Antrozous pallidus*. Journal of Mammalogy 57:19–42.

1159. Veilleux, J. P. 2008. Current status of white-nose syndrome in the northeastern United States. Bat Research News 49: 15–17.

1160. Veilleux, J. P., and S. L. Veilleux. 2004. Intra-annual and interannual fidelity to summer roost areas by female eastern pipistrelles, *Pipistrellus subflavus*. American Midland Naturalist 152:196–200.

1161. Veilleux, J. P., J. O. Whitaker Jr., and S. L. Veilleux. 2003. Tree-roosting ecology of reproductive female eastern pipistrelles, *Pipistrellus subflavus*, in Indiana. Journal of Mammalogy 84:1068–1075.

1162. Veilleux, J. P., J. O. Whitaker Jr., and S. L. Veilleux. 2004. Reproductive stage influences roost use by tree roosting female eastern pipistrelles, *Pipistrellus subflavus*. Ecoscience 11:249–256.

1163. Villa-R., B. 1955. Una extraña y severa mortandad de murciélagos *Mormoops megalophylla* en el norte de México. Anales de Instituto de Biología, Universidad Nacional Autónoma de México 26:547–552.

1164. Villa-R., B. 1966. Los murciélagos de México. Universidad Nacional Autónoma de México, Instituto de Biología, Mexico City.

1165. Villa-R., B., and E. L. Cockrum. 1962. Migration in the guano bat *Tadarida brasiliensis mexicana* (Sassure). Journal of Mammalogy 43:43–64.

1166. Villa-R., B., and M. A. A. Jiménez Guzmán. 1960. Acerca de la posición taxonómica de *Mormoops megalophylla senicula* Rehn, y la presencia de virus rabico en estos murciélagos insectivoros. Anales de Instituto de Biología, Universidad Nacional Autónoma de México 31:501–509.

1167. Villegas-Guzman, G. A., C. López-Gonzáles, and M. Vargas. 2005. Ectoparasites associated to [*sic*] two species of *Corynorhinus* (Chiroptera: Vespertilionidae) from the Guanaceví mining region, Durango, Mexico. Journal of Medical Entomology 42:125–127.

1168. Voigt, C. C., and O. von Helversen. 1999. Storage and display of odour by male *Saccopteryx bilineata* (Chiroptera, Emballonuridae). Behavioral Ecology and Sociobiology 47:29–40.

1169. Vonhof, M. J. 1996. Roost-site preferences of big brown bats (*Eptesicus fuscus*) and silver-haired bats (*Lasionycteris noctivagans*) in the Pend d'Oreille Valley in southern British Columbia. Pp. 62–80 *in* Bats and forests symposium (R. M. R. Barclay and R. M. Brigham, eds.). British Columbia Ministry of Forestry, Victoria, British Columbia.

1170. Vonhof, M. J., D. Barber, and M. B. Fenton. 2006. A tale of two siblings: multiple paternity in big brown bats (*Eptesicus fuscus*) demonstrated using microsatellite markers. Molecular Ecology 15:241–247.

1171. Vonhof, M. J., and R. M. R. Barclay. 1996. Roost-site selection and roosting ecology of forest-dwelling bats in southern British Columbia. Canadian Journal of Zoology 74: 1797–1805.

1172. Vonhof, M. J., C. Strobeck, and M. B. Fenton. 2008. Genetic variation and population structure in big brown bats (*Eptesicus fuscus*): is female dispersal important? Journal of Mammalogy 89: 1411–1420.

1173. Wahl, R. 1993. Important Mexican free-tailed bat colonies in Texas. Pp. 47–50 *in* 1989 National Cave Management: Proceedings of the Symposium (J. Jordan and R. Obele, eds.). Texas Parks and Wildlife Department, Austin.

1174. Wai-Ping, V., and M. B. Fenton. 1989. Ecology of spotted bat (*Euderma maculatum*) roosting and foraging behavior. Journal of Mammalogy 70:617–622.

1175. Walker, C. W., J. K. Sandel, R. L. Honeycutt, and C. Adams. 1996. Winter utilization of box culverts by vespertilionid bats in southeast Texas. Texas Journal of Science 48:166–168.

1176. Walton, D. W., and J. D. Kimbrough. 1970. *Eumops perotis* from Black Gap Wildlife Refuge. Southwestern Naturalist 15:131–143.

1177. Walton, D. W., and N. J. Siegel. 1966. The histology of the pararhinal glands of the pallid bat, *Antrozous pallidus*. Journal of Mammalogy 47:357–360.

1178. Warner, J. W., J. L. Patton, A. L. Gardner, and R. J. Baker. 1974. Karyotypic analysis of twenty-one species of molossid bats (Molossidae: Chiroptera). Canadian Journal of Genetics and Cytology 16:165–176.

1179. Warner, R. M. 1985. Interspecific and temporal dietary variation in an Arizona bat community. Journal of Mammalogy 66: 45–51.

1180. Watkins, L. C. 1969. Observations on the distribution and natural history of the evening bat (*Nycticeius humeralis*) in northwestern Missouri and adjacent Iowa. Transactions of the Kansas Academy of Science 72:330–336.

1181. Watkins, L. C., and K. A. Shump Jr. 1981. Behavior of the evening bat, *Nycticeius humeralis*, at a nursery roost. American Midland Naturalist 105:258–268.

1182. Watkins, S. 1956. The old man of Frio Cave. Texas Game and Fish 14:26–27.

1183. Webster, F. A., and D. R. Griffin. 1962. The role of the flight membranes in insect capture by bats. Animal Behaviour 10:332–340.

1184. Weinbeer, M., C. F. J. Meyer, and E. K. V. Kalko. 2006. Activity pattern of the trawling phyllostomid bat, *Macrophyllum macrophyllum*, in Panama. Biotropica 38: 69–76.

1185. Weller, T. J., S. A. Scott, T. J. Rodhouse, P. C. Ormsbee, and J. M. Zinck. 2007. Field identification of the cryptic vespertilionid

bats, *Myotis lucifugus* and *M. yumanensis*. Acta Chiropterologica 9:133–147.

1186. Weller, T. J., and C. J. Zabel. 2001. Characteristics of fringed myotis day roosts in northern California. Journal of Wildlife Management 65:489–497.

1187. Werner, H. J. 1966. Observations of the facial glands of the guano bat *Tadarida brasiliensis mexicana* (Sassure). Proceedings of the Louisiana Academy of Science 29: 156–160.

1188. Wethington, T. A., D. M. Leslie Jr., M. S. Gregory, and M. K. Wethington. 1996. Prehibernation habitat use and foraging activity by endangered Ozark big-eared bats (*Plecotus townsendii ingens*). American Midland Naturalist 135:218–230.

1189. Wethington, T. A., D. M. Leslie Jr., M. S. Gregory, and M. K. Wethington. 1997. Vegetative structure and land use relative to cave selection by endangered Ozark big-eared bats (*Corynorhinus townsendii ingens*). Southwestern Naturalist 42:177–181.

1190. Weyandt, S. E., T. E. Lee Jr., and J. C. Patton. 2001. Noteworthy record of the western yellow bat, *Lasiurus xanthinus* (Chiroptera: Vespertilionidae), and a report on the bats of Eagle Nest Canyon, Val Verde County, Texas. Texas Journal of Science 53: 289–292.

1191. Weyandt, S. E., and R. A. Van Den Bussche. 2007. Phylogeographic structuring and volant mammals: the case of the pallid bat (*Antrozous pallidus*). Journal of Biogeography 34:1233–1245.

1192. Weyandt, S. E., R. A. Van Den Bussche, M. J. Hamilton, and D. M. Leslie Jr. 2005. Unraveling the effects of sex and dispersal: Ozark big-eared bat (*Corynorhinus townsendii ingens*) conservation genetics. Journal of Mammalogy 86:1136–1143.

1193. Whitaker, J. O., Jr. 1967. Hoary bat apparently hibernating in Indiana. Journal of Mammalogy 48:663.

1194. Whitaker, J. O., Jr. 1972. Food habits of bats from Indiana. Canadian Journal of Zoology 50:877–883.

1195. Whitaker, J. O., Jr. 1995. Food of the big brown bat *Eptesicus fuscus* from maternity colonies in Indiana and Illinois. American Midland Naturalist 134:346–360.

1196. Whitaker, J. O., Jr. 1998. Life history and roost switching in six summer colonies of eastern pipistrelles in buildings. Journal of Mammalogy 79:651–659.

1197. Whitaker, J. O., Jr. 2004. Prey selection in a temperate zone insectivorous bat community. Journal of Mammalogy 85: 460–469.

1198. Whitaker, J. O., Jr., V. Brack, and J. B. Cope. 2006. Are bats in Indiana declining? Proceedings of the Indiana Academy of Science 111:95–96.

1199. Whitaker, J. O., Jr., and P. Clem. 1992. Food of the evening bat, *Nycticeius humeralis*, from Indiana. American Midland Naturalist 127:211–214.

1200. Whitaker, J. O., Jr., and D. A. Easterla. 1975. Ectoparasites of bats from Big Bend National Park, Texas. Southwestern Naturalist 20:241–254.

1201. Whitaker, J. O., Jr., D. A. Easterla, and A. Fain. 1987. First record of the mite, *Ewingana* (Doreyana) *doreyae* Dusbabek 1968 (Acarina, Myobiidae) from the United States with notes on streblid flies from Big Bend National Park, Texas. Southwestern Naturalist 32:505.

1202. Whitaker, J. O., Jr., and S. L. Gummer. 1992. Hibernation of the big brown bat, *Eptesicus fuscus*, in buildings. Journal of Mammalogy 73:312–316.

1203. Whitaker, J. O., Jr., and S. L. Gummer. 2000. Population structure and dynamics of big brown bats (*Eptesicus fuscus*) hibernating in buildings in Indiana. American Midland Naturalist 143:389–396.

1204. Whitaker, J. O., Jr., and W. J. Hamilton Jr. 1998. Mammals of the eastern United States. Cornell University Press, Ithaca.

1205. Whitaker, J. O., Jr., C. Maser, and S. P. Cross. 1981a. Foods of Oregon silver-haired bats, *Lasionycteris noctivagans*. Northwest Science 55:75–77.

1206. Whitaker, J. O., Jr., C. Maser, and S. P. Cross. 1981b. Food habits of eastern Oregon bats, based on stomach and scat analysis. Northwest Science 55:281–292.

1207. Whitaker, J. O., Jr., C. Maser, and L. E. Keller. 1977. Food habits of bats of western Oregon. Northwest Science 51:46–55.

1208. Whitaker, J. O., Jr., C. Neefus, and T. H. Kunz. 1996. Dietary variation in the Mexican free-tailed bat (*Tadarida brasiliensis mexicana*). Journal of Mammalogy 77: 716–724.

1209. Whitaker, J. O., Jr., and L. J. Rissler. 1992. Winter activity of bats at a mine entrance in Vermillion County, Indiana. American Midland Naturalist 127:52–59.

1210. Whitaker, J. O., Jr., C. M. Ritzi, and C. W. Dick. 2009. Collecting and preserving ectoparasites for ecological study. Pp. 806–827 *in* Ecological and behavioral methods for the study of bats (T. H. Kunz and S. Parsons, eds.). 2nd ed. Johns Hopkins University Press, Baltimore.

1211. Whitaker, J. O., Jr., R. K. Rose, and T. M. Padgett. 1997. Food of the red bat *Lasiurus borealis* in winter in the Great Dismal Swamp, North Carolina and Virginia. American Midland Naturalist 137:408–411.

1212. Whitaker, J. O., Jr., B. L. Walters, L. K.

Castor, C. M. Ritzi, and N. Wilson. 2007. Host and distribution lists of mites (Acari), parasitic and phoretic, in the hair of wild mammals of North America: records since 1974. Faculty Publications from the Harlod W. Manter Laboratory of Parasitology, University of Nebraska, Lincoln.

1213. Whitaker, J. O., Jr., and N. Wilson. 1974. Host and distribution lists of mites (Acari), parasitic and phoretic, in the hair of wild mammals of North America. American Midland Naturalist 91:1–67.

1214. White, J. A., P. R. Moosman Jr., C. H. Kilgore, and T. L. Best. 2006. First record of the eastern pipistrelle (*Pipistrellus subflavus*) from southern New Mexico. Southwestern Naturalist 51:420–422.

1215. Wibbelt, G., A. Kruth, D. Hellmann, W. Weishaar, A. Barlow, M. Veith, et al. 2010. White-nose syndrome fungus (*Geomyces destructans*) in bats, Europe. Emerging Infectious Diseases 16:1237–1242.

1216. Wilhide, J. D., M. J. Harvey, V. R. McDaniel, and V. E. Hoffman. 1998a. Highland pond utilization by bats in the Ozark National Forest, Arkansas. Journal of the Arkansas Academy of Science 52: 110–112.

1217. Wilhide, J. D., R. V. McDaniel, M. J. Harvey, and D. R. White. 1998b. Telemetric observations of foraging Ozark big-eared bats in Arkansas. Journal of the Arkansas Academy of Science 52:113–116.

1218. Wilkins, K. T., W. J. Boeer, D. S. Rogers, and W. S. Modi. 1979. Records for eight Texas mammals. Florida Science 42:59–60.

1219. Wilkinson, G. S. 1992a. Information transfer at evening bat colonies. Animal Behaviour 44:501–518.

1220. Wilkinson, G. S. 1992b. Communal nursing in the evening bat, *Nycticeius humeralis*. Behavioral Ecology and Sociobiology 31:225–235.

1221. Wilkinson, G. S., and G. F. McCracken. 2003. Bats and balls: sexual selection and sperm competition in the Chiroptera. Pp. 128–155 *in* Bat Ecology (T. H. Kunz and M. B. Fenton, eds.). University of Chicago Press, Chicago.

1222. Wilkinson, L. C., and R. M. R. Barclay. 1997. Differences in the foraging behaviour of male and female big brown bats (*Eptesicus fuscus*) during the reproductive period. Ecoscience 4:279–285.

1223. Wilks, B. J., and H. E. Laughlin. 1961. Roadrunner preys on a bat. Journal of Mammalogy 49:98.

1224. Williams, J. A., M. J. O'Farrell, and B. R. Riddle. 2006. Habitat use by bats in a riparian corridor of the Mojave Desert in southern Nevada. Journal of Mammalogy 87: 1145–1153.

1225. Williams, L. M., and M. C. Brittingham. 1997. Selection of maternity roosts by big brown bats. Journal of Wildlife Management 61:359–368.

1226. Williams, T. C., L. C. Ireland, and J. M. Williams. 1973. High altitude flights of the free-tailed bat, *Tadarida brasiliensis*, observed with radar. Journal of Mammalogy 54: 807–821.

1227. Willis, C. K. R., and R. M. Brigham. 2004. Roost switching, roost sharing, and social cohesion: forest-dwelling big brown bat, *Eptesicus fuscus,* conform to the fission-fusion model. Animal Behaviour 68:495–505.

1228. Willis, C. K. R., and R. M. Brigham. 2005. Physiological and ecological aspects of roost selection by reproductive female hoary bats (*Lasiurus cinereus*). Journal of Mammalogy 86:85–94.

1229. Willis, C. K. R., R. M. Brigham, and

F. Geiser. 2006a. Deep, prolonged torpor by pregnant free-ranging bats. Natural Sciences 93:80–83.

1230. Willis, C. K. R., K. A. Kolar, A. L. Karst, M. C. Kalcounis-Rueppell, and R. M. Brigham. 2003. Medium- and long-term reuse of trembling aspen cavities as roosts by big brown bats (*Eptesicus fuscus*). Acta Chiropterologica 5:85–90.

1231. Willis, C. K. R., C. M. Voss, and R. M. Brigham. 2006b. Roost selection by forest-living female big brown bats (*Eptesicus fuscus*). Journal of Mammalogy 87:345–350.

1232. Wilson, D. E. 1979. Reproductive patterns. Pp. 317–378 *in* Biology of bats of the New World family Phyllostomatidae, Part III (R. J. Baker, J. J. K. Jones, and D. C. Carter, eds.). Special Publications, Museum of Texas Tech University, Vol. 16.

1233. Wilson, D. E. 1985. Status report: *Leptonycteris nivalis* (Sassure), Mexican long-nosed bat. P. 33 in Unpublished report prepared for Office of Endangered Species, U.S. Fish and Wildlife Service.

1234. Wilson, D. E., R. A. Medellín, D. V. Lanning, and H. T. Arita. 1985. Los murciélagos del noreste de México. Acta Zoologica Mexicana nueva serie 8:1–26.

1235. Wilson, D. E., and S. Ruff, eds. 1999. The Smithsonian book of North American mammals. Smithsonian Institution Press, Washington, D.C.

1236. Wimsatt, W. A. 1945. Notes on breeding behavior, pregnancy, and parturition in some vespertilionid bats of the eastern United States. Journal of Mammalogy 26:23–33.

1237. Winchell, J. M., and T. H. Kunz. 1996. Day-roosting activity budgets of the eastern pipistrelle bat, *Pipistrellus subflavus* (Chiroptera: Vespertilionidae). Canadian Journal of Zoology 74:431–441.

1238. Winhold, L., A. Kurta, and R. Foster. 2008. Long-term change in an assemblage of North American bats: are eastern red bats declining? Acta Chiropterologica 10: 359–366.

1239. Wiseman, J. S. 1963. Predation by the Texas rat snake on the hoary bat. Journal of Mammalogy 44:581.

1240. Wiseman, J. S., B. L. Davis, and J. E. Grimes. 1962. Rabies infection in the red bat, *Lasiurus borealis borealis* (Muller), in Texas. Journal of Mammalogy 43:279–280.

1241. Wolf, W. W., J. K. Westbrook, J. R. Raulston, S. D. Pair, and P. D. Lingren. 1995. Radar observations of orientation of noctuids migrating from corn fields in the lower Rio-Grande Valley. Southwestern Entomologist 18:45–61.

1242. Wood, W. F., and J. M. Szewczak. 2007. Volatile antimicrobial compounds in the pelage of the Mexican free-tailed bat, *Tadarida brasiliensis mexicana*. Biochemical Systematics and Ecology 35:566–568.

1243. Woodsworth, G. C. 1981. Spatial partitioning by two species of sympatric bats, *Myotis californicus* and *Myotis leibii*. M.S. Thesis, Carleton University, Ottawa.

1244. Woodsworth, G. C., G. P. Bell, and M. B. Fenton. 1981. Observations of the echolocation, feeding behaviour, and habitat use of *Euderma maculatum* (Chiroptera: Vespertilionidae) in southcentral British Columbia. Canadian Journal of Zoology 59: 1099–1102.

1245. Wotton, J. M., T. Haresign, and J. A. Simmons. 1995. Spatially dependent acoustic cues generated by the external ear of the big brown bat, *Eptesicus fuscus*. Journal of the Acoustical Society of America 98:1423.

1246. Yancey, F. D., II. 1997. The mammals of Big Bend Ranch State Park, Texas. Special

Publications, Museum of Texas Tech University 39:1–210.

1247. Yancey, F. D., II, J. R. Goetze, B. M. Gharaibeh, and C. Jones. 1995a. Distributional records of small mammals from the southwestern rolling plains of Texas. Texas Journal of Science 47:101–105.

1248. Yancey, F. D., II, J. R. Goetze, C. Jones, and R. W. Manning. 1996. The mammals of Justiceburg Wildlife Management Area and adjacent areas, Garza and Kent counties, Texas. Occasional Papers, Museum of Texas Tech University 161:1–26.

1249. Yancey, F. D., II, and C. Jones. 1996a. Notes on three species of small mammals from the Big Bend region of Texas. Texas Journal of Science 48:247–250.

1250. Yancey, F. D., II, and C. Jones. 1996b. New county records for ten species of bats (Vespertilionidae and Molossidae) from Texas. Texas Journal of Science 48:137–142.

1251. Yancey, F. D., II, and C. Jones. 1997. Rafinesque's big-eared bat, *Plecotus rafinesquii* (Chiroptera: Vespertilionidae), from Shelby County, Texas. Texas Journal of Science 49:166–167.

1252. Yancey, F. D., II, C. Jones, and R. W. Manning. 1995b. The eastern pipistrelle, *Pipistrellus subflavus* (Chiroptera: Vespertilionidae), from the Big Bend region of Texas. Texas Journal of Science 47:229–231.

1253. Yancey, F. D., II, R. W. Manning, J. R. Goetze, and C. Jones. 1998. The mammals of Lake Meredith National Recreation Area and adjacent areas, Hutchinson, Moore, and Potter counties, Texas. Occasional Papers, Museum of Texas Tech University 174:1–20.

1254. Yancey, F. D., II, R. W. Manning, and C. Jones. 2006. Mammals of the Harte Ranch area of Big Bend National Park, Brewster County, Texas. Occasional Papers, Museum of Texas Tech University 253:1–15.

1255. Yancey, F. D., II, P. Raj, S. U. Neill, and C. Jones. 1997. Survey of rabies among free-flying bats from the Big Bend region of Texas. Occasional Papers, Museum of Texas Tech University 165:1–5.

1256. Young, D. B., and J. F. Scudday. 1975. An incidence of winter activity in *Myotis californicus*. Southwestern Naturalist 19:452.

1257. Zehner, W. 1985. First record of *Pipistrellus subflavus* (Chiroptera: Vespertilionidae) on Padre Island, Texas. Southwestern Naturalist 30:306.

1258. Zielinski, W. J., M. J. Mazurek, and J. Zinck. 2007. Identifying the species of bats roosting in redwood basal hollows using genetic methods. Northwest Science 81:155–162.

1259. Zimmerman, R. 2009. Biologists struggle to solve bat deaths. Science 324:1134–1135.

1260. Zinn, T. L., and W. W. Baker. 1979. Seasonal migration of the hoary bat, *Lasiurus cinereus*, through Florida. Journal of Mammalogy 60:634–635.

1261. Zinn, T. L., and S. R. Humphrey. 1981. Seasonal food resources and prey selection of the southeastern brown bat (*Myotis austroriparius*) in Florida. Florida Scientist 4481–4490.

1262. Zubaid, A., G. F. McCracken, and T. H. Kunz. 2006. Functional and evolutionary ecology of bats. Oxford University Press, New York.

Index

*Letters and abbreviations following a page number denote: figure (fig.), note (n), plate (pl), table (t)

roost sites, 26, 97 /cohabitation sites, 26, 193; specimens examined, 99; status, 99; subspecies (*M. c. californicus*), 96

Campbell, Charles A. R., 47, 48fig.7; *Bats, Mosquitoes, and Dollars*, 47

canyon bat. *See* American parastrelle

Capote Canyon (Presidio County), 220, 221

Caprock Canyons State Park (Floyd County), 207

capture (of bats), 31, 33; and harp traps, 31; and mist nets 31, 33, 79, 152, 211

carbon (stable) isotope analysis, 30, 34, 35, 181, 195

Carlsbad Caverns (NM): and Brazilian free-tailed bat emergence, 203, 206; and guano carbon-isotope analysis, 30; and pesticide-residue studies, 200, 201; and pocketed free-tailed bat, 211; and thermal-imaging studies, 200

Carter, D. C., 201, 202

cave myotis (*Myotis velifer*), 112pl; dental formula, 10t3, 113; description, 4–5fig.3g, 57, 64, 93t8, 113 /trenchant characteristics, 93t8; distribution, 113map7, 113 /seasonal, 12t5, 114; etymology, 111; food, water, and foraging, 114–15 /emergence habits, 114–15; fossil specimen, 14t6; hibernation and migration habits, 24, 114; population, 26, 114; predators, 116; and rabies, 38t7 (*see also main entry:* rabies); reproduction and breeding, 115 /and delayed ovulation and fertilization, 115; and roost sites, 24, 49 /cohabitation sites, 78, 82, 105, 110, 114, 203 /decline of, 26 /maternity sites, 114 /and disturbance, 116 /and thermoregulation through clustering, 21; specimens examined, 116; status, 116; subspecies (*M. v. incautus*), 112–13, (*M. v. magnamolaris*), 112–13; and white-nose syndrome (WNS), 43, 44

Central Great Plains ecoregion (9), xi(fig.1), bat distribution by scientific name, 11t4; distribution by common name: American parastrelle, 158, cave myotis, 112, northern yellow bat, 144, pallid bat, 192, Seminole bat, 147, Townsend's big-eared bat, 185, 186, 187

Chaparral Wildlife Management Area, 141

Chihuahua, Nuevo León (Mexico), 224

Chihuahuan Desert (Trans-Pecos) ecoregion (12), xi(fig.1), 9; bat distribution by scientific name, 11t4; distribution by common name: American parastrelle, 158, big free-tailed bat, 213, California myotis, 96, cave myotis, 112, eastern red bat, 127, fringed myotis, 109, ghost-faced bat, 76, long-legged myotis, 117, pallid bat, 192, Townsend's big-eared bat, 185, 186/ western red bat, 123, western small-footed myotis, 99, western yellow bat, 150

Chinati mountains, 76, 85, 117, 167, 224

Chiroptera (order), 9, 13; molecular study of, 9, 13. *See also main entry for individual species*

chiroptorium (artifical bat cave), 48, 49fig.8

Chisos Mountains. *See* Big Bend National Park. *See also* Emory Peak Cave

Choeronycteris mexicana. See Mexican long-tongued bat

Clarity Tunnel (Briscoe County,) 207, 209

classification, xiii, 9, 12–13

climate: as determiner of bat distribution, 41, 90, 119; and effects of global climate; change, x, 41–42; and hibernation and migration, 9; and microclimates, 42, 130, 216; and parturition, 22; ranges of, x

clusters/clustering, 85, 95, 194 defined, 21; maternity/nursery clusters, 23, 95, 165, 205; and census counts, 26, 114. *See also* thermoregulation

Cockrum, E. L., 24

common vampire bat (*Desmodus rotundus*), 89, 90, 224. *See also* hairy-legged vampire bat

Congress Avenue Bridge (Austin), 19, 50, 203, 209

conservation, 40–52; bat protection efforts, 46–47, 50, 52; and legislation, 46, 48; and the role of research, 52. *See also* population (of bats)

Corynorhinus (genus): and classification, 185; and delayed ovulation and fertilization, 22; and echolocation, 17; interfemoral membrane, 91; lower teeth, 71fig.34b; olfaction, 19; upper teeth, 68fig.29b

Corynorhinus rafinesquii. See Rafinesque's big-eared bat

Corynorhinus townsendii. See Townsend's big-eared bat